Geographies of Health

for Caroline, Anna and Emma (ACG)
for Bob, Rebekah and Naomi (SJE)

Geographies of Health: An Introduction

Second Edition

Anthony C. Gatrell and
Susan J. Elliott

WILEY-BLACKWELL

A John Wiley & Sons, Ltd., Publication

This second edition first published 2009
© 2009 Anthony C. Gatrell and Susan J. Elliott

Edition history: Blackwell Publishing Ltd (1e, 2002)

Blackwell Publishing was acquired by John Wiley & Sons in February 2007. Blackwell's publishing program has been merged with Wiley's global Scientific, Technical, and Medical business to form Wiley-Blackwell.

Registered Office
John Wiley & Sons Ltd, The Atrium, Southern Gate, Chichester, West Sussex, PO19 8SQ, United Kingdom

Editorial Offices
350 Main Street, Malden, MA 02148-5020, USA
9600 Garsington Road, Oxford, OX4 2DQ, UK
The Atrium, Southern Gate, Chichester, West Sussex, PO19 8SQ, UK

For details of our global editorial offices, for customer services, and for information about how to apply for permission to reuse the copyright material in this book please see our website at www.wiley.com/wiley-blackwell.

Library of Congress Cataloging-in-Publication Data

Gatrell, Anthony C.
    Geographies of health : an introduction / Anthony C. Gatrell and Susan J. Elliott. – 2nd ed.
        p.   cm.
    Includes bibliographical references and index.
    ISBN 978-1-4051-7576-0 (hardcover : alk. paper) – ISBN 978-1-4051-7575-3
(pbk. : alk. paper)   1. Medical geography.   2. Environmental health.
    I. Elliott, Susan J.   II. Title.
    RA792.G38 2009
    614.4′2–dc22

                                                        2008030384

A catalogue record for this book is available from the British Library.

Set in 10 on 12.5 pt Minion by SNP Best-set Typesetter Ltd., Hong Kong
Printed in Singapore by Markono Print Media Pte Ltd

1   2009

Contents

List of Figures

List of Tables

Figure Sources

Preface

In the seven years since a first edition of *Geographies of Health* was published, much has changed to warrant the publication of a revised edition. For one thing, the sub-discipline has moved on considerably, and this has demanded the addition of new material. Recent and emerging research (including both methodological issues and substantive topics), as well as an intensification of research on topics considered in the first edition, have thereby required the updating of some examples and illustrations. We have preserved the original chapter structure, but have added a brief concluding chapter that looks at current and future health concerns and at the shifting landscape that is health geography. We hope that a careful comparison with the first edition, with which some readers will be familiar, reveals this new material.

Anyone who writes a book waits anxiously for the reviews to emerge. The first edition was well received, and has been quite widely adopted as a course text, much to the great pleasure of the first author, whose book it was. Given the changes to the intellectual and health landscapes identified above, Tony Gatrell welcomed the opportunity to undertake a revised version of the book. However, in the years since publication of the first edition his career has moved in a different direction, away from a conventional research and teaching environment and into an academic leadership role. The role of Dean of a Faculty leaves precious little time for tasks other than management of budgets, people and other resources, and so the thought of undertaking a revision of the book seemed quite daunting. Fortunately, and with the encouragement of the publisher, Tony had the good sense to find a co-author; indeed, he was sufficiently astute to ask Susan Elliott to collaborate – a well-known health geographer who brought to the task other skills and intellectual interests, as well as a different regional perspective on health. Further, since she was herself a Dean, Tony had the added reassurance of knowing that his co-author would understand the "day job!"

This, then, explains the origins of a second edition of *Geographies of Health*. In helping us to undertake this task we would both like to thank Justin Vaughan at Wiley-Blackwell, who has been a constant source of encouragement and advice. Dan Harrington at

McMaster University provided several of the new illustrations, and also updated the list of web-based resources in the Appendix. We thank Michelle Vine (also at McMaster) for her painstaking work on the reference list. Thanks also to Ric Hamilton (School of Geography and Earth Sciences, McMaster University) for his cartographic skills in preparing some new illustrations.

Tony has particular thanks to give to two colleagues in his former Faculty of Arts & Social Sciences at Lancaster University for their wonderful support as this second edition took shape. They are Heather Moyes (now at Cardiff University) and Lesley Waite. He would also like to thank – as before – Caroline Gatrell for being the best possible intellectual companion and wonderful life-partner, and to dedicate his contributions to her and to their lovely daughters, Anna and Emma. Any health and well-being that he enjoys is due in large part to their presence in his life.

For Susan, working on the book while also attending to her role as Dean has been made possible because of the outstanding staff in the Faculty office who do their jobs so well; she would especially like to thank Eleanor Frank, Linda Norek, and Susan Todd-Morris. Of course, family trumps all and Susan dedicates her work to her darling and intellectually challenging daughters, Rebekah and Naomi. Without the support of her husband, Robert Park, this work would never have been completed.

Tony Gatrell (Lancaster, UK)
Susan Elliott (Hamilton, Ontario, Canada)

Part I

Describing and Explaining
Health in Geographical Settings

Chapter 1

Introducing Geographies of Health

It is difficult to pick up a newspaper or magazine, or to listen to a television or radio broadcast without being alerted to a health "problem," whether this be concerned with poor quality screening for a disease, timely access to care, a possible risk arising from environmental contamination, or lack of access to the basic resources needed to live a healthy life. We may find it hard to define what we mean when we say we are healthy – though perhaps easier to describe our feelings of being unwell – but "health" is something we all have, or have had. At the same time, we all live somewhere on the earth's surface; most of us have homes, some have workplaces, others spend time at school or engaged in retirement activities. We occupy locations and, in the course of our lives, move from place to place. We all have our own "geographies" as well as our biographies.

We aim to convince you in the course of this book that our "health" and our "geographies" are inextricably linked. The screening we get for diseases will be available differentially from country to country or from one health region to another. Where you live affects the treatment you get. The risks arising from environmental contamination, be this poor air quality or polluted groundwater, are not uniform over space. If you live on a busy main road, very close to a source of electromagnetic radiation, or near a site disposing of hazardous waste, you may be more at risk of illness than others who do not. Where you live affects your risk of disease or ill-health and therefore your well-being. Access to basic resources, such as nutritious food, clean water, decent housing, and rewarding (and properly rewarded) employment is also geographically differentiated. Where you live affects how accessible or available are such resources. These relationships are further complicated if you experience any type of disability; typically, access to resources to enhance health and well-being are greatly hindered.

If you approach this book with a geographical imagination already well developed via other books or courses you will, we hope, find the statements above uncontroversial, though if you are new to geographies of health we intend to persuade you that the subject of "health" is a rich source of material that bears study by the geographer. If we can stimulate more of you to take up "geographies of health" as an area for further enquiry or research we will be delighted. But we hope, too, to interest other readers; those who come to the subject with a background in health research – perhaps public health, general practice, or nursing – or another social or environmental science, and who are intrigued by what geographers might have to say on the subject.

To set the scene, we need to explain some basic ideas and concepts. For some readers (perhaps some geographers) we need to say something about health, illness, disease, and disability. These are high-level concepts which are far from unproblematic, but we will endeavor to say something about these in this first chapter so that some of the early material makes sense. For other readers (primarily non-geographers), we need to introduce some fundamental geographical principles and concepts.

Having laid some of this groundwork we want then to consider five case studies: examples of work we consider as part of geographies of health. Our purpose in describing these studies is two-fold. First, we wish at a very early stage to introduce some pieces of research in the geographies of health in order to capture the imagination. Second, we intend to use them in order to show something of the richness and diversity of geographical research on health. There are several contrasting approaches to "doing" geographies of health, and there is no single style of enquiry accepted by those working within this field. We shall not say very much in this chapter about these different styles, though we hope it will be clear from what we do say that the differences exist. Instead, we shall leave until Chapter 2 the task of explaining how these five studies differ, and we shall use other examples to show how there are, broadly speaking, five alternative modes of explanation within the geographies of health. It will emerge later that this classification is far from clear-cut, but it serves as a useful organizing device. It will also become clear from the rest of the book that by no means all approaches are widely used in studying the geographies of health. Nonetheless, we think it is essential at an early stage to set out different styles and approaches to understanding the geographies of health.

Health and Geography: Some Fundamental Concepts

This section considers some concepts needed for a basic understanding of "health" and of "geography." They are examined separately – and in a real sense the remainder of the book represents an attempt to show the intimate connections between the two. Clearly, entire books are devoted to each; our aim is simply to provide some ideas that will aid the grasp of later material.

Concepts of health

Health has been defined in many ways. In 1957, the World Health Organization invited us to see health as more than simply the absence of disease; rather, "a state of complete physical, mental and social well-being" (World Health Organization, 1957). This ideal state does not, however, assist us very much since, according to the definition, most if not all of us are unhealthy at all times! We could instead take health to mean the availability of resources, both personal and societal, that help us achieve our individual potential. This is consistent with a more recent definition of health (Epp, 1986), stemming from the Ottawa Charter, where health is defined as a resource for everyday living that allows us to cope with, manage, and even change our environments. Alternatively, we might think of health as being physically and mentally "fit" and capable of functioning effectively for the good of the wider society. Linked to this is the idea of health as personal or mental "strength," fitness, or energy, or engaging in what we might think of as healthy behaviors or lifestyles (drinking alcohol in moderation, getting regular physical exercise). Further still, we could think of health as a commodity, to be given or lost, bought or sold; we "invest" in health perhaps by taking out private health care insurance, and lose it when we break a leg or become ill. Clearly, then, there are many ways to construct "health."

Consider how you behave if you feel unwell. This might take the form of a headache or a sore throat. In the first instance you would possibly decide to manage this symptom yourself, perhaps by taking to bed or by self-medication using an over-the-counter remedy. If the symptoms persisted, or took a different form, you might consult a health professional: perhaps a nurse in a clinic, or a general physician. You do this because the symptoms represent a departure from your usual healthy state. You may be examined and tested for signs of some underlying pathology or disturbance in the body's functioning. You experience some discomfort, some pain perhaps; you feel ill. *Illness*, then, is a subjective experience. The health professional, however, is concerned to offer a diagnosis; to "identify the specific underlying pathology in the patient's body that is producing the signs and symptoms, distinguish it reliably from other possible diagnoses, and label it correctly with the name of a medically recognized *disease*" (Davey and Seale, 1996: 9). Put simply, people suffer illnesses, while doctors diagnose diseases. The doctor or physician wishes, then, to cure the patient of the disease; the patient will, of course, wish to be cured of any disease, but also wants to be freed from feeling ill.

Disease and illness may or may not be associated, in that it is perfectly possible to feel ill without there being any detectable biological abnormality, while the person who has been diagnosed with such an abnormality might feel quite well. For example, those who are debilitated by a feeling of complete lethargy may find that a health professional is unable to detect any obvious "cause" (and hence conditions such as myalgic encephalomyelitis, or ME – also known as chronic fatigue syndrome – may go unrecognized by doctors; it may be a condition they are not prepared to diagnose; see Clauw et al., 2003 and Krieger et al., 2005 for an extended discussion). Equally, a middle-aged man who visits his general physician for a health check-up may be feeling well but is diagnosed with high blood pressure; he arrives as a healthy person and leaves as a patient (Seale and Pattison, 1994: 16).

Since the absence of health is perhaps easier to grasp than health itself it is no surprise that we find it easier to collect data on disease and illness. Further, we find it easier in principle to "measure" disease, since we can observe and record numbers of people with a particular cancer or heart disease while, for example, illness, as a subjective experience may need recording in other ways, as we see later. We call the study of disease in populations *epidemiology*, and a substantial body of material in this book could be labeled "geographical epidemiology:" the study of how disease is distributed in geographical space. Epidemiologists focus on *mortality* (death) and its causes, or on *morbidity* (sickness, which can include both disease and illness). Almost always we find it sensible to calculate mortality or morbidity rates, since this allows for comparisons between populations. We also usually compute *age-standardized rates*, thus controlling for the age structure of a population; knowing that the crude (not age-adjusted) death rate in one place is twice as high as elsewhere carries little information if we also know that there are many more older people living there. Adjustment for age structure yields so-called *Standardized Mortality Ratios*, allowing for comparisons between places that have contrasting age compositions (see Box 1.1).

Box 1.1 Standardized mortality ratios

Frequently, we wish to make comparisons of death rates between different places, but we need to allow for the fact that one area (such as a retirement community) might have an older population than another. Since older people are more likely to die than younger people we need to ask whether the pattern of deaths in one area is independent of its age composition. We therefore "standardize" mortality rates. Such standardization is usually done by agegroup where, for each areal unit, we predict or "expect" a particular number of deaths in different age categories, based on the distribution of the population by age in that area and the age-specific rates of death or disease in some wider (or "standard") population. We then calculate the ratio of observed to expected deaths, multiplying by 100. A value of 100 indicates that the death rate in an area is the same as that over a wider region (perhaps the country as a whole), a value greater than 100 suggests mortality is elevated even allowing for population structure, and a value less than 100 suggests it is reduced. Formulae are available that allow us to assess the extent to which values in excess of 100 are significantly elevated.

This method is known as *indirect* standardization and is useful in making comparisons between areas. The same principles apply to morbidity data, in which case we speak of standardized incidence ratios. We may then map standardized mortality (or morbidity) rates across the study area, though this is not without problems, as we see in Chapter 3.

Some authors use *direct* standardization, essentially the reverse of indirect standardization. The age-specific death rates for an area of interest are applied to the numbers of people in age groups of a standard population, giving the number of deaths expected in the standard population if the death rates in the area had applied. An age-standardized death rate for the area of interest may then be calculated. See McConway (1994: 90–1) for further details, including a worked example of an indirectly standardized SMR.

Mortality data in the developed world come from death certificates, which also specify cause of death; this may be far from easy to establish, particularly among the elderly. Moreover, mortality is a drastic measure of ill-health! Many illnesses and diseases cause a burden to the sufferer, as well as impacts on health care systems, without leading to death. As a result, health researchers often collect morbidity data, via a number of possible routes. These can include one-off patient surveys, or data from hospital consultations or emergency room visits. We shall encounter studies based on these sources at various points during the course of this book. Such data permit the estimation of an *incidence rate*, the number of new cases occurring within a given time interval expressed as a proportion of the number of people at risk from the disease. Alternatively, we can estimate *prevalence*, the number of people with the disease or illness at any one point in time.

Without attempting here a comprehensive classification of disease, we need to draw a distinction between different broad categories. In particular, we need to distinguish between *chronic* and *acute* diseases or disease episodes. Chronic diseases are those such as heart disease and diabetes, which may be long-lasting and even life-long, while acute diseases are those such as myocardial infarction (heart attack), sudden stroke, or appendicitis: conditions that start abruptly, last perhaps for only a few days and then settle, though perhaps developing into chronic conditions or leading to death. Those suffering from a disease such as asthma may experience it in both a chronic and acute form, able to manage it themselves on a long-term basis but perhaps requiring hospital admission if they have a sudden acute attack. *Infectious* diseases (such as measles, influenza, and tuberculosis) are those caused by organisms that can spread directly from one person to another.

If someone is restricted in some way from general physical or mental functioning, we can speak of *impairment*. For example, chronic respiratory disease may limit one's ability to negotiate stairs, while visual impairment varies from the quite mild (short-sightedness) to the most severe (blindness). Others whose impairment confines them to a wheelchair are disadvantaged by social attitudes or poorly designed environments and buildings as well as by the cause of their impairment. Some authors (for example, Gleeson, 1999; Thomas, 1999) prefer to make a formal distinction between "impairment" and "disability," arguing that the former refers to some defective or missing body part while the latter is a socially or culturally constructed form of exclusion. What this means is that at different periods in human history, or in different geographical settings, the same physical or mental impairments might be regarded quite differently by the wider societies. For example, Gleeson (1999) suggests that in feudal societies impairment was not uncommon, but that the treatment of such individuals as "disabled" only emerged with the rise of capitalism: "Within the complex, layered dependencies which constituted feudal village life, physically impaired people were not isolated as 'social dependants' – this abject identity was a construction of the capitalist social order" (Gleeson, 1999: 97). Even in contemporary Western society many of those who are impaired may be oppressed in just the same way as other minority groups whose faces (or bodies) do not "fit." Chouinard and Crooks (2003), for example, profile the hardships associated with being a "disabled" academic. Wilton (2004a) articulates how newly emerging flexible work environments impact differentially on disabled people engaged in paid work. Health geographers (for example, Driedger et al., 2004) have contributed to our understanding of how space and place shape the experiences of those experiencing impairment as a result of chronic illness, while at the same time reminding academics of their obligations to their research participants, ensuring that we are sensitive to, and empower, those participating in our research endeavors (Valentine, 2003).

We saw earlier how illness may trigger a visit to a general physician or other health professional engaged in "primary" care. The diagnosis may call for a referral to other, more specialized health professionals, usually in a hospital-based setting (the "secondary" sector), or even for very specialized care (perhaps complex surgery) in the "tertiary" sector. But this possible sequence of care is very much a traditional western model. *Complementary* (or alternative) medicine has grown rapidly in some western countries, however, and some health care will be delivered by practitioners such as osteopaths, acupuncturists, and homeopaths; for some of these groups, treating the person rather than the disease takes priority. Geographers are beginning to map how the use of such complementary medicine varies from place to place, and why (Verheij et al., 1999; Wiles and Rosenberg, 2001). In the non-western world, "*traditional*" medicine is the norm. Like complementary medicine, this emphasizes the links between mind and body in a holistic approach to illness and disease. Among a variety of such perspectives is Ayurvedic medicine, practiced among Hindus in India for over 2000 years. In common with other traditional forms, this emphasizes the necessary balance between three primary "humors," with therapies involving changes in lifestyle and diet, use of herbal remedies, and meditation (Seale and Pattison, 1994: 18–21). The integration of traditional and western health care among Aboriginal peoples of the world has also been studied (Hunter et al., 2006).

Geographical concepts

We want to begin this brief exploration of some geographical concepts by considering *location*. We shall take location to mean a fixed point or geographic area on the earth's surface, somewhere that can be pinpointed by using a pair of locational coordinates. These coordinates are often latitude and longitude; for example, 51.17°N, 30.15°E refers to a location about 51 degrees north of the equator and 30 degrees east of the Greenwich meridian, while the location 23.16°N, 77.24°E is closer to the equator and further east. At a more local or regional scale it is more conventional to use a coordinate-referencing scheme specified by a national mapping agency. For example, the Ordnance Survey in Britain uses a pair of coordinates (or "grid reference") known as "eastings" and "northings." The grid reference (348211, 458826) thus identifies a unique location in Britain, and has a precision of 1 metre. Of course, few of us use such locational identifiers in our daily lives; instead, we refer to a location using an address. Such an address, when containing a postal code of some sort (known as a postcode in the UK and Canada, or a zip code in the USA, for example), turns out to be extremely valuable to a geographer, since there may be computer-based "look-up" tables that link such postal codes to the coordinates described earlier. The uses to which this may be put are considered in Chapter 3.

What we mean by a "location" does, of course, depend on whether we are looking at health or disease in the world as a whole, or perhaps within a region. For example, if we wished to look at the way disease might be spread via air travel it would be useful to define a set of locations as the set of major cities connected via air networks. But at a regional level it would make little sense to treat Beijing or Paris as a single "point" location.

Locations, then, are points or areas on the earth's surface; they do not seem to *mean* very much. However, your own home, which has an address (a location), may mean a great deal to you. A favorite theatre, or sports stadium, or woodland, or town, all identifiable locations, may well mean something to you. And if we tell you that the locations 23.16°N, 77.24°E and 51.17°N, 30.15°E, referred to above, identify Bhopal and Chernobyl, respectively, then both were etched forever on our collective consciousness as the sites of devastating industrial explosions in 1984 and 1986. Once named or labeled, these locations become *places*. Locations become places when they are charged with meaning. Until the explosions, these places were, for many of us, simply dots on a map – locations – though for those living there they had always been places. We consider the Bhopal explosion further in Chapter 7.

Places, like locations, can refer to very small areas, or be of quite vast extent. Your grandfather's favorite chair, positioned to observe daily life outside the living room window, may mean a great deal to him and contribute significantly to his mental well-being; for him, it is a place. For others, particular buildings will be of enormous importance. Some stories appearing in the media, to which we referred earlier, focus on the possible closure of hospitals, and while some will object to a consequent reduction in access to health care, for others it is the symbolic quality of the institution or building that matters as much. For others still, small neighborhoods are foci of meaning, and proposals or action to alter their character, for example by locating a noxious facility in their midst, will provoke opposition among those who fear it carries a health risk. More broadly, people develop attachments to places that may be cities or regions, as well as to nation states.

Places may be good or bad for health – indeed, this is a major theme running through this book. In some cases, as in the opposition to a noxious facility, the public perception of risk may be as important as any measurable impact on morbidity. As we shall see in Chapter 4, those places which are impoverished in terms of access to health-promoting resources, such as leisure and recreation facilities, will not be associated with good health to the extent that resource-rich neighborhoods are. Moreover, what happens in one place may have negative, even drastic, consequences for those living both nearby and at a considerable distance. Such "externality effects" are illustrated by the impact on health of those living downwind of a major chemical installation or in "Cancer Alley," the 87-mile section of the Mississippi River between Baton Rouge and New Orleans, the location of what is arguably the largest concentration of petrochemical industries in the world (Rosner and Markowitz, 2002), or, more dramatically and over much greater distances, by the Chernobyl explosion. Conversely, other places and landscapes are considered to be beneficial to health; they are *therapeutic landscapes* (see Box 1.2). For ordinary, "lay" people free of any disease, however, there will be other, much more anonymous places where we feel close to nature, where we feel secure, and with which we identify; in short, where we feel "well."

Box 1.2 Therapeutic landscapes

"Therapeutic landscapes are places that have achieved lasting reputations for providing physical, mental, and spiritual healing" (Kearns and Gesler, 1998: 8). According to Wil Gesler, the geographer who first developed the concept, these reputations might be built on the qualities of the physical environment, such as a source of water or a distinctive piece of topography. Or, they may rest on the qualities of buildings, such as temples. But such therapeutic properties are socially or culturally constructed. People might seek out such places in order to be "cured" of a chronic disease, perhaps, or to hope for an improvement in well-being.

Gesler's own early examples include: the sanctuaries first established in classical Greece (at Epidauros, for example: see Gesler, 1993); spa towns (such as Bath in England: Gesler, 1998) based on the perceived healing qualities of local springs; or sites of deep religious significance (such as Lourdes for Catholics and other Christians: Gesler, 1996). Different people "see" these sites in different ways and at different times; for example, some claim genuine healing powers for spa waters, while for others a spa town may be simply a brief stop on a vacation. Even those who are inspired by a piece of landscape, but do not necessarily gain therapeutically from it, may notice an improvement in their mental health. Further, depending on the circumstances and our mood, even rather ordinary or everyday landscapes may contribute to well-being. For example, Milligan and her colleagues (2004) found the concept of value in interpreting the use of allotments (community gardens) by older adults. Wilson (2003) has sought to extend the concept to an understanding of health and identity among First Nations peoples in northern Ontario. As she points out, the relationship between health and place for such people is an intimate one. Moreover, therapeutic landscapes are not necessarily physically located; rather, the way of life of the Anishinabek peoples she studied – their use of the land for hunting and fishing and its direct contribution to health (nutrition) – is connected spiritually as well as materially to the land. Other researchers, such as Wilton and DeVerteuil (2006) take a critical look at the concept of therapeutic landscapes, arguing that in some circumstances we need to recognize the differential power relations that shape health behaviors in particular settings.

For further details, examples, and overviews, see Curtis (2004: Chapter 2) and Williams (2007).

Attachment to places for some may mean separation for others. As Cornwell (whose work we examine more fully in the next chapter) notes, "where there is belonging, there is also not belonging, and where there is inclusion, there is also exclusion" (Cornwell, 1984: 53). Those "attached" to a place may object if others wish to attach themselves to it, especially if they are a different color. At the extreme, this can have severe "health" consequences for those on the receiving end of violence directed towards the "other," those whose faces are "out of place." What this means is that places, and how we identify with them, are not simply a matter of subjective experience; "rather, such feelings and meanings are shaped in large part by the social, cultural and economic circumstances in which individuals find themselves" (Rose, 1995). This indicates that there is a danger in romanticizing the notion of a sense of place. As Mohan (1998: 120) observes, "sense of place has most often and most strongly been associated with economic adversity – for instance, the instinctive collectivism of communities suffering the excesses of capitalist industrialization, such as mining settlements." Is, he asks, "the implication really that one should celebrate such conditions?" For plenty of people places are health-damaging (often in the sense of being a locus of unemployment) rather than health-promoting, and typically those with adequate resources are more likely to find themselves in the latter.

As we shall see later, geographers may choose to study health in a particular place, or they may want to make comparisons between places and study health events and outcomes in a set of places. If the latter is important they will frequently want to consider and measure the *distance* that separates places. How far are people from those facilities delivering health care? How far are people from a possible source of pollution, such as a smelter? Over what distances do diseases spread? We have already seen that locations can be pinpointed in an absolute sense; but, as these examples make clear, we will often need to look at where places are located in relation to other places ("relative location"). Distance, then, is something which relates one place (or location) to another. It is perhaps *the* fundamental concept in geography. How is it measured? An obvious and important measure is the physical distance separating one location from another on the earth's surface. If measuring the distance from Chernobyl to Kiruna in northern Sweden we shall need to take into account the earth's curvature to do this; however, if looking at distances between the home locations of those stricken in a city by an infectious disease we could safely ignore this curvature and measure straight-line ("Euclidean") distance. It is worth pointing out that there are other concepts of distance that may be significant in a health context. We can think of spatial separation in terms of travel time, for example, or travel cost, or people's estimates of such separation ("cognitive distance"), or the social distance that separates them (in terms of class, income or lifestyle, perhaps) from their neighbors. Measuring distances between areal units (such as health regions, counties, or catchment areas, for example) creates special problems. Sometimes we can measure straight-line distances between the centers of such zones; often we are only interested in whether one zone is adjacent, or connected, to another zone and in this case simple contiguity or adjacency serves as a measure of distance.

We have already made oblique reference to it, but we want now to introduce formally the concept of spatial *scale*. This, too, is quite fundamental to what follows, since while health is a property of the individual we can aggregate health events for those living in a neighborhood, city, region, or country, in order to estimate disease rates for a set of such units. We might then choose to study disease incidence in a city, or make comparisons between rates for all countries in Europe, for example. We may find that factors that explain variations in disease incidence at one scale may be quite unimportant at another. For instance, international variations may be a function of how much expenditure on health care is committed by governments, or even by differences in diagnostic or recording methods, while variations from place to place within a small region may be explicable by an environmental factor. Equally, the kinds of events that impact upon our health may operate at different scales. Local contamination of a groundwater source may have consequences for those living in a quite restricted area, while (as noted already) a catastrophic nuclear explosion may have an impact across continents. On a very different scale the quality of social relations within the home will have impacts on the health of its occupants. We shall see in what follows, therefore, that geographers span a wide range of spatial scales in studying health and disease.

Although geographers concern themselves fundamentally with spatial concepts we should not neglect to mention the importance of *time*. This is because while locations remain fixed over time (if we ignore the continental drift that takes place over geological time!) places do not. Those which are inhabited may gain or lose population, with possible health impacts. Chernobyl has been a place for as long as humans have inhabited it, though one could argue that for most of us it carried little or no significance until 1986. And time-scale matters as much as spatial scale; we may choose to study the health consequences of catastrophic, extreme events such as earthquakes or the tidal waves or *tsunami* that devastate Pacific coasts, most dramatically in late 2004 (Wickrama and Kaspar, 2007), or major floods, such as that which devastated New Orleans in 2005. Alternatively, we may concern ourselves more with the impacts of longer-term changes, such as global environmental change (discussed in Chapter 9). Climate affects health over different time-scales; daily hospital admissions for asthma may be elevated by climate events such as thunderstorms the day before, while seasonal change brings marked mood swings in some people ("seasonal affective disorder").

Places may be good or bad for health at different times and over different time-scales. We may, for example, be exposed to particular sources of environmental contamination at different periods of our lives, depending perhaps upon where we live and the work we do. Our "life courses" will have a major bearing on our health, and we cannot neglect the influence of our migration histories on health outcomes; indeed, Chapter 6 is devoted to this theme. Further, our health is affected by what happens to us as we move around during our daily lives, perhaps being exposed to air pollution from motor vehicles as we commute to work, to a risk of accidents in some occupations, or maybe to overt or verbal violence in domestic settings. We shall consider one way of conceptualizing the role of time in the next chapter (see pp. 40–1).

While places may change over time in observable, measurable ways, it is important to note, too, that our *experience* of, or *beliefs* about, them may change too. The marshlands of south-east England and other parts of Britain were considered very unhealthy 200 years ago, because of the risk of malaria (literally, "bad air;" see Dobson, 1997) while south-east England now carries a reputation as one of the healthier parts of the country. Moreover, our experience of time changes through our lives. This, too, may affect our psychological well-being. But this experience of place also changes over very short time-scales; for example, we might feel entirely safe walking through a park during the day, yet quite threatened at night-time. The location remains the same, but the "place" changes character dramatically during 24 hours. One's fear of crime can have a very real impact on health and well-being. Further, our access to health services changes over time. We may live next door to a health center, but if it is closed for the weekend we may well have to travel much further for immediate attention, while if it closes permanently because of health service restructuring there will be longer-term impacts on access to services.

Places may also be thought of as social settings or social environments; we are literally surrounded, or "environed" by other people and features of the landscape. However, we also think of *environment* in the sense of the physical world and how it impacts upon us. Climate affects health, in both a direct and indirect sense, as we shall see in Chapter 9, while in earlier chapters we look at the impacts of environmental degradation, both of air and water, on health. Even the local geology can have health impacts. For example, goitre (an enlarged thyroid, resulting in severe swelling of the neck) in areas such as south-central Sri Lanka is thought to be due in part to low levels of iodine in water and soils (Dissanayake and Chandrajith, 1996).

The physical environment figures prominently in a branch of the geography of health known as *disease ecology*. Here, the argument is that one cannot understand the distribution of a disease, particularly an infectious or parasitic disease, without knowing about its relationship to local and regional ecologies – the interactions between topography, climate, water, soils, plants, and animals. Various examples of an ecological approach figure in this book, of which malaria is a good example, since we require data on particular configurations of rainfall and temperature, as well as knowledge of animal (mosquito) and human behaviors, to predict its spatial distribution. Yet the environment has impacts on health in much more subtle ways. A case can be made for suggesting that loss of biodiversity, and the despoiling of landscapes, has a negative impact on well-being. Those who derive mental health from the enjoyment of particular landscapes may well find that others' modification of such landscapes in an environmentally insensitive way causes genuine, albeit hard-to-measure, ill-health. Modern public health sees the environment as social and psychological, not merely as physical. In this sense, then, "environment" and "place" converge to provide a spatial context for health that transcends the individual's own behavior and health outcomes.

Geographies of Health: Five Case Studies

Having set out some key ideas and concepts, we want to add some color by illustrating some work that we take to be representative of the rich variety of the geographies of health.

Asthma in New York

Consider, first, a study of the geographical epidemiology of asthma among children in New York City (Corburn et al., 2006). Asthma is a chronic disease characterized by a narrowing of the airways that results in mild to severe wheezing and/or loss of breath, causing death in its most severe form. It is a relatively common disease among children, particularly in the developed world (prevalence is around 10%). In New York City, one in six children are thought to show symptoms of asthma and the condition is the major cause of school absenteeism, hospitalization, and visits to emergency rooms. Corburn and his colleagues sought in this study both to *describe* how patterns of asthma hospitalization vary across the city, and to *explain* or interpret such patterns in terms of the social and environmental characteristics of neighborhoods within the city.

The authors had data for 2016 census tracts in New York City. Such data included: asthma hospitalization rates (between 1997 and 2000); median household income; the percentage of the population classified as Latino and African-American; and the percentage of housing classified as dilapidated (all data from the 2000 census). The authors also included data on the density of air polluting facilities as well as truck routes, again for each of the census tracts.

Asthma "hotspots" in this study were defined as contiguous (adjacent) census tracts where the hospitalization rate was significantly greater than expected, given the population of children. To define such hotspots the authors used a "spatial scan statistic," a method of identifying clusters that has become widely adopted; we describe the method in Chapter 3 (pp. 54–5). These hotspots can then be mapped; in so doing, we observe four such clusters, with particularly high rates in central and east Harlem (see Figure 1.1). The hospitalization rate here is 18.4 per 1000 children, about three times the rate elsewhere in the city. Residents in hotspots are five times as likely to be living in public housing compared with those outside hotspots, and three times as likely to be in dilapidated housing. The relations between asthma rates and other variables were examined using correlation coefficients (see Box 1.3). For example, within census tracts in east/central Harlem there is a strong negative correlation (−0.81) between asthma rates and median household income, and a strong positive correlation (0.73) with dilapidated housing, but a lower correlation (0.12) with the density of truck routes.

This study is a good example of scientific investigation in health geography. It begins by assembling a set of numeric data, both on the health problem and the variables (covariates) hypothesized to be related to that problem. It maps the health data, but goes beyond simple mapping by identifying areas with elevated disease rates, using statistical methods of spatial data analysis. The research then examines the environmental and social characteristics of those areas with higher than expected rates (hotspots), using simple correlation analysis to see whether hospitalization rates are associated with particular socio-economic variables and markers of outdoor air pollution. Last, like many similar studies, it has in mind potential applications to public (health) policy; specifically, the aspiration is to use geographical intelligence as a basis for directing limited public health resources at particular neighborhoods. Rather than distribute scarce resources (people and money) across New York City, are we not better off targeting those neighborhoods most at risk, and hence most in need?

This study makes a contribution to our understanding of asthma epidemiology. But, like any piece of research, there are limitations. First, it is frequently the case that we have to rely on data that are already available, unless we have the means to collect our own. Second, hospitalization data represent only the most extreme cases of asthma. There will be a sizeable proportion of the child population who self-medicate, or are managed by a family doctor (physician). Next, given this study was undertaken in the USA, we also have to take into account that there is not universal access to health care. So we have to assume – perhaps heroically – that the data we have are an unbiased subset of overall asthma incidence.

The study relies on aggregated, rather than individual or household, data. As the authors acknowledge, there is plenty of evidence linking asthma to more direct markers of indoor

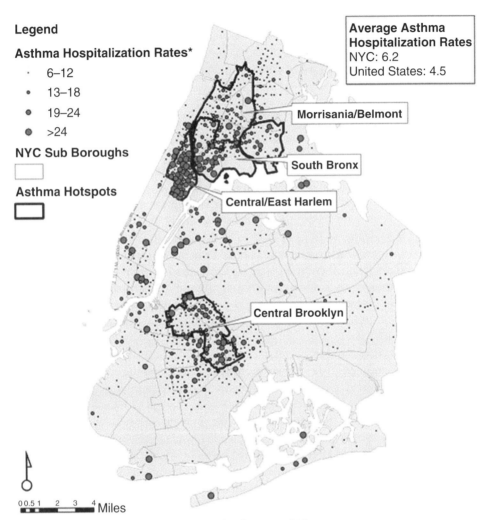

Legend

Asthma Hospitalization Rates*

·   6–12

●   13–18

●   19–24

●   >24

NYC Sub Boroughs

Asthma Hotspots

Average Asthma Hospitalization Rates
NYC: 6.2
United States: 4.5

Morrisania/Belmont

South Bronx

Central/East Harlem

Central Brooklyn

0 0.5 1   2   3   4 Miles

* Hospitalizations per 1000 persons under the age of 14

Sources: New York State Department of Health (1996–2000), Statewide Planning and Research Cooperative as cited in infoshare (http:\\www.infoshare.org).

Figure 1.1   Neighborhood asthma hotspots in New York City

Box 1.3 Correlation and regression analysis

Correlation analysis looks at the association (correlation) between a pair of variables. Following Corburn and colleagues' example, we have a table or data matrix of variables, as follows. For example, X_{21} represents the known median household income in census tract 2.

Census tract	Y (Asthma hospitalization)	X_1 (Median household income)	X_2 (Dilapidated housing)
1	Y_1	X_{11}	X_{12}
2	Y_2	X_{21}	X_{22}.

Each observation, a census tract in the New York City example, may be represented as a point on a graph (the set of points forming a "scatterplot"), as the diagram overleaf suggests. We may then compute a correlation coefficient which represents the degree of association between the two variables. As noted in the text, the correlation between Y (asthma) and X_1 (income) is negative; simply put, in those census tracts with higher income, the asthma hospitalization rate is lower. The correlation between Y and X_2 is positive; census tracts with higher proportions of dilapidated housing tend to have higher asthma rates. A correlation coefficient varies from −1.0 (perfect negative correlation), through 0 (no correlation) to +1.0 (perfect positive correlation).

Regression analysis (not used by Corburn and his colleagues) is a widely used statistical method that seeks to use one or more explanatory variables (sometimes called "independent" variables or "covariates") to account for variation in a response variable ("dependent" variable). It seeks to draw a line through the scatterplot that passes as close to the points as possible. The slope of this line (the regression coefficient) shows how Y varies with X; specifically, how a unit change in X produces a unit increase in Y. This might be positive (as X increases, Y increases too) or negative (as X increases, Y decreases). The extent to which the data points cluster round the regression line is measured by the correlation coefficient described above. The square of the correlation coefficient represents the proportion of the variation in Y that is "explained" by X.

The regression line has two parameters, its slope and intercept (the value of Y that is predicted when X is zero). For any observed value of X we can consider two accompanying values of Y: the observed value (Y_i), and that predicted by the regression model ($\hat{Y}_i$). As the graph makes clear, some data points are poorly predicted by the regression line; that is, the observed and predicted values of Y are very different. The difference between the observed Y value and its predicted value is called a regression "residual" (e_i). Observations lying above the regression line are positive residuals (observed minus expected Y is greater than zero), while those below the line are negative residuals.

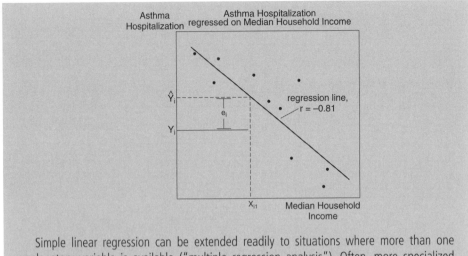

Simple linear regression can be extended readily to situations where more than one explanatory variable is available ("multiple regression analysis"). Often, more specialized forms of regression analysis are required, known as *generalized linear modeling*. For example, disease rates are formed by dividing a count of disease cases by some suitable denominator. Rather than use the rates as a dependent variable it is more appropriate to use a so-called *Poisson regression model* (McNeil, 1996).

air quality, whether exposure to environmental tobacco smoke or to allergens associated with mold, dust, and dampness. We often encounter studies that do not, or cannot, collect direct measures of exposure; instead, they rely on surrogate information. Here, for example, data on the condition of public housing were used as a marker for the likely presence of factors that trigger asthma. Many classical epidemiological studies collect individual, rather than aggregated (sometimes called "ecological") data; in that situation we have the ability to assess directly what factors distinguish individuals with the disease ("cases") from those without ("controls"). In Chapter 3 we shall discuss a method that allows us to look both at individual effects *and* at neighborhood-level factors.

Area effects on smoking in disadvantaged communities in Glasgow

Smoking is a primary risk factor for chronic diseases such as lung cancer, diabetes and cardiovascular disease. Health geographers are interested in how smoking behaviors and their determinants vary from area to area. Stead et al. (2001) have conducted fieldwork in disadvantaged communities in Glasgow, Scotland, to explore possible geographical influences on smoking patterns. As we see in Chapter 4, there is increasing evidence that residence in a disadvantaged area can have effects on ill-health, independent of individual characteristics such as income, education or employment. There are several reasons for this – not least of which is the fact that disadvantaged communities have fewer amenities that are beneficial to health, such as good housing, access to shops with healthy foods, leisure and recreation facilities, and employment opportunities. Further, living in a disadvantaged area also tends to bring increased exposure to stresses produced by higher crime rates, violence and general incivilities. All of this is underscored by the feelings of social exclusion that often accompany living in a disadvantaged community. Data indicate that while, nationally, smoking rates in Canada, the USA and the UK are decreasing, levels of smoking among lower income groups remain constant, as do the amounts of tobacco consumed. The literature indicates several explanations for this pattern of tobacco use: economic insecurity; education; isolation and stress; poorer psychological health; and deliberate targeting of low income groups by the tobacco industry. Stead and colleagues are interested "particularly in the ways in which smoking might be fostered (and smoking cessation hindered) by residence in communities excluded economically, culturally and physically from mainstream society" (Stead et al., 2001: 334). But instead of collecting quantitative data the authors collect qualitative data from focus groups (described in Chapter 3), comprising individuals living in disadvantaged communities. In their respondents' own words (and preserving the distinctive dialect):

Ah'd take my kids an' get away if ah could. Because o' the drugs an that kind o' thing (male respondent, 18–24 years).

Lost a job, and you're unemployed, and you're just sitting there – it's [smoking] somethin' tae do (male, 18–24 years).

Other themes emerging from the focus group discussions include pro-smoking community norms, limited experience of environments that encourage cessation, and cumulative barriers to successful cessation. Again as their data suggest:

Ma Maw used to buy me fags so ah woudnae smoke a' hers (female, 18–24 years).

When we were at school there were plenty o' jobs, then Maggie [reference to a former Prime Minister of Britain, Margaret Thatcher] come along and took all the steel works away . . . We would all like to stop [smoking] but we know we cannae. No incentive. No will power (male, 25–44 years).

It is ma only pleasure. Aye, an' it is hard [to stop] (female, 25–44 years).

The final set of themes that emerge relate to isolation from wider social norms and the role that smoking in these communities plays in fostering social participation and belonging:

I think when people ask you where you come from and you say Possilpark, they think you are going to stab them . . . they think you are a pure criminal, or you're scum or a junkie or something (female, 19–24 years).

You [can] always tap money for a packet of fags (female, 25–44 years).

Stead and colleagues conclude that in these disadvantaged communities "smoking appears to act not only as a means of coping with stress and exclusion but also as a means of expressing identification and belonging. The collective aspects of smoking – sharing, lending and borrowing cigarettes – provide a means of giving and receiving support, and arguably help bind people together" (Stead et al., 2001: 341). Their research illustrates not only the role of place on health (and unhealthy behaviors), but also the relevance of the beliefs of ordinary people living in particular places. The aim is to understand a health issue from the standpoint of those directly involved, out of which may emerge more effective health promotion strategies. By not opting for a large-scale survey of numerous families, but rather for in-depth discussions with a small sample, their view is that this method is likely to be productive in uncovering the psychosocial factors involved in shaping health in these particular neighborhoods.

The changing political economy of sex in South Africa

AIDS (Acquired Immune Deficiency Syndrome) results from infection by the Human Immunodeficiency Virus (HIV) and has been an epidemic almost unparalleled in human history, such that over 33 million people were living with HIV/AIDS across the globe in 2007, and 2.1 million deaths from AIDS were recorded that year. Yet this global figure masks huge variations from place to place, with the vast majority of those infected living in sub-Saharan Africa. Recent estimates (from avert.org, an international AIDS charity) indicate that the prevalence rate of HIV infection among women in their late twenties in South Africa is 33.3%, while for men aged 30–39 years it is 23.3%; in other words, over a third of women under the age of 30 are thought to be infected, and almost a quarter of men in their thirties. What factors explain these high rates, and how have both the rates and the causal factors changed in the years since the end of apartheid? These are questions posed by Hunter (2007). Hunter's explanations lie in three sets of factors: "rising unemployment and social inequalities, dramatically reduced marital rates, and the extensive geographical movement of women as well as men in contemporary South Africa" (Hunter, 2007: 690). He points to very high HIV rates in informal settlements, characterized by a population that is black African, young, unmarried, and lacking in secure employment. These informal settlements are "spaces of poverty and sex exchanges" (Hunter, 2007: 696).

Unpacking the three explanatory factors further, Hunter suggests that the number of economically active (potentially employable) women searching for work increased by 2 million between 1995 and 1999; those women who did find jobs were "pushed into poorly remunerated and highly unstable informal work; consequently, women's median income fell sharply in the post-apartheid period" (Hunter, 2007: 694). Social inequalities have increased markedly. Second, far fewer women are marrying and relationships are more unstable. In 1970 about half of all Africans over the age of 15 years were married, but this had dropped to 30% in 2001; marriage is now, Hunter argues, seen as a middle-class institution. Third, evidence suggests that rural women are moving more frequently than men. Hunter rejects earlier arguments that suggested African women are left in their rural homes and are infected by male migrant partners returning from the cities wherein they are employed. Instead, the informal nature of work "can propel women into the sexual economy, a scenario ultimately driven in some instances by World Bank/IMF sponsored structural adjustment programmes" (Hunter, 2007: 691).

Hunter's theoretical arguments are supported by empirical research during a lengthy period of fieldwork in KwaZulu-Natal, where he spent five years living in an informal settlement (Hunter, 2007: 690). How, then, do the three structural factors help explain increasing rates of HIV infection? Hunter suggests that there is an informal sexual economy and complex sexual networks within the settlements. It is not that prostitution is "rife" in these informal settlements. But, he suggests, it is not uncommon "to hear stories of women having material relationships – 'one for money, one for food, and one for clothes' – but also common to hear about love letters and signs of affection" (Hunter, 2007: 697). Condom use may well be higher in "commodified" relationships (prostitution) but lower when there is an element of perceived (but possibly misplaced) trust between partners. The latter can drive infection rates.

This is an example of research undertaken by a geographer on one of the most serious and intractable public health issues, HIV/AIDS, that has continued into the twenty-first century. The author sees the pandemic as a symptom of structural inequalities, such inequalities having roots in the apartheid, but also post-apartheid, era. Hunter's research is in a particular spatial setting – part of rural South Africa – and comprises detailed empirical observation that endorses his theoretical arguments. We will elaborate on this kind of theoretical explanation in the next chapter, illustrating it with other examples.

The personal significance of home

Our fourth example involves the recognition of the multiple contexts within which our daily lives are played out, contexts that are nonetheless constrained by societal structures that shape our access to resources or "capital:" material, cultural, social. The example (Angus et al., 2005) draws on a sample of individuals receiving long-term home care. Specifically, these researchers examine "the ways that recipients of long-term home care are positioned and active within altered and shifting relational patterns that enter into the construction, dismantling and reconstruction of their experiences of home" (Angus et al., 2005: 162). In so doing, they draw upon the concept of "habitus."

Habitus describes the durable pattern of dispositions, deeply inculcated and reflective of the social position of the individual or the material conditions to which that individual is accustomed. "Dispositions are corporeal in that they are embodied and converted into motor schemas and bodily automatisms that materialize in practice as postures, gestures and movements" – a sort of social biomechanics, if you will – "and as aesthetic performances or tastes" (for food, clothing, culture; Bourdieu 1977 as cited in Angus et al., 2005).

Practices do not stem directly and wholly from habitus; they depend on contexts or fields and those fields are in turn shaped by the individual's access to material, cultural and social capital. While it is usually the case that our practices are pre-reflexive, when new factors/variables/aspects enter our life world (e.g., chronic illness and the need for long term home care), those practices necessarily change (because our access to capital – material, cultural, social – has changed): "the corporeal sedimentation of habitus into mannerisms, tasks and skilled movements seems to rely on the continued facility of the body to reproduce patterns of dispositions associated with a particular social position" (Angus et al., 2005: 166).

The data for this research stem from a larger research program in Ontario, Canada called the Hitting Home project, designed to investigate issues related to society, culture and health. Several data collection techniques were employed, including in-depth inter-views, field notes, videotapes, and (non-participant) observation. The sample consisted of 16 detailed case studies of individuals (both adults and children) receiving long-term home care in urban, rural, and remote parts of the province. These data were applied to the following problem: if, as Bourdieu posits, social fields are imbued with practices influenced by habitus and constrained by different forms of capital, then a transposition of those practices will naturally take place when fields (i.e. home and health care) "collide." The result is improvisation.

Three primary themes emerged from analysis of the data. The first is *the politics of aesthetics*. This refers to the disruption between care recipients' bodies and the objective conditions provided by the material spaces of the home. For instance, a previously well-appointed bedroom for a middle aged woman becomes shifted to a habitus of disability or dependency when a commode is introduced to that space. Alternatively, a child's bedroom – previously unkempt and untidy with toys – takes on a very different habitus when the space is (necessarily) converted to a home hospital room with spaces allocated to necessary health care equipment, space where toys once took precedence.

The second primary theme that emerged from the data relates to the *maintenance of order and cleanliness*. Meanings of home shifted shape as more and more of their inhabitants had to rely on paid home care workers to keep up the order and maintenance of things, which may or may not (typically the latter) meet the participants' expectations. This shift in habitus was a major source of frustration for some, as "the state of homes in these cases bore visible testimony to the physical deterioration of their occupants, demonstrating the co-vulnerability of home and body" (Angus et al., 2005).

The final theme relates to *transcending the limitations of the home*. In order for long-term care clients to access their communities, they were required to undertake improvisations. These included the acquisition of devices (especially those related to communication and connection with extra-local sites), the delegation of tasks to service providers as well as family caregivers, and occasional subversive disregard by providers for the institutional logic of health care (for example, being accompanied by a home health care worker on an errand in the community, which was strictly against the rules).

In short, Bourdieu's concept of habitus essentially mediates the dichotomy of structure/agency. In the context of the study by Angus and colleagues, the improvisations of the research participants in the conditions and arrangements of their living spaces signified their own social repositioning: as chronically ill, dependent users of long-term home care.

Embodied spaces of health and medical information on the Internet

The volume of information, about health and well-being, available on the Internet has exploded in the past few years. At the time of writing a Google search of "online health information" reveals over 420 million sites! Among others, Hester Parr (2002a) has explored the ways in which these sites provide alternative forms of access to health care, as well as providing new communities and forms of belonging; communities that are not defined in geographical space but, rather, virtual space.

Of considerable interest to geographers and other social scientists is "whether a 'real' social world is being supplanted or supplemented with multiple "virtual" social worlds through which the usual boundaries of human action, cohesion, embodiment, and knowledge are being expanded, broken down, and reconstituted" (Parr, 2002a: 74). Parr suggests that we are encountering a new form of "spatialization," a term used by the French historian Michel Foucault to denote alternative conceptions of space. While earlier, eighteenth century, "spatializations" (Parr, 2002a: 79) sought to look at ("gaze upon," as Foucault would say) the body on the dissecting table, then into the three-dimensional body itself, before focusing on the institutions of health care (hospitals and clinics), Parr argues convincingly that the Internet may be thought of as a new form of spatialization "which facilitates a specialist medical gaze entering into the fabric of 'lay' everyday life (for those who are connected)" (Parr, 2002a: 79). The parenthetical comment is important, since although we live in an increasingly globalized world, not everyone is "connected."

Web-sites providing health information or promoting good health therefore offer a new form of medical gaze; they allow the lay user to access knowledge, though of course questions can be asked about the quality of information imparted. As Parr notes, we can interpret such sites, particularly those produced by government agencies, in Foucauldian terms, as vehicles for "disciplining" the body; in effect, ensuring that people look after themselves, for their own and the common good. But these sites are not necessarily interactive, in the sense that they permit groups of users to ask questions or solicit advice. Here, "new geographies of medical knowledge/power, health, and technological embodiment are argued to be emergent" (Parr, 2002a: 76). In order to explore these geographies, Parr entered chat rooms used for communication (virtual travel) by people with multiple sclerosis (MS). Here, users share experiences, talk about their bodies, and gain both formal and informal advice and support online. Her data indicate that the "social isolation which many of those with MS feel in 'real space' is compensated for in virtual space" (Parr, 2002a: 88). Further, these uses may be "a form of resistance to medicalization and medical power where subjective, embodied experiences rather than conventional medical knowledges are privileged" (Parr, 2002a: 89).

Since Parr undertook her research, Web 2.0 social networking technologies (such as MySpace and Facebook, as well as virtual worlds such as Second Life) have developed that permit new forms of social connections and architectures to emerge, expand, and dissolve (see also Box 2.1 below). Exploring the geographical dimensions of multiple social networks, as well as their consequences for health, will prove a rich area of research in the years ahead. For example, such technologies can permit otherwise excluded groups, such as Aboriginal women (Hoffman-Goetz and Donelle, 2007), to operate in online neighborhoods and thereby access health-related social support that might not otherwise be available.

Concluding Remarks

We hope it is clear already from this first chapter that there are "geographies" of health. The five vignettes have been chosen to illustrate a variety of approaches to the subject, a set of different perspectives that can be brought to bear on the study of health and place. Some look to be more obviously geographical, in that they produce mappable patterns, whether of historical or more contemporary disease or illness. The geographical content of others may appear less obvious; nonetheless, location, and space and place figure prominently in all. In the next chapter we shall set out in more detail what these different perspectives entail. We do this by laying out some of their characteristics and by describing some further case studies. In so doing, we hope to persuade the reader of the richness of approaches to the subject, as well as laying some groundwork for considering particular themes in subsequent chapters.

Further Reading

There are several relevant journals that anyone interested in geographies of health can usefully consult. Of these, we draw attention to: *Health & Place*, *Social Science & Medicine*, *Journal of Epidemiology and Community Health*, and *American Journal of Public Health*. All of these have good international coverage. In addition, other epidemiological and more "mainstream" health/medical and geographical journals carry relevant papers from time to time: the *British Medical Journal*, *The Lancet*, *Public Health*, *Journal of Public Health Medicine*, *New England Journal of Medicine*, and the *American Journal of Epidemiology*.

As far as the present chapter is concerned, an excellent discussion of some of the conceptual issues underlying health research may be found in Aggleton (1990). Also very highly recommended is the series of books on Health and Disease produced by the UK Open University; the introductory chapters in Davey and Seale (1996) and Seale and Pattison (1994) are worth reading, while the volume edited by McConway (1994) provides a superb accompaniment to both the present and the following chapter.

Jones and Moon (1987) is the classic text on the geography of health; its first chapter covers some of the introductory material dealt with here, while the second chapter looks in detail at epidemiological principles and sources of data. Other key introductory texts and collections of essays are: Curtis (2004), Meade and Earickson (2000), Kearns and Gesler (1998), and Butler and Parr (1999).

If new to geography, you could usefully start with Haggett (2001). Good overviews of contemporary human geography are: Cloke et al. (1999), and Daniels et al. (2004).

For further, detailed, examination of concepts of distance and space see an earlier book by one of the authors: Gatrell (1983).

On "place" see: Cresswell (2004), or the integrated collection of essays in Massey and Jess (1995), especially Gillian Rose's chapter. For a detailed study of the impact of the Chernobyl disaster see Gould (1990).

Chapter 2

Explaining Geographies of Health

In the first chapter we described in some detail five case studies or "vignettes," examples of work that can be bracketed under the broad heading of the geography of health. Their subject matter differs from one to another, but the difference we wish now to highlight is that of approach to explanation. Put another way, underlying each is a different philosophical stance, and while these positions were implicit rather than overt we want now to render them explicit. It is these different philosophical approaches that lead us to speak of "geographies" of health.

What the studies had in common, however, was a "problem" deemed worthy of investigation, an issue that required some understanding or reflection. An explanation of why there was place-to-place variation in disease incidence was sought in one case; an understanding of how people managed long-term care in the home was required in another. Our task in this chapter is to say more about these different kinds of explanation. We do this by setting out the key features of each, but will also introduce other examples. This will serve to fix ideas, but we believe that all of the examples considered here are of interest, and will thus, we hope, serve to further draw the reader into the study of the subject, in all its variety.

We want to end these introductory remarks by saying – if it were needed – that there is no single correct philosophical stance or mode of explanation. We shall comment on both the strengths and weaknesses of each. And while we want to discourage readers from subsequently pigeon-holing into one or other category each and every piece of geographical research in this field that they subsequently encounter, we do believe it is a useful exercise to reflect upon the theoretical framework, either explicit or implicit, that is employed.

Positivist Approaches to the Geography of Health

In the first chapter we used research on asthma in New York City because it encapsulates very neatly the way in which many geographers approach the geography of health. The investigation began, in effect, with a map of hospitalization rates for childhood asthma.

The map is a visual representation of something that had been medically defined; its value depends on an accurate recording of all cases of the disease in question. It then becomes, in this and similar studies, a tool for asking questions. Is the spatial arrangement of rates random or not? What explains the set of above average rates that seem to cluster in parts of the study area? Additional data are sought in order to examine if the rates are spatially associated with one or more variables (covariates). Such associations are tested statistically, by using correlation analysis to test hypotheses and account for variability in incidence.

Positivist explanation

The study of asthma by Corburn and his colleagues is a "scientific" study, in the sense that it adopts the methods of natural science, looking for order or spatial patterning in a set of data. The study relies on accurate measurement and recording and searches for statistical regularities and associations. It emphasizes, via mapping and spatial analysis, what is observable and measurable; it has many of the characteristics of a *positivist* or *naturalistic* approach to investigation. In a health context, such approaches seek to uncover causes or "aetiological" factors, though usually the best that can be established is strong association rather than "cause." A classically positivist account would have, as its end goal, a search for laws, though weaker versions strive simply to make generalizations.

The concern in positivist approaches to the geography of health is usually to detect areal pattern or to model the way in which disease incidence varies spatially; people with the disease only appear as numbers that compose spatially varying disease rates. Location and spatial arrangement matter – indeed, these are the crucial variables in *medical* geography – but "place" is incidental. Certainly, the individuals do not speak to us; how children cope with asthma in daily life, and what they think might cause them, but not their friends, to have the illness is not a matter for "scientific" investigation. We emphasize that this does not mean that a positivist does not collect individual-level data. Far from it. Individual-level data may well be collected, on both the disease and other characteristics of such individuals (perhaps their age, sex, residential history, family income, and so on), and used to sort out what factors discriminate those with the disease from those without. This may be done either via survey work – what we can refer to as "social positivism" – or in the laboratory, where "bio-medical positivists look for causes in micro-organisms and in anatomical and physiological abnormalities" (Aggleton, 1990: 75). But a positivist medical geography typically involves mapping disease data and then striving to describe and explain the spatial distribution.

In health research in general, a positivist approach also involves frequently adopting a biomedical perspective. Here, the body is seen as a machine that may not be in good working order and needs "mending." What matters is to investigate – and, for the geographer, to investigate in a spatial setting – specific diseases that have one or more specific causes. The individual is a rather anonymous person whose features and characteristics can be ticked off on a check-list. Critics of this perspective argue that it is reductionist (the individual is "reduced" to a collection of body parts and behaviors); even if "lifestyle" factors (such as diet, stress, and social support) or the effects of air and water pollution are recognized, these are attributes to be attached to individuals and used in a statistical analysis. Positivist approaches, in general, rely on the use of quantitative, usually statistical, methods, often sampling from a wider population and seeking to generalize from the sample to the population. A corollary of this is, where possible, to have the sample as large as possible, since this strengthens the conclusions.

That location and distance are key variables in a positivist geographical approach is seen not only in the study of health and disease but also in the study of health care and its delivery. Rational planning demands the organization of services on an efficient basis, such that the costs of overcoming distance are, in the aggregate, minimized. Here, too, the search for order is paramount, a feature which some writers have traced back to the philosophy of the Enlightenment in the late-eighteenth and early-nineteenth centuries, where the Enlightenment or modernist "project" was to impose order and to eliminate disorder.

Further examples of positivist approaches

To illustrate further the broad range of essentially positivist accounts we draw on a number of other studies. First, we examine some research on AIDS conducted in the developed, rather than the developing, world. We then consider some of the large body of work on disease spread or diffusion. All are studies drawn from a potentially large set that adopt essentially positivist approaches. We then examine one study that is concerned with health care planning in a geographical context.

Gould and Wallace (1994) have researched the spread of the human immunodeficiency virus (HIV) through mapping the cumulative incidence of AIDS cases across the USA and also analyzing the number of cases reported in the New York region. At the national scale, data for 1982 (when AIDS was becoming recognized as a real problem) show concentration in the major urban agglomerations, followed in later years by spread to smaller cities via a process known as *hierarchical diffusion* in which the patterns appearing in subsequent maps are structured by the flows of people, along the major transport arteries, initially among the largest urban centers and then to smaller towns. Later, more local processes of *contagious diffusion* appear to operate, in which physical proximity matters more than the leapfrogging of geographical space and spread down the urban hierarchy. This contagious diffusion was "driven in large part by the extensive daily commuting fields" (Gould and Wallace, 1994: 107). The argument is that since the main form of HIV transmission is via sexual or needle contact, these personal contacts (for which data are rare, or impossible to obtain) "reflect the massive flows of interaction at much greater geographic scales" (Gould and Wallace, 1994: 107), although later they speak of the spread as being "forced" by flows of commuters. The importance of commuting flows is confirmed by plots of AIDS rates against an index of commuting for the 24 boroughs and counties of the New York Metropolitan Region (see Figure 2.1).

Gould and Wallace anticipate one criticism of their research – the neglect of the human actor or voice – when they acknowledge that some researchers might prefer more autobiographical accounts of disease spread and that others might feel that poetry and other mediums of expression can be more illuminating. "But in the same way that the scientist, *qua* scientist, can say little about the individual, so the poet can write little that will help us formulate public policies . . . Sonnets, no matter how empathetically wrenching, do not stop viruses, and do not help us to look ahead" (Gould and Wallace, 1994: 105). "[B]ecause we pose questions ultimately of cause and effect, or in other words questions of mechanism,

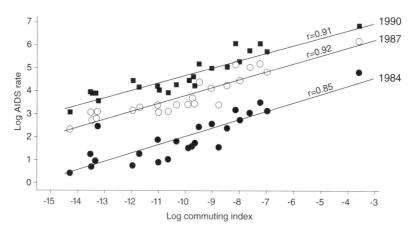

Figure 2.1   Relationship between AIDS rates and commuting in New York City region

'social physics' is inevitably at the heart of any scientific investigation" (Gould and Wallace, 1994: 105). In their avowedly naturalistic approach there is no need for those affected by the disease to speak out. Nor is there much space in their account for consideration of the broad structural societal forces that may also cause disease spread, though they do refer to the social marginalization of those intravenous drug users living in unrelieved poverty. The study is a classic example of a positivist geography of health, one that has established a convincing association between disease incidence and an aggregate social variable. Yet, although in this large metropolitan region flows of commuters are very highly correlated with AIDS rates, the causal mechanisms are missing. While the authors refer to a newspaper article which suggests that 70% of the clients of Manhattan prostitutes are suburban commuters, there needs to be much more empirical work – not necessarily all positivist in conception – that illuminates this and other possible causes.

Other geographers have worked more on some of these possible causal mechanisms, via both detailed historical research and more formal modeling of the disease diffusion process, as instanced by the series of monographs and papers produced by Peter Haggett, Andrew Cliff and their colleagues over many years. As one example, consider the spread of influenza in Iceland (Cliff et al., 1986). Iceland is a quite self-contained island setting, with superb medical records going back to the late-nineteenth century, all of which allow for detailed geographical description and analysis. At least 30 well-defined influenza epidemics, some involving several thousand cases, others a few hundred, have affected the country since 1900, the population having grown from about 78,000 to 210,000 in 1970 (when the authors terminate their study). In early decades the role of shipping is highlighted, with the 1918 epidemic triggered in particular by one trawler arriving in Reykjavik from Copenhagen. Communication bans and the quarantining of "infectives" were soon imposed, with the result that the epidemic was contained successfully in south-west Iceland. In the 1937 epidemic, spread was via both hierarchical and contagious processes. From Reykjavik it spread down the urban hierarchy, via local shipping movements and, once established,

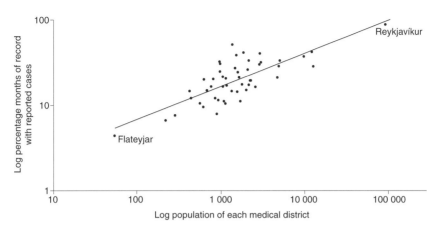

Figure 2.2 Relationship between number of months with reported influenza and population size in Iceland, 1945–70

moved outwards from towns into adjacent rural areas, facilitated by social and business gatherings (such as a farmers' conference). In 1957–8 a new strain of the virus ("Asian influenza") arrived from overseas via air as well as sea routes, this time reaching the eastern coast. For the authors, a mixed hierarchical-contagious diffusion model seems generally applicable to all the major epidemics they consider.

Cliff et al. (1986: 168–205) identify a number of "statistical regularities" in both the temporal and spatial spread. One, for example, concerns why influenza appears in Iceland as a series of regular epidemics rather than being endemic (occurring continuously). The explanation lies in the sparsity of population. If we plot the percentage (logarithmically transformed) of months between 1945 and 1970 in which medical districts report influenza, against population size (also logarithmically transformed), we see a quite strong linear relationship (see Figure 2.2). This is described by the following regression model:

$$\log T_i = 0.85 + 0.381 \log P_i$$

where, for the i'th district, T_i is the percentage of months with reported cases and P_i is population size. If we give T the value of 100% (that is, a district has cases every month) and solve the equation for P, we obtain a value of about 110,000. This is then the *threshold* for endemicity. Even the capital of Iceland, Reykjavik, has too small a population to sustain the disease continuously.

Cliff et al. (1986: 207–59) go beyond these statistical descriptions to build mathematical models of the diffusion process, with the longer-term view of forecasting the future spread of the disease. Several models are explored, the simplest of which are based on an accounting framework, expressing the number of people infected at any point in time as a function of: the number of infectives in the previous time period; newly emerging cases; those removed by being no longer infective; those "arriving" (via birth or immigration); and those "departing" (via death or emigration). This can be expressed graphically (see Figure 2.3) where, if we plot the proportion of the population that is susceptible (S)

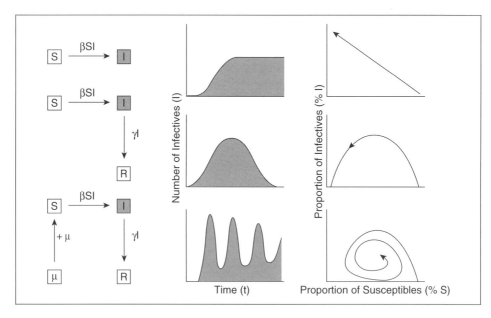

Figure 2.3   Relationship between susceptible and infected population in a simple diffusion model

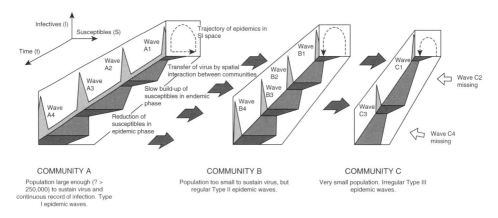

Figure 2.4   Spread of a communicable disease in communities of different sizes

to the disease (that is, capable of being infected) against the proportion that is infected (I) we can trace the simple dynamics of disease spread according to different modeling assumptions. These dynamics can be visualized in another way (see Figure 2.4). Above the population threshold for endemicity there is a continuous persistence of the disease, but at a low level. This provides a "reservoir" of infection that triggers an epidemic when the susceptible population has grown (new births, immigration, or lack of people immunized). As the proportion of the population that is infected increases so the proportion that is susceptible diminishes; this gives rise to the relationship between S and I shown in Figure 2.3. Where the size of the population is too small and the disease is not endemic, the epidemics become less frequent and less intense (Haggett, 2000: 21–25).

Diffusion models for more than one district assume that the probability of contact between an infective in one district and a susceptible in another depends upon the distance separating them. However, the models are "data-hungry," ideally requiring daily data since measles has a serial interval (the time between the onset of symptoms in one person and in another person directly infected by the first) of only 4–5 days; typically, only monthly data are available.

These studies of spatial diffusion form, along with others, a substantial body of knowledge relating to the spatial and temporal spread of infectious disease. They involve careful recording of cases, statistical description of the data, and subsequently mathematical and statistical modeling of the process. Attempts may then be made to forecast disease spread, with the longer-term goal of devising strategies to control this spread. But those infected are anonymous individuals, aggregated into areal recording units, and it is left to other approaches to let those who are infected by, or affected by, disease spread have a voice.

There are positivist approaches to examining the provision and delivery of health services. Given limited resources, where should we locate health services? How should patients be allocated to particular health centers or hospitals? The set of tools used here are known as *location-allocation* methods (see Thomas, 1992 and Bailey and Gatrell, 1995, Chapter 9

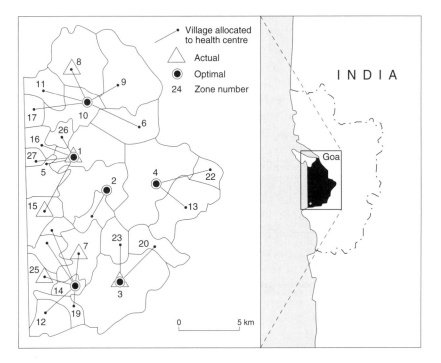

Figure 2.5　Optimal location of health centers in West Goa

for an introduction). In a developed world context, Smallman-Raynor and his colleagues (1998) have demonstrated the usefulness of location-allocation modeling in planning the location of cancer units in a part of central England. In the developing world Hodgson (1988) has emphasized the hierarchical nature of the planning problem; in other words, the need to locate both basic health services, but also centers offering a wider range of health and medical care. Particular location-allocation models propose different location criteria, typically minimizing the aggregate distance traveled from place of residence to the health center. The sizes of the centers are important, however, in that they may be taken to represent, albeit in a rather crude way, their "attractiveness;" other things being equal, people will travel further to a larger center. Models are therefore needed that maximize net benefit to the patient, locating a set of health centers in such a way as to ensure that they are as close as possible to such patients. Hodgson applies a hierarchical model to the location of health centers in a region of Goa, India (see Figure 2.5).

Social Interactionist Approaches to the Geography of Health

We have seen how positivist accounts focus on the observable, the measurable, and the generalizable. Individual characteristics might be recorded, but nowhere is there a concern for individual *meaning*. In a health care planning context, the attractiveness of health centers is represented by easily available data on bed spaces, or number of health care workers, for example, rather than by the more intangible features (such as friendly staff) that are likely to be important to people. Consequently, other researchers emphasize what is less readily measured and quantified; the subjective experience of health and illness. We shall, following Aggleton (1990), refer to these approaches to the study of health and illness as *social interactionist* (some use the term *social constructionist*). They are so called, because meanings are *constructed* out of the interactions (which may be conversations or encounters) that we have with each other in everyday life. The theory of social interaction stems from the seminal work of sociologist Erving Goffman (1959). Geographers have sometimes referred to these perspectives as "humanistic," since they address implicitly human beliefs, values, meanings, and intentions. In a health context, such approaches see people not simply as collections of (possibly) diseased body parts. Nor do they see people as merely passive recipients of knowledge about health and health care; rather, people are continually engaged in the construction of such knowledge. A corollary of this is that the views of ordinary people (referred to as "lay" or commonsense views) have as much status as those of the health professional. These embodied knowledges add substantially to our understanding of the meanings of health and illness ascribed by individuals.

Social interactionist explanation

In social interactionist accounts, therefore, the emphasis is on the meaning of the illness or disease to the individual and the task for the researcher is to uncover and *interpret* these meanings that make it "rational" to act in a particular way; in other words, to see things from *their* point of view. For example, not going to have a child immunized may be irrational to the health professional yet perfectly logical to the parent.

By extension, social interactionist approaches study small numbers of people; geographical versions will typically study small communities or neighborhoods, rather than studying large areas. This is because the experience of place is more important than the accurate recording of large numbers of locations, or the pigeon-holing of people into a fixed set of areal units. The methods used are typically qualitative rather than quantitative, and the ultimate goal is empathetic understanding and explanation rooted in the social, rather than the natural, world.

Such approaches are, as one might expect, subject to criticism from positivists, who argue that the verifiability of results is hard to obtain, and any conclusions drawn from small numbers are difficult to credit; for a positivist, the objective matters more than the subjective. For others, the emphasis on "human agency" means that, like positivist accounts, the social interactionist neglects wider structural influences on health.

In the example used in the previous chapter on area effects on smoking, for instance, Stead and colleagues (2001) did not focus on smoking rates, or models to predict smoking; rather, they wanted to understand the role of *social exclusion* in smoking behavior, particularly as linked to socio-economic status. For them, a limitation of a positivist approach to science is that it tends to merely *describe* rather than *explain*. They want a greater *understanding* of the social processes involved in shaping health-related behaviors and outcomes.

Further examples of social interactionist approaches

Another example of research that can be bracketed under this heading is that on the experiences of local residents living near waste disposal sites; specifically, landfill sites and an incinerator in southern Ontario (Eyles et al., 1993). This work parallels a more classical epidemiological survey (Elliott et al., 1993), uncovering the psychosocial effects via a series of in-depth interviews and focus groups (see Chapter 3 for further discussion of these techniques). Individuals were selected from among those who had, and had not, expressed concern about the waste disposal facilities, giving a total of 39 interviewees over the three sites. Only one interviewer was used (thereby eliminating one source of possible bias), and the interviews (lasting between 20 and 90 minutes) were taped and transcribed. A checklist of topics to be covered included: attitudes to home, environmental contamination and quality of life, and more specific feelings about the local facility. The results depend on whether the site is a landfill or incinerator. Those living near the latter complained about air pollution, with some worrying about short-term health impacts and others about longer-term damage, though since the incinerator is located in a highly industrialized area it is difficult to separate out the effects of the facility from those of other sources of environmental contamination. Those living near one landfill site expressed few concerns about its impact on water quality or human health; for those close by the problems were of debris and noise, while for those further away it was the additional traffic generated that caused concern. At the second (at the time, planned) landfill site there were more concerns about possible contamination of water supplies. But, to the authors' surprise, associations between environment and health were not so very prominent. Rather, anxieties were latent and concerns were more about the lack of control over siting decisions than about likely health impacts.

The authors justify their qualitative study as being complementary to the epidemiological companion paper (Elliott et al., 1993), arguing that it enriches the survey results (Eyles et al., 1993: 810). A positivist would question whether a very small number of interviewees is representative of the local populations, and worry about whether their concerns can be pinned down as due exclusively to the waste sites, rather than to some other possible sources of environmental pollution. This is countered by the argument that qualitatively based research such as this does not set out to make generalizations.

Cornwell (1984) offers a detailed case study of ordinary people's accounts of health and illness, and their experience of health services, in Bethnal Green, part of East London. Cornwell's approach is anthropological and ethnographic, and she does not see "health" as a field around which one can put sharp boundaries. To that end, she is as much concerned with understanding aspects of daily lives that are, on the surface, less directly health-related: housing, work, and family life, for example. Her method is to conduct in-depth interviews with 24 people, drawn through a "snowball" process whereby the first contact spawned subsequent names to be recruited to the study. The interviews, based around a core schedule of topics and conducted over numerous sessions with the same people, provide a level of detail that no survey could ever hope to match. Cornwell conducted multiple interviews with her informants, since a single one might reveal only "public" accounts (in a sense, what the respondent thinks the interviewer might wish to hear), whereas a succession of interviews begins to tease out the "private" accounts, the deeper thoughts and feelings emerging from the respondents. The interviews were, as far as possible, led by respondents' own concerns, since one of the dangers of survey research (and some interviews) is that the relationship between interviewer and interviewee is very hierarchical, with the "expert" merely gathering as quickly as possible the data needed to advance the research. As Cornwell says, the sample is not statistically representative, but is "typical" in the everyday sense that the people's lives "faithfully reflect the history of social and economic life in East London over the past eighty or more years" (Cornwell, 1984: 1). As with Eyles' work, this marks a very clear separation from a positivist approach.

The distinction between public and private accounts is important, and often conflicting. Public accounts emerge from answering questions; private accounts emerge from the telling of stories about personal experiences. For example, public accounts stressed the almost romantic sense of place – the strong sense of community and high quality of social relationships. By contrast, private accounts emphasized the negative aspects – the fights, the arguments, and lack of concern for others. Similarly, public accounts of family life stressed the "unity" of family life, the virtues of hard work and the devotion of women to family welfare; private accounts uncovered the rifts, the disharmony, and the stresses of living together. This emphasis on family and working life is important, since it acts as the context for accounts of health and illness.

As far as public accounts of health and illness were concerned, the respondents wanted to assure Cornwell that illnesses were "real." They also made light of their own ill-health, being reluctant to be labeled as someone with poor health. Good health is morally worthy, while illness is not, unless it bears the stamp of official approval by diagnosis, in which case it is beyond the power of the individual to control it. Private accounts implicated poor working environments, such as the problems of deafness and discomfort caused by driving older lorries (trucks) and the back problems caused by handling heavy goods. Public theories of disease causation tended to be the interpretations of "medical" opinion by ordinary people; they frequently suggested that disease and illness were beyond individual control and that the person affected was not to blame. Private theories were more complex, suggesting multiple causes of disease and illness drawn from events throughout life.

As a final example of this style of research, Krenichyn (2006) explores the meaning of an urban park for women in Brooklyn, NYC, in the context of investigating facilitators and barriers to physical activity for women living in a densely populated urban area in the USA. "Physical activity is becoming a growing public health concern: the current numbers of people in the USA engaging in regular exercise are so small that our sedentary lifestyles and related health problems are now characterized as epidemic" (Krenichyn, 2006: 632). Women are particularly underrepresented among those who are physically active. The benefits of physical activity to the prevention of chronic disease (cardiovascular disease, stroke, diabetes, some cancers) are well documented. So, too, are the benefits of exercise for mental health (e.g. anxiety, depression). What Krenichyn discovers in her interviews with women who undertook physical activity in an urban park was the meaning that place had for them.

Prospect Park is a 526 acre urban park in Brooklyn. It was designed and built in the 1860s by the well-known landscape architects, Frederick Law Olmstead and Calvin Vaux. The primary purpose of the park was "to preserve nature within the industrialized city and provide opportunities for healthy recreation, socialization, and spiritual elevation of the masses" (Krenichyn, 2006: 631). Krenichyn found that women did indeed experience these feelings in using this urban park for exercise. It was a place to "get away" right in the middle of the city. It was free of traffic and people other than those also seeking respite and exercise. They felt safer in the park than in the streets of their dense urban borough of New York and reflected on the park as a thing of beauty, expressing their appreciation for the green space and the sensory experiences of the change of seasons. As one respondent puts it:

> The park is, you know, just being outside and surrounded by trees. And the smell – you know, the greenery, and flowers when the flowers are in bloom, but even just grass. And trees. And, you know, all that stuff is so great compared to being . . . in the urban setting that we're in most of the time. It's just so wonderful, you can actually kind of pretend that you're not in the city (Krenichyn, 2006: 636).

Women in this study spoke not just of the aesthetic aspects of the park, but also its therapeutic and spiritual qualities (see Box 1.2 above); indeed, Olmstead's original goal in designing the park was to create an escape from everyday stimuli that would be a "place for the restoration of mental capacities." The women interviewed also described the park as a socially intimate place where their activities were enriched by the presence of others, given the park was a place for others – family, meeting friends, encountering strangers (e.g. other runners) on a regular basis.

Structuralist Approaches to the Geography of Health

The work by Hunter that we considered in Chapter 1 sought to identify some of the political and social factors that explain high rates of HIV in South Africa. As we show in this section, related studies suggest that the underlying causes of disease are embedded in political and economic systems. Explanations are not to be sought at the individual level – for example, the kinds of "unhealthy" behaviors they adopt. Instead, it is the broader social context that matters. As Turshen (1984: 11), has it, sickness lies not in the body but the body politic. Because of the stress on these macro-scale social, political, and economic structures, this style of approach is often referred to as *structuralist*, or alternatively as a *political economy* perspective.

Structuralist explanation

Structuralist approaches derive much of their impetus from Marxist theories of oppression, domination, and class conflict, where inequalities are embedded in society. Such theories take a variety of forms but, put simply, they propose that economic relations and structures underpin all areas of human activity, including health and access to health care, and further that the economic "determines" the social. It is clear from this, therefore, that such a perspective is antithetical to that of social interactionism; human agency is absent from classically structuralist accounts.

Some authors offer an explicitly Marxist interpretation of health and health care, especially in the context of late-nineteenth- and early-twentieth-century colonialism (Ferguson, 1979). As we shall see below, the emphasis in these contexts was very much on curative medicine rather than on preventive (public) health. There are two reasons for this, according to Ferguson. One is that the expertise developed by the doctor over several years' training produces a commodity (medical skill) that has an exchange value which is realized by treating those able to afford the private care on offer. The second point he makes is that since poverty is the main cause of ill-health, and poverty results from capitalism, there is little incentive among those controlling, and working in, the health care system to attack these root causes, since more money is to be made from providing medical cures than reducing poverty and preventing disease and ill-health in the first place. Whether in the developing or the developed world, biomedicine emphasizes cure rather than prevention and is keener to promote "high technology" approaches to health care, even in quite rural developing countries where this is wholly inappropriate. Medicine thus serves to perpetuate social inequalities and to widen the gap between the rich and poor; it does nothing, according to the political economist, to reduce these disparities. Health improvements are contingent on overcoming such dependency and on transforming the economic system from capitalism to socialism. Unfortunately, there is scant evidence that those living in many former socialist states had very different health outcomes, or that inequality was less than in capitalist societies.

There are, however, other deep structures embedded in society that are based on conflict and power relations, of which the most obvious is the role played by male power (patriarchy) in structuring women's health. A broader reading of structuralist explanations would therefore want to see a wide-ranging emphasis on conflict, or power relations, whether this be between social or ethnic groups, between men and women, between people with different sexual orientations, between those owning the means of production and those employed as laboring classes, or between societies. Some writers (Gerhardt, 1989; Stainton Rogers, 1991) refer to *conflict* or *dominance* theories, embracing all forms of economic, imperialist, and patriarchal domination. As a result, the studies examined in this section look at links between imperialism and both the spread of disease and the nature of health care delivered under colonial rule, before turning briefly to work that is shaped by gender-based conflict.

Further examples of structuralist or conflict-based approaches

In tracing the historical roots of Marxist approaches to understanding the social dimensions of health, Waitzkin (2005) reminds readers of the contributions made by Friedrich Engels in the nineteenth century, and Salvador Allende in the twentieth century. As he puts it, while "numerous studies have described occupational and environmental diseases, they generally have not considered the structural contradictions in the capitalist organization of production that foster illness and early death" (Waitzkin, 2003: 21); these writers did, either explicitly or implicitly. In Engels' famous work *The Condition of the Working Class in England*, published in 1844, his "theoretical position was unambiguous. For working-class people, the roots of illness and early death lay in the organization of economic production and in the social environment" (Waitzkin, 2003: 23). Engels described the effects of environmental pollution, poor housing conditions, poor nutrition, and poor working conditions, outlining the consequences for tuberculosis (then known as "consumption"), malnutrition, industrial accidents, disability and eye problems. All these issues remain of concern to epidemiologists over 150 years later; the critical point here is that, for Engels, "the solution to these health problems required basic social change – limited medical interventions would never yield the improvements that were most needed" (Waitzkin, 2003: 27).

Allende is best remembered as the President of Chile between 1970 and 1973, overthrown and murdered in a political coup orchestrated by General Pinochet. Yet he was also a trained doctor, deeply interested in politics and social justice from an early age, and the author of a book on public health published in the 1930s. As Waitzkin describes, the book argues that piecemeal health reforms are useless without broad structural political change and, specifically, the end of capitalist and imperialist exploitation by multinational corporations. In an echo of much contemporary thinking in public health, he argued that "health policy must transcend the health sector alone" (Waitzkin, 2003: 42) and that (as Waitzkin's translation has it) "the solution of the medical-social problems of the country would require precisely the solution of the economic problems that affect the proletarian classes."

We saw earlier how one kind of approach to the spread or diffusion of disease was based on an essentially positivist approach. However, to explain the impact of diseases such as smallpox on native South and Central American populations in the sixteenth century we need to look at the practices of Spanish and British colonialists.

Many epidemic diseases (such as smallpox, measles, yellow fever, typhus, and influenza) can be associated with the Spanish conquests of Peru and Mexico in the early-sixteenth century, and some authors (Brothwell, 1993) consider that they did at least as much as the Spanish military to dismantle the Aztec and Inca empires (see Figure 2.6). Such populations had no natural immunity to the diseases. One set of population estimates for central America suggests that while there were 25.2 million living in 1519, the slaughters wreaked by conquistadors in 1521 and, later, the 1531–2 and 1545–7 smallpox epidemics, had reduced this to 6.3 million by mid-century. The population had dropped to just one million

Figure 2.6   Sources of epidemics in Americas during the sixteenth and seventeenth centuries

by the early 1600s. After the collapse of the Aztec state following Cortes' 1521 campaign, smallpox diffused along trading networks down into South America, devastating the Inca population in 1524–5. One hundred years later, the Andean population had dropped to just 7% of its level in 1524 (Watts, 1997: 91). Smallpox and other viruses were spread widely by those recruiting slave labor to work in silver and other mines, where appalling living conditions encouraged disease spread.

Later, and much further north, the Pilgrims arriving in 1620 were able to observe that "the good hand of God favoured our beginnings . . . in sweeping away the great multitudes of the Natives by the Small Pox" (in Watts, 1997: 93), while in 1763, in order to wipe out further resistance among the Ojibway native Americans around Lake Superior, the British Army commander (General Amherst) oversaw the distribution among them of blankets infected with smallpox. The settlers and military in North America, like the Spanish conquistadors, were less affected, having acquired some exposure, and therefore lifelong immunity, to milder endemic forms of the virus back home.

It would be naive to imagine that native populations in the Americas lived in idyllic conditions, free from any illness. We simply have little or no record of what epidemics swept these populations before the colonialists arrived (Arnold, 1988). But what is certain is that the infrastructure imposed by colonialism, new trade routes, and communication networks aided the spread of micro-organisms and disease vectors that transmitted other diseases. Until the emergence of this infrastructure, poor communications and the "friction" of distance served to quarantine the diseases. And since rates of spread of infectious disease are greater in more densely populated areas, the diffusion of diseases such as smallpox among native Americans was aided not only by trade and labor recruitment, but also by forced resettlement into larger population groupings – whether mining compounds, plantations, or missions – for the purposes of more effective administration and use and control of the labor force. Arnold (1988) speaks of colonialism in this context as "biological warfare" or, at the very least, a major health hazard for the indigenous peoples of native America.

In a detailed study of the *political ecology* of disease and ill-health in a single East African country, Tanzania (formerly Tanganyika), Turshen (1984), too, shows how the health of the indigenous population deteriorated in the presence of colonial rule and how the economic transformations brought about by colonialism and capitalism impacted on disease. For Turshen, as for others (Ferguson, 1979), the causes of disease are rooted in political and economic systems rather than individual lifestyles and behaviors. However, we dwell less on these aspects – some of which we have already examined – and more on the health care system that emerged under colonial rule (initially, under the Germans between 1884 and 1918, then under Britain until 1961) and, following independence in 1961, under the new socialist government led by Julius Nyerere. This provides a useful contrast with more positivist geographical analyses of health care systems.

Before the "scramble for Africa" in the later-nineteenth century, Christian missions had been established in East Africa, many of which introduced western medicine. This emphasized the cure of disease rather than preventive medicine, and there developed a clear association between medicine and conversion. Archdeacon Walker wrote in 1897 that: "I regard the medical work from its missionary aspect . . . I consider how far it is likely to aid our work, not how much suffering will be relieved" (quoted in Ferguson, 1979: 319). After the partition of Africa, Germany took control of what was known as German East Africa (later, Tanganyika) and established army hospitals to serve local garrisons. Under both German and early British control separate facilities were created (either separate hospitals or separate wards) for Africans, Asians, and Europeans, again with the emphasis on curative and surgical medicine. The contrasts are apparent from a description by the Principal Medical Officer in 1920. The European Hospital in Dar es Salaam held 50 beds and had "a separate maternity section, well-fitted X-ray room . . . and room for the examination of eye cases, spacious operating theatre, outpatient department and quarters for the nursing staff and Medical Officer." The Sewa Haji Hospital, built for Africans and Asians, and also with 50 beds, was "a curious rambling collection of buildings of which the administrative block is the outstanding feature" (in Ferguson, 1979: 326). The ratio of beds to population served was 1 to 10 in the former and 1 to 400–500 in the latter.

These health services were superimposed on the traditional health care system, essentially a group of practitioners whom the earlier missionaries had sought to discredit, since their traditions ran "counter to the belief that Victorian civilization was the acme of human achievement" (Turshen, 1984: 145). Turshen sees the active participation of local communities in the healing process and prevention of illness as a possible political threat to colonial autonomy and the social control that western medicine provided; this was another reason for discouraging traditional practices. Until independence in 1961 the emphasis remained on cure rather than prevention, especially among the urban labor force. Among women and children living in rural areas little attention was given to meeting nutritional requirements, though this was not the case among laborers; an official government report in 1936 confirmed that "under-nourished natives cannot be expected to, and are not capable of, working really efficiently" (in Turshen, 1984: 151). The emphasis given to an urban-based curative system thus produced a very uneven distribution of health care.

Paul Farmer's research on AIDS in Haiti uses a structuralist or political economy perspective (Farmer, 1992). He attempts to relate large-scale (both historical and contemporary) events to the lived experience of ordinary village people and their families and neighbors. His research has involved spending several years among people living in one village, Do Kay, a community of about 1000 people that had originally comprised refugees displaced in the 1950s by a hydroelectric dam. The experiences of three people living with (and dying of) AIDS are examined in depth. One, for example, a man called Dieudonné, has a structuralist explanation of the illness: "What I see is that poor people catch it more easily. They say the rich get [AIDS]; I don't see that. But what I do see is that one poor person sends it on another poor person" (Farmer, 1992: 12). Quoting others, he suggests that "the poor, in their popular wisdom, in fact 'know' much more about poverty than does any economist. Or rather, they know in another way, in much greater depth" (Farmer, 1992: 262–3). But the misfortunes of ordinary "lay" people are linked to those of the communities they live in as well as the appalling economic position of Haiti itself.

Haiti's place within a West Atlantic economic system whose hub is the United States has played a major role in shaping the epidemiology of HIV and AIDS in the country. These economic linkages have included the tourist industry, and other qualitative work has demonstrated that impoverished "beach boys" have been infected by, and in turn infect, those with whom they have sexual contact. Economically driven male prostitution, involving a largely North American clientele, has played a pivotal role in the spread of HIV to Haiti (rather than, as some have claimed, being spread from a Haitian "source" to North America). Indeed, Farmer (1992: 261) goes so far as to suggest: "Given that unequal relations between the Caribbean and North America have contributed to the current epidemiology of HIV, an analogous exercise leads to a somewhat analogous observation: the map of HIV in the New World reflects to an important degree the geography of U.S. neocolonialism." HIV has diffused "along the fault lines of economic structures long in the making" (Farmer, 1992: 9). These deeper structures cannot be observed but are nonetheless real. Farmer's work is, therefore, couched firmly within a structuralist framework.

Other structuralist or political economy-based research indicates the role played in HIV risk by the structural adjustment programmes imposed on non-industrialized countries by the International Monetary Fund and World Bank (Lock et al., 2006: 70). Such countries had sought loan agreements in order to pay off debt, but such agreements required them to adopt economic measures that created new problems. For example, in Malawi (Craddock, 2001) fertilizer subsidies to farmers were withdrawn in the late 1980s and some farmers were unable to grow enough maize to sell, thus reducing family income; the poorer female heads of household may be forced to turn to commercial sex work. The "inequitable power relations within which much commercial sex takes place leaves women with little leverage to insist upon the use of condoms" (Craddock, 2001: 50).

Another "conflict-based" approach is a *feminist* perspective which argues that historical relations between men and women are basic to an understanding of society; power is divided unequally, with men the dominant group. Health research in general has made considerable use of a feminist perspective (for instance, in looking at women's experiences of antenatal care), and feminist perspectives in health geography are increasingly important. But such research cannot possibly be "forced" into a simple box; it draws as much on social interactionist and post-structuralist approaches as on the structuralist perspectives considered in this section.

Chouinard (1999) has drawn attention to the marginalization of disabled women within some feminist agendas as well as within dominant male interest groups. She explores, with particular reference to Canada, the "barriers to a body politics of empowerment that disabled women have encountered within both the disability rights and the women's movements"(Chouinard,1999:270).Economically,disabledwomenarestructurallydisadvantaged, with low incomes and high rates of unemployment. But in seeking to be heard, they have also suffered by being marginalized either by uncommitted governments (withdrawal of funding) or by male domination within disability politics and organizations. These structural disadvantages are then compounded by barriers of physical access and (perhaps well-intentioned) "ableist" attitudes. However, political activism at the international scale is, Chouinard suggests (1999: 290), helping to "unsettle oppressive regimes of power," not least by forging alliances with other marginalized groups. She herself (1999: 292) links the "devaluation" of disabled women in Canada to a "general and diverse site of oppressions in late patriarchal capitalism," including (for example) the exploitation of female workers in branch plants operating in industrializing economies. Like other researchers, Chouinard has a strong sense of responsibility to contribute to the real political struggles of those with disabilities; to force societal improvements and not merely study the injustices.

Glassman (2001) draws attention to the considerable risks faced by women workers in industrializing countries. He describes a factory fire outside Bangkok, Thailand, that killed nearly 200 workers, all but fourteen of whom were women, arguing that "women bear an increasing burden of occupational illness and mortality, rooted in their selective recruitment to sweatshops which offer wages otherwise unavailable to the rural poor" (Glassman, 2001: 62). Factory labor in general is under-regulated, women wield little influence in the trade unions and suffer from low wages. Efforts to tackle HIV/AIDS, encouraged by international organizations, are not matched by the investment needed to improve working environments; indeed, Glassman claims that the burden of occupational disease is greater than that of HIV. He points to the lack of resources available for resisting dangerous work environments, and argues that, with the increasing global mobility of capital, there is a need for women workers in the industrial periphery to operate collectively to improve their social position and health status.

Structurationist Approaches to the Geography of Health

We have seen how one set of alternatives to positivist explanation in the geography of health is to give much greater weight to the lives of real people; what has become known as "human agency." We have seen further that another is to invoke the broader social, economic, and political structures that mold – even determine – health and health care provision. It is in the search for a middle ground that a third alternative to positivism has come to the fore. This is known as structuration, and is most closely identified with the British social theorist Anthony Giddens. Some geographers have drawn extensively on his theoretical work (see Cloke et al., 1991, for an overview, as well as Gregson, 2005 for applications). Although Bourdieu (encountered in Chapter 1) does not use the term "structuration," his ideas are broadly similar.

Structurationist explanation

Structurationism recognizes the duality of structure and agency. That is to say, it acknowledges that structures shape social practices and actions, but that, in turn, such practices and actions can create and recreate social structures. One possible "language" in which this can be expressed is that of time geography, first outlined by the Swedish geographer Hägerstrand in the mid-1960s. Consider a time-geographic diagram (see Figure 2.7), in which members of an imaginary family engage in daily activities. For example, the female partner takes her young child to a pre-school center, before going to work, stopping off to shop on her way back to collect the child, and then returning home. Her engagement in various activities consumes time at particular locations, while travel between such places also takes time. Although not shown in the diagram, her activity (and that of other family members) in particular settings brings her into contact with others. While the purpose of this particular example is fundamentally epidemiological, it was used by the Swedish geographer Schærstrom (1996) to make the point that time-varying exposure to various environmental problems and social stresses has health consequences; the diagram also serves to give weight to the interaction of structure and agency. Social structures require that particular activities can only be carried out at particular times and in particular settings, but, equally, such structures may themselves be transformed by social action. For example, the opening times of clinics or health centers, together with workplace commitments, dictate when it might be possible to take the young child to see a doctor or nurse; the structure of health care delivery constrains agency (health-seeking behavior). On the other hand, the difficulties parents might have in taking a child to be immunized (for example) and the consequent lack of uptake of health care in particular places may ensure that the patterning of health care resources in space is re-fashioned; in other words, agency may transform structure. Time-geography has not gone uncriticized, notably by feminist researchers (Rose, 1993: 19–40) who consider that it refers mostly to masculine bodies and

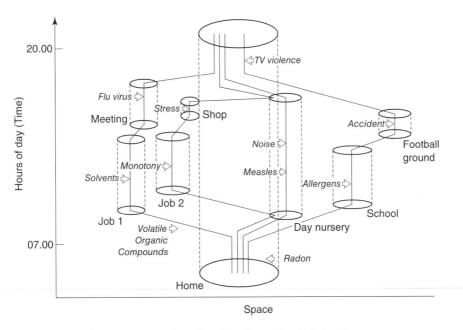

Figure 2.7 The time geography of an imaginary Swedish family

spaces. A renewed interest in time-space geographies, linked with Geographic Information Systems (GIS: see pp. 65–75 below) has also engendered interest in feminist analyses of the structural constraints on women's agency (Kwan, 1999).

Further examples of structurationist approaches

In the end, humans make their own health, but not in the conditions of their choosing. For example, Dyck and McLaren (2004) undertook a qualitative exploration of 17 immigrant and refugee women in Vancouver, Canada in order to investigate the impacts of immigration and settlement on women's work, health and gender relations in the household. The topic is informed by feminist theoretical and methodological debate that understands accounts of "everyday life . . . to be located within broader relations and distributions of power that play out unevenly within the particularities of time and place" (Dyck and MacLaren, 2004: 513). In so doing, these authors saw their original research objectives modified by the women's own agency; that is, while the original focus of the interviews was on practicing femininity, motherhood and the family as a problem of identity performance informed by structural immigration policies and practices, the women inserted themselves in the interview process by reframing the problematic as one closely related to the everyday struggles of immigrant resettlement; finding a job, raising children, "fitting in," impacts on health, both physical and psychosocial. They further inserted themselves into the research process by using the researchers/interviewers as knowledge brokers for leads on jobs and other questions related to education, economics and health.

A classic example of the use of structuration theory within health geography comes from Dear and Wolch (1987) in their exploration of the social processes involved in "deinstitutionalizing" the mentally ill, that is, taking them out of the psychiatric institution and into the community. These processes – which occurred in the 1970s and 1980s in much of North America and the UK – resulted in changes in spatial form, creating "landscapes of despair" or urban ghettoes of this and other "service-dependent" groups (including the dependent elderly, substance abusers, and ex-prisoners). Those discharged from institutions gravitated towards inner urban areas, where they found agencies willing to help and house them; as the numbers swelled so, too, have more services appeared, leading to further in-migration as a self-perpetuating cycle of ghettoization emerged. For Dear and Wolch the analysis recognized both structure and agency in a reciprocal way. Long-lasting "macro-level" political and economic forces worked as these structures got translated into specific policies and programs by the state. Alongside these structures are the shorter-term routine behaviors of individual human agents involved in the social welfare system. As the authors put it, the "service-dependent ghetto has been created by skilled and knowledgeable actors (or agents) operating within a social context (or structure), which both limits and enables their actions" (Dear and Wolch, 1987: 10). The reflexive or reciprocal relationship between social process and urban form arises in several ways. For example, inner-city improvement in recent years ("gentrification") has limited the possibilities for housing in the inner city; social processes influence spatial form. Yet the existence of the ghetto (urban form) shapes community awareness (social process) of the service-dependent community.

A more recent example comes from Watson's (2006) study of the health impacts of the stress experienced in a neighborhood of Krakow (Nowa Huta) in the south west of Poland during the period of transition after the fall of communism in this country. Essentially, democratization in Poland resulted in substantially increased rates of mortality for both males and females, as well as increased levels of mental distress. Watson looks for an explanation through a qualitative investigation of the psychosocial influences shaping experiences of change and therefore health. Nowa Huta was a purpose-built community, an exemplary undertaking for the socialist project. The local economy was based in steel production. While questionable emission and disposal practices created health concerns for some, age standardized mortality rates for both men and women in this "socialist utopia" were actually substantially lower than in the rest of Krakow. However, with the fall of communism, from the mid-1990s onwards, the welfare functions of the state were shed, while pay and pensions remained at state socialist levels which presupposed that certain social services (education, health care, etc.) were still in place. A steady decline in health status was seen from this point on. Watson explores the impacts of stress as a result of the transition on the health status of this local population. Through focus group and archival research, she finds that indeed stress played a major factor in her respondents' lives. Three major themes pervaded the interviews: financial strain, health care reform, and social unequalization. As Watson points out (2006: 374): "the subjectivity of an individual person can be seen . . . as constituted by the social environment, a social environment of which that person represents a constitutive element at the same time."

In the end, humans make their own health but not in the conditions of their choosing.

Post-structuralist Approaches to the Geography of Health

As we have seen, some of those engaged in the geography of health approach it from a positivist epidemiological stand-point, where measurable covariates are employed to shed light on disease incidence and key variables such as distance are used to explain disease spread. We shall encounter other examples later in the book. Others, we have noted, adopt interactionist approaches, exploring and interpreting the "authentic," lived experience of ill-health in particular places. Structuralists criticize both groups for their neglect of the broad social and economic forces, while structurationists acknowledge that both structure and agency matter.

Post-structuralist explanation

In the last twenty years, however, some geographers, in common with other social scientists (and including many health researchers), have begun to engage with other theoretical developments, which may be labeled *post-structuralist*. This is something of a "catch-all" term but, in essence, these perspectives are concerned with how knowledge and experience are constructed in the context of power relations. Such perspectives have illuminated work on health "risk," on representations of the body (see Box 2.1) and of social groups, and on what it means to be a healthy citizen. They question the rationalist assumptions on which much public health research is based (the so-called Enlightenment or *modernist* tradition in which scientific "truth" reigns supreme). Since there are close links between public health and geography it is important to say something about post-structuralism. We should say at the outset that what some social scientists call post-structuralist others refer to as *post-modern*, but that – as with positivist and other approaches – there is no simple box into which we can place particular studies.

Further examples of post-structuralist approaches

For writers such as Petersen and Lupton (1996) the "new public health" (which exhorts us to adopt healthy lifestyles – to eat well, exercise regularly, and cut down on smoking – as well as to play our part in creating healthy and sustainable environments) is a "modernist project." Having largely "solved" the problems of most infectious disease in the developed world, attention has shifted to the prevention of heart disease and cancers (and, latterly, AIDS), by promoting "good" health behavior. In doing so, control or power is exercised, not through repression but, as Michel Foucault has demonstrated, "through the creation of expert knowledges about human beings and societies, which serve to channel or constrain thinking and action" (Petersen and Lupton, 1996: xii) and, indeed, by self-governance. Petersen and Lupton see the new public health "as but the most recent of a series of regimes of power and knowledge that are oriented to the regulation and surveillance of individual bodies and the social body as a whole" (Petersen and Lupton, 1996: 3).

"Surveillance," of course, calls to mind the explicit monitoring of asylum populations in the nineteenth century (Philo, 1989). But the Foucauldian "gaze" is more subtle. Spatial

Box 2.1 The body

In a classically positivist account one might take the body, or one of its organs, to provide a set of "facts" that help to inform the study of disease. But the past 30 years has spawned a huge literature, primarily in sociology and cultural studies, and influenced very substantially by feminist writers, on "the body." This literature, including that by geographers, adopts a mix of very different theoretical stances, though structurationist and post-structuralist perspectives have tended to dominate.

Let us suggest, uncontroversially we hope, that people help to create places, and in turn people are shaped by the places they inhabit. The feminist writer Adrienne Rich (quoted in Nast and Pile, 1998: 2) puts it like this: "I need to understand how a place on the map is also a place in history within which as a woman, a Jew, a lesbian, a feminist I am created and trying to create." We can apply these same ideas to the body, for example the way it is dressed, looked after, and worked on. Bodies, it can be argued, are not "given," they are made. A structurationist would argue that they are "made" both by the individual who asserts her/his agency by the practices s/he adopts, but also by the ways in which society as a whole structures that behavior. For example, bodies that smoke, or exercise, or adorn themselves with tatoos or piercings, do so deliberately but what is acceptable, or possible, or impossible, is shaped by societal rules or conventions or, geographically, for example, by access to fitness centers or restricting, spatially, the locations in which smoking is permitted.

A post-structuralist would take a different tack and would wish to foreground bodies that do not "fit" or are in some sense "othered." They (for example, Petersen and Lupton, 1996) would argue that the primary focus in medical sociology (and indeed, medical geography) has been on white, male, heterosexual and able bodies. They are united by a "political concern with the subjugation of bodily diversity and creativity in the contemporary era" (Shilling, 2005: 17). But they tend to downplay human agency; yet, as we saw in Chapter 1, some bodies put *themselves* under surveillance (Parr, 2002a); there is "agency" here that sits comfortably in a structurationist perspective.

Others, notably Longhurst (2001), have been critical of a lack of attention, by both geographers and poststructuralist writers, to the materiality of bodies, and especially to their lack of engagement with "messy bodily fluids." Her empirical work involves the ways in which male and female workers "manage" their bodies in the workplace, the usually private spaces of men's bathrooms, and the pregnant body in public spaces. On the latter, her argument is that pregnant women "are thought to threaten and disrupt a social system that requires them to remain largely confined to private space during pregnancy" (Longhurst, 2001: 33). One might extend the argument to look at debates about *where* to give birth, and also breastfeeding practices, which may be seen as policed and confined to certain places in which the practice is seen as socially acceptable (Gatrell, 2007); the pregnant and breastfeeding bodies can be regarded as "othered" in post-structuralist terms.

We can extend a focus on materiality with particular reference to disability, since people with disabilities are among the most "othered" of all bodies in society, in the sense of being physically and socially excluded. Ironically, people with disabilities have often been neglected by post-structuralists, who "ignore institutional sedimentations in the built environment, assuming instead an apparently obstacle-less, frictionless plain of social interaction" (Dorn, 1998).

Existing and emerging technologies are opening up new possibilities – both positive and negative – for bodies and how they interact with places. Some of you reading this will have your own identities in virtual landscapes such as "Second Life." How are these bodies ("avatars") shaped by interactions with virtual places? Indeed, what are the meanings of "care," or "health," or "illness" in cyberspace (Shilling, 2005: Chapter 8)? Such interactions in cyberspace may be good for well-being, opening up the possibilities of new friendships and social support; this may be particularly the case for those who are "othered" and lack the freedom to develop healthy lives in real places. But virtual spaces can also be less welcome, as when they lead to bullying behavior or invite meetings in "real space" that expose potentially vulnerable people to dangerous situations.

There are rich veins of material for geographers to mine here. For further discussion see the collections edited by Nast and Pile (1998) and Butler and Parr (1999), the brief review by Parr (2002b), but also the monographs by Longhurst (2001), Crossley (2001) and Shilling (2005).

controls of quarantines and "cordons sanitaires," and of isolation hospitals (and, to some extent, long-stay hospitals for those with mental illness and learning disabilities: the asylums studied by Philo), have given way to non-spatial controls implemented via legislation (such as the mandatory wearing of seat belts or controls on smoking in public spaces: Poland, 1998), inspection, and large-scale population surveys of those factors deemed to increase the risk of ill-health and disease. The direct "gaze" of the medical professional on the individual's body thus gives way to the social survey, whereby the health landscape of the population at large is "surveyed" and the "policing" of health is more subtle.

To illustrate this argument further Peterson and Lupton (1996: 120–45) consider the concept of the "healthy city," a public health project emerging from the World Health Organization for Europe (and since spawning other "healthy" social settings, including hospitals, schools, universities, and prisons). The basic aim was to bring together various organizations, associations, community leaders, and local citizens to promote and achieve better health for all in particular urban settings. The idea (Ashton, 1992) was that such cities would become models of good practice and that this would inspire other cities outside Europe to follow this lead. The emphasis is on a holistic approach to health, with as much focus on housing quality, control of pollution, and "community spirit" as on measuring and monitoring of health status. The "sustainability" of urban life, the role of active citizenship, and the identification, monitoring, and reduction of environmental risk are all key features in the "healthy cities" movement. Cities are seen as ecosystems requiring management, but management that ideally involves active participation by all their citizens. If the city remains "unhealthy," rational, modern planning will ensure that the "organism" is restored to full health. Petersen and Lupton (1996: 127) are rather scathing in their attack on this fundamentally modernist project, which "obscures the power relations, uncertainties and ambiguities that underlie the development and implementation of policies, and conveys the impression that national, cultural and local differences, competing interests and inequitable access to resources are irrelevant to policy outcomes."

The emphasis on *difference* or *otherness* is a key feature of post-structuralist accounts. These draw attention "to the fact that the assumed or constructed human subject of Western modernist discourse is an exclusive subject in that it is predominantly male, European, heterosexual, middle aged, and middle class" (Petersen and Lupton, 1996: 10). This rather sweeping generalization (somewhat ironic coming from postmodernists!) ignores the quite substantial body of public health research concerned with inequality, whether according to social class, gender, or ethnicity (see Chapter 4). But, having said this, let us consider a good example of how the "other" has come under scrutiny from a health geographer, namely Craddock's study of the treatment of the Chinese community in San Francisco during the smallpox epidemics in the latter half of the nineteenth century (Craddock, 1995).

Craddock argues, in implicitly Foucauldian terms, that the analysis of disease in a geographical context "needs to include not just the visible configurations of spatial exclusion of the diseased, but also the many other ways the diseased are defined, disempowered, and controlled through metaphoric associations of place and affliction, inscriptions of contagious space, and the restructuring of purportedly diseased environments" (Craddock, 1995: 958). Smallpox probably diffused into San Francisco not with the Chinese but with other immigrants arriving from the eastern seaboard and the Midwest. But given the need to find a scapegoat for the epidemics, and that the mechanisms by which the disease spread were poorly understood, the Chinese were blamed for bringing the disease into the city, not least by public health officials. As one of them observed in 1881, the Chinese, "coming in contact with our people generally as no other class of our inhabitants do . . . are a constant source of danger to the health and prosperity of the entire community" (quoted in Craddock, 1995: 963). The appalling conditions in which they lived were seen as due to natural "depravity" rather than economic circumstances and the lack of jobs. Chinatown came almost to be equated with smallpox, and certainly stood as a metaphor for disease. It was a bounded area of the city that represented the "headquarters of disease." In Craddock's (1995: 966) memorable phrase, the Chinese "represented the most 'other' of all others" and the "threatening" urban space in which they were confined "was reproduced in the image of a threatening of disease" (Craddock, 1995: 967).

In other research, on HIV/AIDS, Craddock (2001) gives explicit attention to the gendered inequity that lies, she argues, at the heart of HIV risk. Gender determines, in part, access to community resources, the degree of control over household income, and sexual exchange. Craddock suggests (2001: 45) that post-structuralism has been significant "in bringing attention to voices of the 'other' not commonly recognized in political economic analyses intent upon illuminating the larger brushstrokes of class and institutional relations." Women are not passive players; there is space for them to "make decisions, negotiate responses, or construct resistant practices and productions of knowledge."

Fox (1999) suggests that the iconic figure in modernism is the detective Sherlock Holmes; he hypothesizes, solves problems and finds the "truth." In contrast (and drawing on the postmodern authors Deleuze and Guattari), the postmodern icon is the *nomad*. She does not put down roots or conform; rather, she celebrates her difference, her otherness. She "wanders in a field in which biology has not constructed a map; there are no landmarks, and no familiar paths" (Fox, 1999: 9). However, the "nomad" is often taken to be a rather romantic, and distinctly "ableist" construction, as Dorn (1998) has noted (see Box 2.1). For many people with disabilities the reflection on "otherness" gives way to a political activism that seeks *enabling* improvements in the physical environment. The disability activist Patty Hayes, whose daily life Dorn observes closely, may thus be considered a "frontier surveyor, clearing or removing hurdles from her way as she recolonizes an urban jungle" (Dorn, 1998: 199).

Concluding Remarks

This chapter has revealed the range and diversity of approaches to the geography of health. We think it is clear from what has been said – and hope it will emerge from a reading of some of the subjects covered subsequently – that there are multiple interpretations of, and possible explanations for, particular health issues in geographical settings. For example, we have seen in this chapter that the spread of disease can be described statistically, or, alternatively, explained in terms of the impacts of colonial structures. We shall see in Chapter 4 the richness of variety in accounting for health inequalities; for some, the effects of place and space are captured by positivist epidemiological approaches, while others seek structuralist, class-based explanations, and yet others give more credence to ordinary people's accounts of such inequalities. We shall see in Chapter 7 that an understanding of the impacts of air pollution on health can be understood by adopting a positivist "risk factor" approach, or a social interactionist account that gives weight to "lay" or popular views of the association, or a structuralist interpretation that the world industrial-economic system is the root cause of the problem.

We think it appropriate to end on a cautionary note: "Any attempt to categorize the theoretical approaches taken in medical geography is surely doomed to be flawed and partial, to illuminate some aspects of the intellectual landscape while obscuring others, and in doing so to be just one possible way of telling the story among many" (Philo, 1996: 36). We do not wish to put an intellectual straightjacket onto the work done by health geographers and allied workers. The "landscape" is complex and does not require bulldozing. It does require a map, however, and this is what we have endeavored to provide. But just as all maps are models of the territory they purport to describe, so is our own simply one, hopefully reasonable, sketch of the recent historical and contemporary terrain.

Further Reading

We have followed others, in both the geographical and health research arenas, in a three-fold classification of work as "positivist," "social interactionist," and "structuralist." See, for example, Aggleton (1990) among the latter and Litva and Eyles (1995) among the former. However, other important bodies of social theory need to be addressed, and we have sought to give due weight to "structurationist" and other "post-structuralist" accounts. Thomas (2007) addresses a range of theoretical perspectives in medical sociology and disability studies and should be consulted.

For a wonderfully clear justification for taking philosophy seriously, as well as a succinct journey through some of the debates, we recommend that the geographer begins with: Graham, "Philosophies underlying human geography research," in Flowerdew, and Martin (2005), before moving on to Cloke et al. (1991). The latter gives an extended treatment of "humanistic" approaches (what we have termed social interactionist), as well as structuralist (Marxist), structurationist, and postmodern approaches.

On broadly "positivist" styles of the geography of health see the wide-ranging analyses of disease records for world cities in the late-nineteenth and early-twentieth centuries (Cliff et al., 1998) as well as for island populations (Cliff et al., 2000). An excellent pedagogic account of classical spatial diffusion models is given in Thomas (1992). For a recent study of spatial diffusion in an historical context (of poliomyelitis in the USA) see Trevelyan et al. (2005).

Several of the essays in the collection edited by Butler and Parr (1999) draw on a social interactionist perspective; moreover, the editors' introduction gives a very good insight into contemporary geographies of health and disability.

On structuralist or political economy approaches, see Arnold (1988) for a very concise summary of the impact of colonialism on the health of native populations, while among a number of comprehensive syntheses we have drawn on Watts (1997). A series of detailed case studies is presented in the collection edited by Hartwig and Patterson (1978). Detailed Marxist-based accounts of the development of health care and the impact of colonialism on disease outbreaks in Tanzania (Tanganyika) are given in Ferguson (1979) and Turshen (1984). For a global political economy approach to AIDS see Lee and Zwi (2003).

On feminist scholarship see Rose (1993), but also the collection of essays edited by Dyck and others (2001). Rose (1993) reflects critically on structurationist approaches, particularly inasmuch as they have tended to ignore women. See also Dyck and Dossa (2007). Crossley (2001) writes very clearly about Bourdieu.

The classic text by Jones and Moon (1987) has a great deal to say about different perspectives on the geography of health. It is particularly valuable on positivist and structuralist (referred to there as "materialist") accounts. Although some of the examples are a little dated, the book is still well worth reading.

For an introduction to postmodern perspectives on health see Fox (1999). An important paper on AIDS, with particular reference to Malawi, and drawing on post-structuralist perspectives, is by Craddock (2000). Petersen and Lupton (1996) cast a wonderfully critical, and implicitly geographical, post-structuralist eye on public health.

Chapter 3

Method and Technique in the Geography of Health

Having set out some broad approaches to studying the geography of health we want now to consider the variety of methods and techniques that can be brought to bear in empirical studies. We have already encountered some of these in the examples used in previous chapters. For example, studies of disease incidence and spatial diffusion tend to use cartographic and statistical methods, while studies of how people with ill-health or disability cope with their environments call usually for qualitative methods.

The present chapter is divided into two substantial sections. The first examines what we call the "mapping" of health. Here, we consider not only techniques for visualizing health outcomes but also some of the tools of quantitative analysis that geographers and others use when studying the spatial distribution of disease and illness. The aim is to get across the flavor and range of methods that are used; readers will have to look at the source material suggested at the end of the chapter for details.

The second section examines a range of qualitative methods; here, the aim is less on measurement and more on the interpretation and understanding of ill-health, disease and disability in the context of place. Again, this material is not sufficient to produce experts in qualitative data analysis, but as with the first section we hope it is enough to allow the reader to appreciate some empirical work considered in this book and elsewhere.

We should also say, by way of introduction, that it is a mistake to equate quantitative methods with positivist accounts, and to assume that other approaches to the geography of health simply require qualitative methods. In practice, this will often be the case. But what matters most is simply whether the methods and techniques shed any light on the problem being studied.

"Mapping" the Geography of Health: Quantitative Approaches

Quantitative spatial data analysis typically involves one or more of three tasks. First, we will usually want to draw a map of our data, in order to see whether there seems to be any spatial patterning or perhaps visual evidence of an association with social or environmental factors. We refer to this as *visualization*. Second, we will almost always want to use both graphical and statistical methods to explore the data; this involves a more rigorous search for spatial pattern and possible associations with other variables and is referred to as *exploratory spatial data analysis*. The dividing line between "visualization" and "exploration" is a fine one, since both rely heavily on graphical methods. Third, we may wish to specify a more formal statistical model of our data; typically, such *modeling* will involve testing a hypothesis (for example, that the incidence of a disease is higher around a suspected source of air pollution). In the geography of health, such models will often involve a "spatial" variable (such as distance from a pollution source, or perhaps a simple indicator to distinguish one area from another). Again, the division between exploration and modeling is fluid, and this three-fold separation is one of convenience. In practice, analysts often use all three simultaneously, using modern statistical software in order to move interactively from a map, to another picture, to the fitting of a model, and then mapping the results (Bailey and Gatrell, 1995).

Visualization

In order to draw a map of health data we need some form of spatial referencing. If data have been collected for a set of individuals we might have some form of postal code (not usually a complete address, in order to protect confidentiality) that can be linked to a "point" on the ground. As noted in Chapter 1, a UK postcode "maps onto" an Ordnance Survey grid reference, for example. We could then map the set of individuals who are ill or have been registered on a database. In practice, this will not be very useful, since it will tend to reflect the distribution of population as a whole. Instead, as we shall see in the next section, we need some data on a "control" population in order to make comparisons.

If, however, we aggregate individuals to a set of areal units, we can draw more useful maps. Commonly, data come in this form anyway and unless one is working with health professionals (or those controlling access to individual health records) the geographer will most probably be using such data. Many web-based resources, some listed in the Appendix (page 269), provide access to data of this form.

In practice, geographers rarely map observed counts of cases by area, since these, too, would simply reflect population distribution or the age structure of that population. Even mapping crude rates (for example, number of cases of heart disease per 10,000 resident people) is usually of little value, since it will not have allowed for the age structure of the population; older people are more likely to suffer from heart disease. Consequently, some form of standardization is adopted before data are mapped (see Box 1.1, page 6).

There are many issues, of both a cartographic and a statistical nature, raised by this kind of mapping. Cartographers have long wrestled with ways of defining class intervals for these so-called *choropleth* maps; how many classes should we use, what should the class intervals be, and what shading or color scheme should be used? These are not trivial issues, since they affect how the map is "read" or perceived (Cromley and McLafferty, 2002: 105–10). A further difficulty is that the areas themselves are of variable physical and population size. One simple, and sensible, cartographic solution is to shade only those areas above a threshold population size, and to place a small shaded symbol in the less densely populated areas. Another is to draw a *cartogram* or *isodemographic* map, which transforms the areal

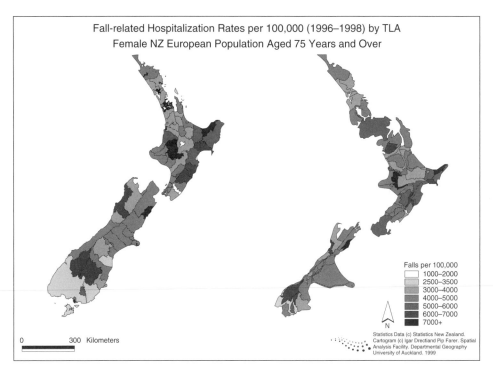

Fall-related Hospitalization Rates per 100,000 (1996–1998) by TLA
Female NZ European Population Aged 75 Years and Over

Falls per 100,000
1000–2000
2500–3500
3000–4000
4000–5000
5000–6000
6000–7000
7000+

0   300   Kilometers

Statistics Data (c) Statistics New Zealand.
Cartogram (c) Igar Drectiand Pip Farer. Spatial
Analysis Facility. Departmental Geography
University of Auckland. 1999

Figure 3.1   Mapping morbidity in New Zealand: "conventional" map and a cartogram

units according to population at risk, such that less densely populated areas appear much smaller than on a conventional map, while more densely populated areas "expand." This idea has a long history in disease mapping (see Shaw et al., 2002: 192–203 for a review), though it has not been widely used, partly because of the technical difficulties in performing the map transformation and partly because the resulting maps, while sensible, are sometimes difficult for lay readers to grasp; however, the latter problem may be lessened when a conventional map is placed alongside (see the example from New Zealand in Figure 3.1). At a global scale it is possible to map a variety of data sets relating to health and disease via the "worldmapper" website (www.worldmapper.org). At a national scale, Vanasse and his colleagues have used cartograms creatively to map adult obesity in Canadian health regions (Vanasse et al., 2006).

There are many examples of more conventional published atlases of disease and ill-health, showing spatial variation at a number of different scales. More recently, public health researchers and epidemiologists are using modern information technology to make these visualizations available over the Internet. For example, the Atlas of United States Mortality, published by the National Center for Health Statistics, provides maps of mortality for 18 causes of death, for nearly 800 Health Service Areas. Given that many such atlases are now in electronic form they allow users to interact with the map and to manipulate the data themselves; some reference sites are given in the Appendix (page 269). The possibilities of interactive data analysis lead us into data "exploration" or "exploratory visualization," which we consider in the next section.

Exploratory spatial data analysis

In addition to problems in mapping health data for a set of areal units, there are statistical issues that need discussing. We highlight two of these. The first of these is the *modifiable areal unit problem*. This is best explained with reference to an imaginary example (see Figure 3.2), showing the residential locations of children born with congenital heart disease. As we shall see shortly, there are ways of exploring this using the point data themselves, but if we wish to analyze how the data are distributed by area we will get very different results depending upon how we configure the areal units. A configuration based on administrative units (such as Census Super-Output Areas in Britain, counties or census tracts in the USA, communes or municipalities in many European countries) will give different estimates of disease incidence than one based on different (and essentially arbitrary) zones or ones that show "bands" of distance from a suspected source of pollution (see Figure 3.2). We say that the areal units are "modifiable" (Openshaw, 1983).

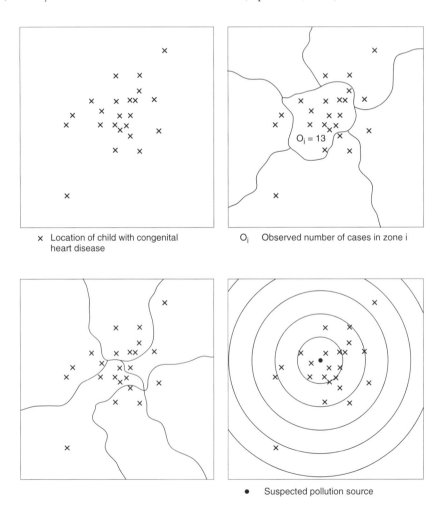

x   Location of child with congenital O_i Observed number of cases in zone i
heart disease

●   Suspected pollution source

Figure 3.2   The modifiable areal unit problem: assigning point data to different sets of areal units

The second issue relates to the numbers of cases used to construct estimates of disease incidence, particularly standardized mortality ratios (SMRs). To some extent this is a problem of scale, since as our areal units become larger the numbers of cases will grow. For example, we will observe more lung cancers at the state rather than county level in the USA, and more cases at the provincial level in Italy than at the municipality level. When mapping and analyzing data for well-populated areal units we can put a considerable degree of confidence in our estimates of the SMRs, much more so than at a fine spatial scale, where the counts may be very small (perhaps one case here, two or three there, none in some areas, and so on). At this finer scale we say that estimates of disease risk are "unstable," since the addition or subtraction of only a single case can greatly alter the estimate. For example, if over a three-year period we expect two deaths in one area, and observe three, we have an SMR of 150. But if we observed none, or one, or two, or five the SMR would change to 0, 50, 100, and 250. This is the so-called *small number problem*.

There are a number of possible solutions. One is to extend the data collection period to cover further years. A second is to work at a coarser scale (though this is of little help if we wish to identify local disease "hotspots"). A third strategy is to place confidence intervals around the estimate of the SMR (so we can tell whether a rate of 130 is significantly different from 100, for example); until recently this was widely used in disease mapping. A fourth alternative is to use *probability mapping*, where we map the probability of observing a count that is more extreme or unusual than that actually observed. Technically (Bailey and Gatrell, 1995: 302; Cromley and McLafferty, 2002: 133–7), we use the Poisson distribution to calculate these probabilities. Low probabilities (say, less than .05) suggest that an area's rate is significantly high if the observed value is greater than the expected, or significantly low if the converse is true. But this approach, too, has been superceded by more sophisticated forms of modeling, which we consider shortly.

Assuming we have constructed a map of disease incidence, what questions might we ask of it? A common and important question is to ask whether areas of high incidence "cluster" together, or if the spatial arrangement of disease rates is essentially random. Non-randomness might suggest clues about disease causation. Such spatial patterning is referred to as *spatial autocorrelation*, and it is a fundamental concept in geography. Using an analogy with ordinary correlation analysis (Box 1.3 above), we speak of positive autocorrelation where similar values tend to occupy adjacent locations on the map, and negative autocorrelation where high values tend to be located next to low ones. If the arrangement is completely random we refer to an absence of spatial autocorrelation.

There are good reasons why we should expect spatial autocorrelation to be a very common feature of geographical data in general and health data in particular. First, as we saw briefly in Chapter 2, the process of spatial diffusion or disease spread means that the incidence of measles or influenza, for example, in one area is likely to affect directly the incidence in nearby areas. Second, there will often be broad regional processes, whether environmental or social, which ensure that disease incidence in one small area is similar to that in neighboring areas. For example, if there is widespread regional pollution it is likely that areas within this wider region will all tend to have high rates of respiratory disease. Or, if this wider region has quite uniform levels of poverty we can expect rates of some chronic diseases to be high in adjacent small areas that make up that region, since poverty is closely linked to many such diseases. Geographers and statisticians have collaborated to devise statistical measures of spatial autocorrelation, and the use of an autocorrelation statistic (Bailey and Gatrell, 1995: 269–74) in characterizing health data is now quite common.

To summarize a whole map pattern by a single number is a little crude, and this has led others to suggest more "local" measures of spatial autocorrelation or spatial association. In a health context, these look at the association between a disease rate in one location and rates in neighboring locations, up to a specified distance. This helps identify disease "hot-spots" (Cromley and McLafferty, 2002: 139–41). Details of the method are given in Getis and Ord (1992) and applications to health data are presented by these authors in that paper and elsewhere (Getis and Ord, 1999). For example, they study the clustering of AIDS cases in the area surrounding San Francisco, between 1989 and 1994. Results suggest that there is a strong tendency for clustering of high rates among the seven counties within about 40 miles (64 km) of San Francisco. Rigby and Gatrell (2000) have used the method to look for clustering of breast cancer in north-west Lancashire, England, while Mather and colleagues (2006) use it to identify clusters of parishes (counties) in Louisiana with high rates of prostate cancer, and Hanchette (2008) to suggest areas of high and low lead poisoning in North Carolina.

We have already drawn attention to the arbitrariness or modifiability of areal units, though in many cases, as when presented with data for administrative units whose boundaries are fixed, we can do little about this. Systems of areal units can be defined using geographical information systems, however, as we see later (Box 3.2, page 67). For example, if we wish to explore the possible association between childhood respiratory disease and proximity to main roads (see Chapter 7) we can create "buffer zones" around the roads and count the number of cases of ill-health within and outside such zones. The same problems of modifiability arise as before (see Figure 3.2), since there will be no real reason why these zones should be 25 meters wide, or 50, or 100; the choice is to some extent arbitrary.

The arbitrariness of areal units, coupled with the increasing availability, in the developed world, of coordinate data derived from individual health records, means that a number of geographers and epidemiologists have been drawn to methods based on spatial point process analysis. Here, we have a map of the distribution of observed health "events" (points in space) and, as with area data, interest centers on whether the map could have arisen by chance. More formally, we might ask whether the distribution is the outcome of a completely random process, such that any sub-region has an equal chance of having an event located there and the occurrence of one event is independent of that of any other. But this is unhelpful, since we know that population is distributed unevenly in the first place, and so the notion of complete randomness is untenable. We need to allow for population distribution in our exploratory analysis. This is done, ideally, by comparing our observed distribution of point events ("cases") with a set of "controls" (similar individuals who do not have the disease).

Numerous statistical methods have been devised to address this issue; for example, *kernel or density estimation* was highlighted in the first edition of this book (Gatrell, 2002: 62–5), while *spatial filtering* was introduced into geographical epidemiology by Rushton and Lolonis (1996); see Cromley and McLafferty (2002: 142–51) for further description and examples. In recent years, however, one particular method has become widely adopted. It is called the *spatial scan statistic*, and was devised by Martin Kulldorff who has produced freely available software ('SaTScan') that can detect disease clusters in space (and, indeed, assess whether disease clusters in both space *and* time; see Box 6.3 below). SaTScan will detect areas of elevated and lowered risk for several different data types. It does so by superimposing a series of circles, of varying radii, across the study region, in effect "scanning" for clusters. For any circle the observed number of cases is compared with that expected by chance, and where this is most extreme a cluster is detected. For example, if one has geo-referenced data for cases and suitable controls, the analysis will conduct a spatial case-control analysis, with suitable adjustment for covariates; or, applied to data already aggregated to areal units, it will detect groups of adjacent zones that have elevated (or reduced) disease rates, again with adjustment for known risk factors.

Among numerous examples that could be drawn upon, consider an application to the prediction of West Nile Virus (WNV) in New York City (Mostashari et al., 2003). WNV (see also Chapter 9) is a mosquito-borne disease that infects animal and human populations, causing encephalitis (inflammation of the brain) in the most severe human cases. Analyzing the spatial distribution of viral activity in dead birds provides a possible early-warning system. Mostashari and his colleagues took reports of dead birds and assigned these to census tracts. The number of recently observed dead birds (cases) within and outside a circular region (a potential cluster) is compared with an expected number of cases, where this is defined on the basis of historical dead-bird data. The study area is scanned in this way, and clusters are defined as sub-regions containing census tracts where the observed counts significantly exceed those expected on a chance basis. For example, on 14 September 2000 (see Figure 3.3) it was evident that there were three major clusters – in all of Staten Island, in large parts of Brooklyn, and in eastern Queen's. Laboratory analysis of mosquitoes or detection of the virus in dead birds is the only sure way to identify the virus; but the authors suggest that "use of routinely collected dead bird reports to detect WNV-related dead bird clusters may facilitate early detection and targeting of scarce surveillance and vector control resources" (Mostashari et al., 2003: 644). Clearly, there are limitations, not least reliance on someone recording the presence of dead birds, and such deaths realistically being ascribed to the virus as opposed to other causes.

Modeling health data in a spatial setting

In exploratory spatial analysis we are not seeking to test explicit hypotheses, other than perhaps trying to assess departures from randomness. For example, when trying to detect disease "hotspots" we do not usually have in mind that there might be clustering around specific pollution sources. This contrasts with situations where we do believe, a priori, that there might be a raised risk of disease near such a source. A well-known example of this kind of "focused" study is the substantial research effort devoted to the suspected elevated risk of childhood leukaemia around plants processing nuclear material. Controversy surrounding the nuclear reprocessing plant at Sellafield on the Cumbrian coast of north west England (see Box 3.1, page 57) has motivated these studies, but there are many other examples.

Again, there is a range of point-based methods that can be put to use on this kind of problem. One (Diggle and Rowlinson, 1994) allows us to test whether the incidence of disease is related to one or more possible sources of pollution, while controlling for the possible effects of other variables allowed for at the individual or household level (such as smoking behavior, or damp in the home). Essentially, this approach takes a set of point

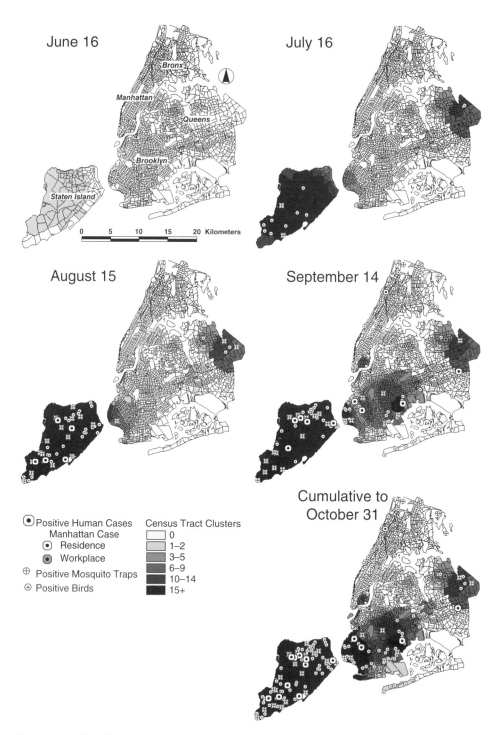

Figure 3.3 Dead bird clusters, West Nile virus (2000)

> ## **Box 3.1** Leukaemia "clusters"
>
> Television publicity was given in 1983 to an unusually large number of cases of childhood leukaemia near the Sellafield nuclear reprocessing plant in West Cumbria, England. Numerous epidemiological studies (including those by geographers: see Openshaw et al., 1988) have confirmed the existence of this "cluster" and indicated that it is restricted to children born in the nearby village of Seascale (6 cases of leukaemia observed in such children, compared with only 0.6 expected).
>
> The Sellafield problem has motivated a considerable volume of sophisticated technical work on the detection of disease clusters. The explanation of the excess number of cases of childhood leukaemia near Sellafield is, however, controversial. For some, it arises as a result of direct exposure to contamination on nearby beaches or via consumption of seafood and locally produced foodstuffs. Some, such as Gardner and his colleagues (1990), suggest that the raised incidence is associated instead with fathers' employment at the Sellafield plant. The high radiation doses given to fathers, before the conception of any offspring, has resulted in leukaemia among their children (a result of a mutation in the sperm). Others contest this as a causal mechanism. In particular, the epidemiologist Leo Kinlen has offered a competing hypothesis based on unusual patterns of migration and population mixing; his work, arguing the role of infectious agents rather than radiation, is considered in Chapter 6.
>
> Other studies have pointed to the existence of leukaemia "clusters," albeit with different possible environmental causes. A well-known example in North America is the excess risk in Woburn, Massachusetts, linked to well-water contamination (see pp. 213–14 below). This gained additional publicity via the movie *A Civil Action* (released in 1998) that drew upon the book of the same name by Jonathan Harr.
>
> For further discussion of the Sellafield cluster see Gardner (1992) and the collection of papers edited by Beral et al. (1993).

locations and asks: how likely is it that any given point is a case rather than a control, given its distance from one or more suspected pollution sources and any additional information we may have about the individual in question? Some researchers have, for example, asked if proximity to sites disposing of hazardous waste, whether via landfill (Dolk et al., 1998) or incineration (Diggle et al., 1990; Elliott et al., 1992), is a risk factor for congenital malformations or cancers of the respiratory system, while others have asked if there are clusters of breast cancer on a Long Island community in New York (Timander and McLafferty, 1998).

Modeling of area data is much more common, since we invariably have census-based data to use as potential explanatory variables or covariates. A standard tool to use here is regression analysis, which seeks to account for variation in a dependent variable (such as disease incidence) in terms of the chosen covariates. Lovett and Gatrell (1988) sought to account for geographical variation in rates of spina bifida (a congenital malformation of the central nervous system) among health authorities in England and Wales. Statistically significant predictors were: percentage of female residents born in the Caribbean (which was inversely related to spina bifida); percentage unemployed; percentage of the workforce employed in agriculture (both positively associated with spina bifida); and, most significantly, an indicator ("dummy") variable that denoted whether or not the health authorities routinely screened pregnant women for the disease. As with all regression models in health research that use spatial units for their analysis, differences between observed and expected incidence can be mapped and the results suggest parts of the country where, in particular, incidence that is higher than expected merits particular attention.

This example serves to make a number of points. First, census data are used frequently in studies of this form simply because the required data do not exist at an individual level. We can hardly conclude from the spina bifida study, for example, that a woman born outside the Caribbean, and either she or her partner is unemployed, is more likely to give birth to a spina bifida baby. The absurdity of this argument is borne out by the fact that employment in agriculture is also a variable in the analysis; it is impossible to be both unemployed and to work in agriculture! Second, there is nothing particularly "spatial" about the analysis. The health authorities could be randomly located anywhere on the map and the regression results would be identical! "Place" enters into the analysis merely as a container for the disease cases. What methods can we use to bring space into the analysis?

To answer this question takes us into some rather difficult areas of applied statistical spatial data analysis, but let us sketch the basic ideas using a real study. Following Middleton et al. (2006), we have a set of observed counts (numbers of cases) of suicides in men aged 15–44 years in small areas of England and Wales, over a seven-year period. For the country as a whole there were 15,710 deaths from suicide, distributed over 9265 small areas. On average, then, there are about 1.7 deaths per area; but this average masks considerable spatial variation. In some areas there were no suicides, while there were 21 suicides in one area. What we would like to map is our very best estimate of risk. Unfortunately, a standardized mortality ratio (observed divided by expected number of suicides) is a poor estimate, since the areas differ greatly in population; in relatively densely populated areas the estimates may be quite stable, but in sparsely populated areas the estimates are highly unstable. In contemporary mapping (technically, referred to as *hierarchical Bayesian modeling*), we seek to smooth the estimates, allowing these smoothed estimates to be determined in part by the average rate for the country as a whole and in part by the tendency for adjacent areas to have similar rates. Put differently, we can model the observed count as a function of the expected number of suicides in a small area, and two additional influences on these counts, referred to as "structured" (or spatially dependent) effects and "unstructured" variation (sometimes called heterogeneity). Middleton and his co-authors (2006: 1041) put it like this: "The less precise the estimated rate in an area is or the stronger the evidence that rates are homogeneous across the study region . . . the higher the amount of smoothing towards the national average. Alternatively, the more similar levels of suicide are in neighboring areas (i.e. the greater the evidence of spatial autocorrelation), the higher the amount of smoothing towards the local mean." Having done this analysis, we can then map the smoothed suicide risk. The unsmoothed map of suicide SMRs shows considerable "noise," but the smoothed map (Middleton et al., 2006: 1043) reveals high rates in urban areas (London, Manchester and Birmingham) and particularly in inner-city areas. In addition, high rates are observed in coastal and remote rural areas.

But we can go beyond this to add in a set of explanatory variables or covariates. Again, specialized software is used to estimate these parameters and the spatially structured and unstructured components for each small area. In the case of suicide in England and Wales important explanatory variables include the proportion of single-person or lone-parent households, and the proportion of the population that is unmarried.

What useful information do we glean from this kind of analysis, and what is missing? On the positive side, we are able to get stable estimates of disease risk from place to place, and – in this example – we can begin to explain such geographical variation. For example we can establish that areas with high proportions of single-person households and unmarried populations (crude measures of social fragmentation, perhaps) are more likely to have higher rates of suicides. But, as the authors stress, such associations over a set of geographical areas do not necessarily imply that these factors apply to individuals. To make inferences about individuals (for example, that single people are more likely to commit suicide) is to commit the so-called *ecological fallacy*. Set against the benefits of a quantitative spatial analysis, however, is the question as to what, precisely, are the gains in explanatory power. We can, demonstrably, quantify statistical associations between suicide and some plausible covariates, but does this really help us to understand the causal factors? Clearly, while we cannot interview the dead we could explore with family members, with carers, with health professionals, and those who are seriously mentally ill, the range of factors in real lives that may predispose people to end such lives. In particular, care needs to be taken in inferring that there are particular features in inner-city areas that "cause" higher-than-average rates of suicide; it could be that those already at increased risk "drift" to such areas.

Suppose that we have a set of data *both* for individuals *and* for the areal units within which they live. What kinds of modeling of health data are possible in this situation? Arguably the most significant contribution made by geographers to quantitative health research has been to demonstrate that data for both places and individuals can be brought together in order to shed light on health outcomes. This body of work is known as *multi-level modeling* (MLM). Such modeling can get quite complicated, and here we simply sketch some basic ideas, relying on graphical rather than mathematical formulations, illustrating the method with examples.

Suppose, hypothetically, we wish to explore the relationship between smoking behavior and age. We have data for a large number of individuals living in several towns; we know their age and how many cigarettes they smoke on an average day, as well as their place (town) of residence. Pooling all individuals and fitting a simple regression model would yield a regression line with a fixed intercept and slope (see Figure 3.4a). But individual smoking behavior might vary from place to place; perhaps there is the same overall relationship between smoking and age, but each town has its own intercept, denoting its varying average cigarette consumption (see Figure 3.4b); here, the intercepts are said to be "random" (meaning, drawn from a probability distribution). In this case, some places simply have higher overall levels of smoking, at all ages. Or, the relationship between smoking and age might itself vary from place to place. In some places there might be no relationship (the slope is horizontal); in others it might be quite steep (as age increases, cigarette consumption rises rapidly); in yet others there might be an inverse relationship (see Figure 3.4c). The slopes and intercepts are both "random." There is, then, a distribution (over the set of places or towns) of possible intercepts and slopes, the relationships between which can be explored. The main point is that these place-varying behaviors are masked by a single-level analysis.

In this example, we are using two levels (individuals and towns), but there might be three or more levels. For instance, and linking back to the study in New York City used as an exemplar in Chapter 1, children with asthma might be affected by age (individual level), by the presence of smoke or damp in the home (a higher-level effect), and by the volume

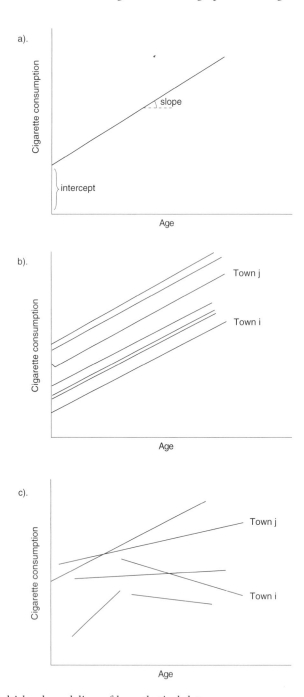

Figure 3.4   Multi-level modeling of hypothetical data

of truck traffic (hence diesel fumes) along their street (a third level). There may be interactions across levels too; for example, the risk of lung cancer will be elevated in those who smoke heavily (an effect at the individual level) *and* who live in areas where there is substantial air pollution (an effect at a higher level). We need, then, to add to our analysis a set of variables that are measured at the higher levels.

Multi-level modeling offers a potentially rich analytical perspective on the geography of health, helping us to establish whether health variations from place to place arise simply because different sorts of people live in different places (so-called *compositional* effects), or because places themselves differ in terms of environmental quality or other attributes (*contextual* effects). If the latter is the case, we can genuinely suggest that there are *ecological* effects on health, in the sense that the environment in which people find themselves has direct impacts on their health. As Jones and Duncan (1995: 30) have it, "individuals and their ecologies must be modeled simultaneously." Further, MLM is an attractive tool for those who work within a structurationist framework, since it suggests that an engagement with both human agency and broader (higher-level) structures is required for a full understanding of health problems. Although it is in no sense a qualitative method, it does offer a means of operationalizing the undoubted connections between individual human behavior and broader environmental and social contexts (Duncan et al., 1996).

In Britain, Kelvyn Jones and his colleagues have applied MLM in a number of different contexts. One example (Jones and Duncan, 1995) concerns respiratory ill-health in England, using the Health and Lifestyle Survey (Cox et al., 1993). Data are recorded for 9003 individuals, living within 396 small areas (electoral wards). As we shall see in Chapter 4, it is possible to create a deprivation score for such areas. Variables measured at the individual level include age, height, gender, smoking and eating behavior, alcohol consumption, exercise, income, tenure, social class, and employment status. The problem is to account for variation in respiratory health ("forced expiratory volume," or FEV) using the individual- and higher-level variables. The authors demonstrate that many of the individual-level variables are significant, for example FEV worsens among those on low incomes and of low social status. But the main point is that there are higher-level effects. Figure 3.5a shows the effect of ward-level deprivation. Lung function worsens in the more deprived areas, regardless of individual social class; but the effect of ward-level deprivation does vary slightly, so that the health risk of living in a deprived area is less if one is of higher social status and greater if working in a manual occupation (social class IV and V: see Box 4.3 below, page 96). Consider, too, the simultaneous impact of smoking behavior at the individual level, and urbanization at the ward level (Figure 3.5b). Here, smokers have poorer FEV than non-smokers; but while urbanization has little additional effect on smokers it seems to have a negative impact on the non-smokers. A tentative conclusion would be that air quality is a risk factor for respiratory health. These results indicate that "health outcomes cannot be simply reduced to individual characteristics" (Jones and Duncan, 1995: 38).

Several studies have used MLM to assess possible area effects on mental health, some indicating that such effects are present, others not. Two studies that have detected such effects are described here. Fone and Dunstan (2006) used data from a large health survey in Wales (almost 30,000 respondents), which asked questions about mental health from which a mental health status score was derived. There were significant associations between mental health status and deprivation after adjustment for individual data, such as age, gender, social class, and unemployment. Mental health status is worse in more deprived areas, and this relationship has largely the same form regardless of employment status; the difference is that the overall level of mental health is higher among those in employment. However, the relationship between mental health and deprivation is much more pronounced among economically inactive people and it varies substantially from place to place.

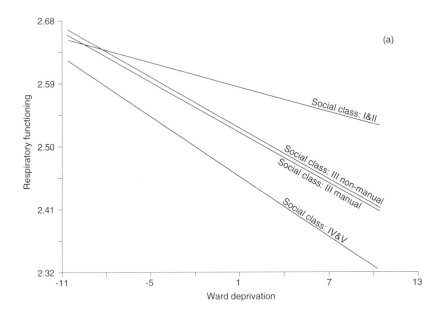

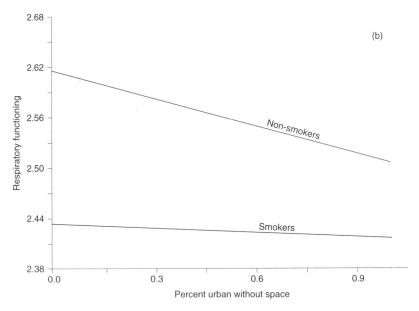

Figure 3.5 Multi-level modeling of respiratory functioning: (a) effects of deprivation and social class and (b) urban index and smoking

Sundquist and Ahlen (2006) looked at admissions to psychiatric hospitals in Sweden. Their results indicated that people living in the poorest neighborhoods had a significantly higher risk of being hospitalized for mental health problems than those in wealthier neighborhoods, even after controlling for individual socio-economic and demographic factors (such as: age; living alone; on low incomes; and with poor education). There did not seem to be place-to-place variation in the *form* of the relationship with the individual-level variables; in other words, the relationship was as depicted in Figure 3.4b rather than 3.4c.

A further set of examples concerns what Moon and his colleagues refer to as "geographies of obesity" (Moon et al., 2007) and we summarize three such studies, all of which have used MLM (see also Box 4.4, page 109 below). Oliver and Hayes (2005) have analyzed the geography of overweight children in Canada. "Overweight" is defined as a body mass index (BMI) of over 25 kg/m^2. Oliver and Hayes hypothesize that there may be differences in weight according to age and gender (individual effects), as well as family effects (a second level) and among neighborhoods (a third level effect). Using data for over 11,000 children in Canada (9000 families and over 5000 neighborhoods) the authors detect statistically significant effects at each level. Children in more affluent families are less likely to be overweight, while those whose primary carers do not have a high school education are more likely to be overweight; these are family-level effects. In addition, there are neighborhood effects; children living in poorer neighborhoods are 29% more likely to be overweight than those in the most affluent area. The suggestion is that "less affluent neighborhoods have social and physical environments less conducive to maintaining a healthy bodyweight" (Oliver and Hayes, 2005: 418). Further, this work has direct policy consequences: to improve the health of children we need to ensure that neighborhoods are adequately provided with the infrastructure necessary for productive physical activity.

In a fascinating study, McLaren and Gauvin (2003) have posed the question: does the average size of women in a neighborhood influence a woman's likelihood of body dissatisfaction? A telephone survey of nearly 900 women in several parts of Canada asked whether women wished to lose weight or were dissatisfied with their weight; over 50% indicated they were dissatisfied. Variation in body dissatisfaction was not explained by individual factors, such as age or affluence, but it was, as expected, influenced strongly by individual body mass index (BMI). Interestingly, however, regardless of whether individual BMI was low or high, a woman living among thinner women within her neighborhood is more likely to feel dissatisfied with her own body than a woman who lives among larger women. This empirical finding is illustrated in Figure 3.6. McLaren and Gauvin (2003: 332) theorize it thus: "living among smaller women provides a more constant reminder of the 'thin ideal' and one's discrepancy from it." Further, if aware (from magazines and television) of the broader cultural "value" placed on thinness this neighborhood effect may add to that low body esteem. Whether these findings generalize to other contexts is an empirical question that a social positivist research perspective could illuminate. Other, more critical, perspectives would seek both to question the whole notion of the "thin ideal" and to worry, quite properly, about the impact this has on the well-being of young women.

Last, Moon and his colleagues (2007) have sought to look at what they call the "distinctive geographies" of obesity in England. The Health Survey for England (HSE) gives data for over 18,000 individuals, living within just under 8000 small areas, 303 health regions (Primary Care Trusts – PCTs) and eight larger regions; thus there are four levels of analysis. What Moon and his colleagues want are good estimates of obesity prevalence (and the

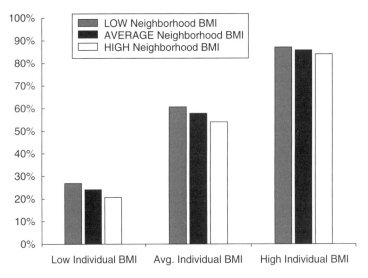

Figure 3.6 Women's body dissatisfaction, by individual and average neighborhood body mass index

prevalence of overweight adults) for PCTs; but with an average of only 60 respondents per PCT, and given that PCTs have populations of over 100,000 we can hardly use the HSE data on their own to provide such estimates! Cleverly, the authors model the data using MLM and then use the model to estimate obesity, disaggregated by age, sex and ethnicity. This allows them to produce maps, by PCT, of (for example) the prevalence of obesity among different sub-groups of the population. In general, obesity appears particularly high in the urban West Midlands (centred on the city of Birmingham).

The number of applications of MLM to health problems has mushroomed in recent years, but alongside the plethora of applications has come some critical reflection on what is meant by neighborhood, and what a theoretically meaningful set of contextual variables might be at a neighborhood level (Rajaratnam et al., 2006). Often, such variables are survey-based or census-derived, and therefore the choice is determined as much by the ready availability of data as by theoretical considerations. "Indeed, the availability of census data drives even the definition of neighborhoods, which are often defined as census tracts or blocks" (Rajaratnam et al., 2006: 554). All the studies outlined above can be criticized on these grounds. For example, the Welsh study of mental health uses "deprivation" as an area-level variable; but the deprivation score is itself estimated from individual data on (among other variables) unemployment, which is already accounted for as an individual-level variable. Similarly, in the Swedish study, since the area-level variable (neighborhood income) is simply an aggregation of income data collected for individuals (itself already in the model), we have to ask if there are not more sophisticated area-level variables that could be considered to impact on individual mental health. Last, the Canadian study of body dissatisfaction used census tracts as neighborhoods, a rather crude reflection of the activity spaces within which the women are likely to circulate.

We saw earlier how it is possible to analyze case-control data in a spatial setting, where the "cases" are individuals with some disease or disability and the "controls" are a group of individuals with the same broad characteristics (age, gender, social class, for example) but without the disease or disabling condition. We used a spatial analytic method – the spatial scan statistic – to see if we could detect "clusters" of cases, relative to controls. Case-control methodology is widely used in quantitative health research, where data for individuals are used to examine the influence of one or more "risk factors," adjusting or "controlling" for the influence of other variables. We shall see in later chapters how these ideas are used in particular research contexts. But here we introduce a standard statistical method, the *logistic regression* model and look briefly at an example.

Logistic regression is appropriate where the variable to be explained takes on one of two values. This contrasts with conventional regression analysis, where the dependent variable is continuous, and with the Poisson regression model, where it comprises a set of counts. Logistic regression is therefore used where we wish to look at factors and variables that predict a binary dependent variable (here, case or control). The regression coefficients that emerge from a logistic regression analysis may be converted into *odds ratios* that reveal the "odds" of disease or illness, given the particular value of an explanatory variable or covariate. For example, suppose we wished to determine whether children living close to a cement works were at greater risk of respiratory disease than a similar group living elsewhere. Here, the cases are children with a respiratory problem, the controls are those without, and the covariate of interest is whether or not the children live near the cement works (itself therefore a binary variable). We can enter into the regression analysis other factors that may "confound" the effects of the cement works. For example, if we found that living near the plant had an adverse effect on health, might this be due to the fact that smoking was much more common in such households, rather than the exposure (external air pollution) itself? A study in East Lancashire, northern England (Ginns and Gatrell, 1996), reveals that the odds of respiratory and other symptoms seem to be increased in the area surrounding a cement works that burns a fuel derived from hazardous wastes (see Table 3.1), although the effect of proximity to the plant ("area" in Table 3.1) is relatively weak. For example, the risk of "sore eyes" is about two and a half times greater in the area near the plant than further away, but the risk is much higher among children suffering from a common allergy ("hay fever"). The confidence intervals allow us to see whether the increased odds are statistically significant or not; an odds ratio of 1.0 means that the covariate in question has no effect on the response variable. Confidence intervals that do not include the value 1.0 therefore suggest that the covariate is significant at a specified level of probability.

Geographical information systems and health

A good deal of the spatial analytical work outlined in the above sections has been paralleled by the development of geographical information systems (GIS). Such systems first emerged in the 1960s, but it is only in the past 25 years that the science and technology have taken off, and only relatively recently that health geographers have begun to exploit these developments.

A GIS is a computer-based system for collecting, editing, integrating, visualizing, and analyzing spatially referenced data. Such data comprise two forms: the locational element

Table 3.1 Results of logistic regression analysis of health survey in part of East Lancashire

Symptom	Covariate	Odds ratio	Confidence interval (95%)
Sore throat	Area	1.83	1.15–2.91
Blocked nose	Hay fever	4.81	2.23–10.36
	Mold in house	2.82	1.25–6.35
	Area	1.58	1.00–2.48
Sore eyes	Hay fever	26.76	11.26–63.60
	Mold	2.04	1.05–3.98
	Area	2.67	1.01–7.09

"Area" takes on two values, according to whether or not a child lives near the cement works
(*Source*: Ginns and Gatrell, 1996: 633)

and associated "attributes." For example, we might have data on the residential addresses (locations) of people suffering from stomach cancer, of different ages and sex (attributes). We might want to link or integrate such data with other databases, such as those concerned with environmental quality. Exploring or modeling the associations between the health and environmental data is complicated, because the data sets will have been collected for different sets of locations. As another example, we might have data on the locations of health centers (which will have attributes such as number and type of staff employed there, and opening times), along with data on patients treated there (who will also have locations and attributes, such as health status); we might wish to model the flows of people to health centers.

Let us consider some examples to illustrate what is possible. Consider first the incidence of malaria in Africa. In the South African province of KwaZulu-Natal the disease has had devastating effects. Malaria control teams equipped with global positioning systems (GPS) receivers have obtained the latitude and longitude of nearly 35,000 family homesteads in parts of the province, and they store "green cards" under the roof eaves to record the application of insecticides as well as results of blood smears for parasite detection; all these data now form part of a geographically based malaria information system (Sharp and le Sueur, 1996; Martin et al., 2002). The system also identifies the locations of clinics to treat affected people, and the GIS allows the user to place *buffer zones* around these clinics in order to determine which homesteads are relatively remote from these sources of health care (see Box 3.2). Maps of malaria prevalence may also be produced, permitting control agencies to direct resources more systematically to areas of need. This approach is now becoming widespread in the developing world, such as in the Tigray region of Ethiopia, while the use of other satellite-based systems (remote sensing) is beginning to be used in disease-monitoring (Thomson et al., 1996). For Africa as a whole, Snow and his colleagues (1999) have sought to produce good estimates of the numbers of people killed each year by malaria. They derive a climate suitability model for the disease, based on climate data for weather stations that are then interpolated onto a 5×5 km grid. Those areas are deemed most suitable for transmission of malaria in which temperature is between 18°C and 22°C (64–70°F) for at least five months, and where rainfall is on average 80 mm (3 ins) per month also for at least five months. Data on population distribution, by age, are obtained for over 4000 administrative areas and also interpolated onto a grid. Last, making some

Box 3.2 Collection and manipulation of digital spatial data

Most paper maps have a system of "geo-referencing" or spatial coordinates that allow us to determine a grid reference, or perhaps a latitude and longitude, for a particular feature. We can also think of linear features (such as roads and rivers) as comprising sequences of coordinates, while areal units comprise ordered sequences of point coordinates that form a boundary. In order to transfer such paper-based information into electronic form we need a means of "capturing" this coordinate information. This is done by a process known as "digitizing," whereby a hand-held cursor on a special table (in which is buried a fine grid of copper wires) tracks the features of interest and transmits the coordinate data to a computer file. Clearly, this can be time-consuming! Fortunately, we do not always have to collect our own digital spatial data, since national mapping agencies do this themselves. In Britain, the Ordnance Survey (OS) makes available digital data at a variety of spatial scales, including large-scale data on the locations of over 27 million individual properties ("ADDRESS-POINT"™) and on road and other transport networks (the Integrated Transport Network™ data).

In the USA the Bureau of the Census provides comprehensive national spatial data coverage. Topologically integrated geographic encoding and referencing (TIGER) files give grid references for street intersections, as well as "topological" data that indicate which blocks lie on which side of each street segment. So-called TIGER/Line® files (revised in 2006) are available publicly that provide data "layers" relating to numerous transport networks and administrative boundaries.

In situations where such databases do not exist, or where paper-based maps are inadequate, we have to rely on other forms of data capture. At broad regional scales we may find remote sensing – the use of satellites – of use, but more locally we use global positioning systems (GPS), as in the South Africa malaria study described in the text.

It is frequently of interest to construct "buffer zones" around point features (such as factories or hospitals). From an epidemiological perspective this allows us to examine disease incidence, not on the basis of fixed areal units but within a given radius of a site. From the viewpoint of health care planning it gives us a simple means of assessing access to health care facilities. We can also construct buffer zones of fixed radius around linear features such as roads or power lines. This, too, could be used to monitor accessibility but a more common use is to assess disease risk along such features; is the incidence of asthma elevated along busy main roads, for example?

A difficulty with buffer zones is that the size of the zone is usually rather arbitrary, and the results depend on how they are defined. This is another manifestation of the modifiable areal unit problem; see pp. 52–3 above.

In order to manipulate all these digital data a proprietary Geographic Information System is required, as described in the text.

See Cromley and McLafferty (2002) and Wilkinson (2006: 16–26) for further details. National mapping agencies, Census Bureaus, and other national organizations have web-sites describing available spatial data.

assumptions about age-specific mortality rates, and applying these to small area population estimates in regions of differential malaria risk (as assessed by climate), Snow and his team suggest that there were likely to have been just under one million deaths from malaria in the continent as a whole during a single year (1995). Others have derived similar estimates, but a GIS "provides a more rational basis for defining disease burden, which is transparent in its inputs, assumptions, limitations and caveats" (Snow et al., 1999: 635). It needs to be borne in mind that the difficulties in collecting data (whether on health or demography) in parts of the developed world are immense.

The possible health effects of living near high-voltage power lines, a source of electromagnetic radiation, has generated considerable controversy in recent years. The issue is difficult to resolve, both because of the problem of measuring the exposure and because the associated health effects (thought to be leukaemia, brain, and other cancers) are rather rare. Some Finnish work (Valjus et al., 1995) has used GIS to assess possible exposure. Finland has a computerized population and buildings register; the locational coordinates of all properties are known, as are residential histories from the 1960s onwards. The authors digitized maps of high voltage power lines (see Box 3.2) and linked these to the buildings and residences files. It was thus possible to calculate the number of people living in buildings located within particular distances or buffer zones of the digitized power lines. Given the locations of the lines, along with average annual currents and distances to the center points of buildings it is possible to estimate the exposure of individuals, over time, to what is known as magnetic flux density. Results (see Figure 3.7) show the distribution of the Finnish population living within 150 meters of power lines in 1989, and their level of exposure. About 15,600 persons were exposed to a field strength greater than the normal background level, and 530 to a significantly greater level. These are small fractions of the total population (about five million persons) and, given the rarity of the possible health events linked to such exposure it would be difficult to draw conclusions about cause (exposure to electromagnetic radiation) and effect (disease). Nonetheless, this is a good example of the value of GIS in a health-related context, made possible only because of the detailed population and buildings database, common throughout Scandinavian countries. More recently, Draper and his colleagues (2005) have sought to link exposure to high voltage power lines to data on childhood leukaemia in England and Wales. Compared with those children who, at birth, lived more than 600 m from such power lines, children who lived within 200 m had a 69% increased risk of leukaemia, and those born between 200 and 600 m had a 23% increased risk. Despite some caveats (not least that the evidence in Figure 3.7 suggests the physical effect of such power lines does not extend beyond about 100 m), this work points to a possible health impact of proximity. Having said this, as one commentator on the Draper study noted, the risk to children is of a much lower order of magnitude than that from (say) road traffic accidents. Both these studies are part of an emerging (and controversial) body of work on non-ionizing radiation. Box 3.3 considers some of the issues involved in assessing possible risk from exposure to other forms of electromagnetic radiation (radiofrequency waves) resulting from mobile phone networks.

A further example of the use of GIS in a health context is drawn from a study of the levels of lead in the blood of children in the city of Syracuse, in up-state New York (Griffith et al., 1998). Here, the authors are concerned to describe and analyze the spatial distribution of such levels; is there an association with transport routes or with residential quality (the age or value of housing, for example)? Lead finds its way into the bloodstream via

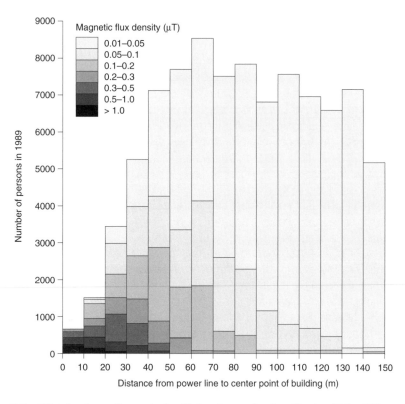

Figure 3.7 Distribution of people in Finland ever having lived within 150 meters of power lines, and their estimated average annual personal magnetic flux densities

emissions from leaded gasoline (petrol), lead-based paint and drinking water (historically transmitted through lead pipes), among other routes. A normal background level is about 0.5 µg/dl, and health experts call for intervention if levels exceed 10 µg/dl; interest therefore centers on levels that exceed this. Using a GIS it is possible to match addresses to streets and census units (tracts and blocks); the dot map of highly elevated levels (see Figure 3.8) shows two areas of concentration, each running from north-west to south-east. The authors place buffer zones around the cases with elevated levels and it is clear from both visual and statistical analysis that there is no association between high levels and proximity to roads.

There is clear evidence of spatial autocorrelation in the blood-lead levels when these are aggregated to either census tracts or census blocks (smaller units); high levels in one census unit tend to be associated with high levels in nearby areas. The authors fit a regression model to the data, in an attempt to find covariates that predict blood-lead levels. Although the findings depend upon the scale – a common feature of geographical analysis – there are striking associations with mean house value (see Figure 3.9) and this proves to be a good predictor. The conclusion is that levels are elevated in the poorer neighborhoods and

Box 3.3   Assessing health risks from exposure to mobile phone masts

With the rapid growth of mobile telephone technology and usage in the last 10–15 years, concerns about the health effects of exposure to radiofrequency (RF waves) have been voiced. This is in part because mobile phone handsets are used close to the human body, but also because large numbers of base station antennas are needed to deliver access to the networks, and anxieties have been expressed about the health of populations living in close proximity to such base stations. A variety of studies have been conducted, including: epidemiological studies of mobile telephone users and workers in RF occupations; experiments on animals exposed to RF waves; and modeling of the spatial distribution of RF field intensity. For example, some studies have detected effects on human cognition and information processing (concentration), on headaches and sleep patterns, and even on rare cancers, including those of the brain. However, a comprehensive WHO report (Valberg et al., 2007) provided little support for adverse health effects arising from RF exposure.

Generalizing from this specific example, there is frequently a tension between two groups of stakeholders: first, experts who, on the basis of scientific studies, and reviews of such studies, draw the conclusion that there is no evidence of health risk; and lay people, often those living close to possible sources of environmental contamination, who claim that there are indeed detectable risks to health. In the presence of such uncertainty some writers would suggest adopting the *precautionary principle*. In general, this principle suggests that if a policy or new technology might cause serious harm to a population, then in the absence of any scientific consensus as to whether or not harm will follow, the burden of proof rests with those who wish to introduce that new technology. Thus, rather than wait for definitive "proof" that RFs do no harm, the precautionary principle suggests that proponents of a new (and potentially harmful) technology ought to have demonstrated that the new technology does no major harm, before it becomes widely employed. An industry response would be that radiofrequency (RF) waves have been used for many years, including for the transmission of radio and television signals, with little or no evidence of health risk. As always, there are competing interests, with claims from lay people and pressure groups that studies funded by industry are likely to be biased, and counter-claims of "scare-mongering" and poor science from industry-sponsored scientists and others.

See www.powerwatch.org.uk for further information but also Wilkinson (2006: Chapter 7) for a fuller account of the health risks from non-ionizing radiation.

that intervention, such as improving the quality of the housing stock, and ensuring it is maintained, needs to be focused here.

A further example of the use of GIS relates to the study of *environmental justice* (Maantay, 2007). Essentially, environmental *in*justice is the disproportionate burden of exposure to environmental pollution or contamination faced by disadvantaged groups (typically, ethnic minority or low income population groups), and the possible effects on the health of such groups. As with the study described in Chapter 1 (pp. 12–16) Maantay looks at the relationships between hospitalization for asthma in part of New York City (the Bronx) and the locations of sources of potential air pollution. She uses the locations of point sources of pollution, drawn from the Toxic Release Inventory (TRI) and the National Emissions

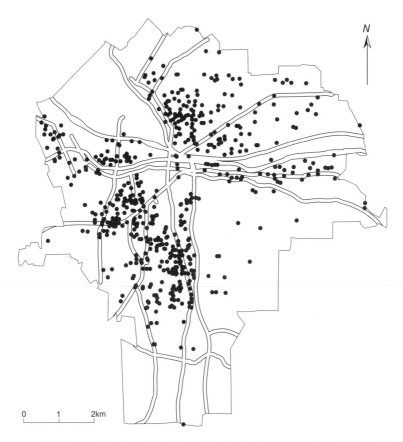

Figure 3.8 Children with high blood-lead levels in Syracuse, New York, with "buffer zones" (150 ft) around roads

Inventory, both compiled by the US Environmental Protection Agency; she also has digital data (TIGER files) on limited access highways and major truck routes. Around these various sources she draws buffer zones (see Box 3.2) and it is clear from census data that more minority and low income groups live within the buffer zones than outside (see Figure 3.10). Further, Maantay links the buffer zones (both separately and as a composite "overlay" map of all possible pollution sources) to the asthma hospitalization data. Unusually, she has privileged access to individual case records, so is able to see whether the residential address of a patient lies within or outside the buffer zones. Her results (see Table 3.2) show that the odds of being hospitalized for asthma are significantly greater for adults and children living within, rather than outside, the buffer zones drawn around most of the putative sources of pollution.

This is a good example of how GIS can assist health geographers and public health researchers to assess relations between environment and health. But a close reading of the paper pays dividends, since the author is careful to draw attention to the shortcomings of

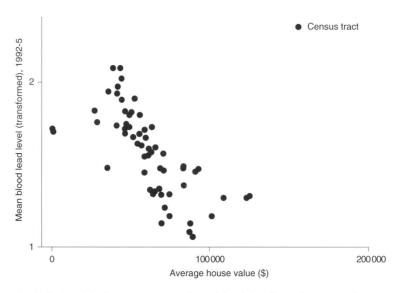

Figure 3.9 Relationship between mean blood-lead levels and average house value in Syracuse, New York

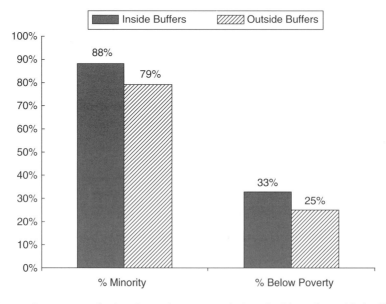

Figure 3.10 Percentage of minority and poor populations inside and outside buffer zones defined by potentially polluting land uses

Table 3.2 Odds ratios for asthma in the Bronx, New York City, with respect to various possible sources of air pollution (1995–99)

Buffer type	Adults	Children	Total population
Toxic Release Inventory	1.29–1.60*	1.14–1.30*	1.33–1.49*
Stationary point sources	1.26–1.66*	1.16–1.30*	1.23–1.32*
Major truck routes	1.07–1.17*	1.00–1.09	1.10–1.15*
Limited access highways	0.90–0.93	0.83–0.99	0.86–0.93
Combined buffers	1.28–1.30*	1.11–1.17*	1.25–1.29*

Notes: Figures are ranges of odds ratios over the five-year period. * denotes statistical significance at the 0.01 level

(*Source*: Maantay, 2007: 49)

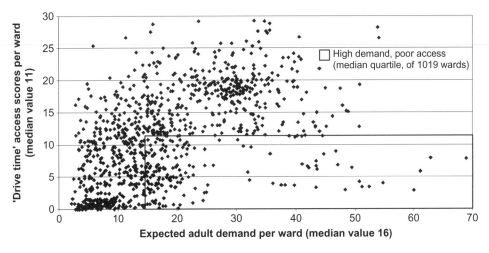

Figure 3.11 Relationship between accessibility to adult in-patient hospice services and expected "demand" for small areas in NW England

both the data and the analysis. Data may be of variable resolution and quality, and doubtless subject to error; merely because data exist in digital form we should not assume automatically that they are "correct." Care, too, must be taken with the analysis; in this specific study, Maantay herself notes (2007: 53) that it "is possible that high asthma hospitalization rates reflect minority and poverty status as much or more than they do high exposures to environmental pollution. [R]egardless of whether the high asthma hospitalization rates are due to environmental causes or result primarily from poverty and other socio-economic factors, the findings of this research point to a health and environmental justice crisis."

Two final examples of the use of GIS relate to a health care planning context, one in north west England (Wood et al., 2004), the other in Canada (Scott et al., 1998). Using data for small areas, Wood and his colleagues examined the relationship between proximity to in-patient hospice services and the possible "demand" for such services. The authors

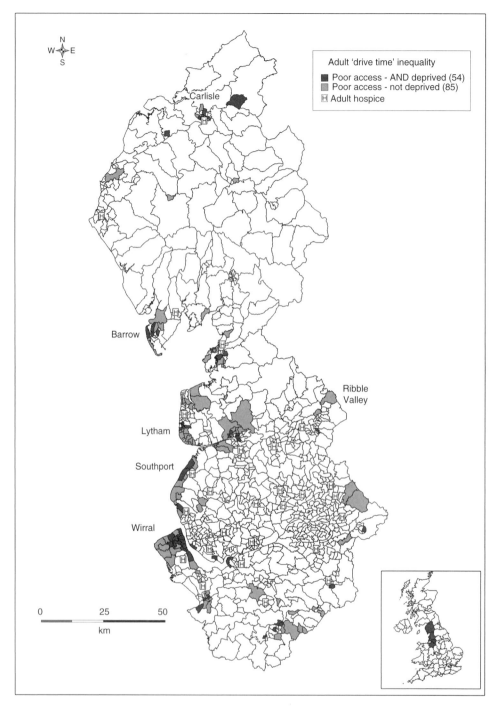

Figure 3.12 Small areas in NW England where accessibility to adult in-patient hospice services is below average and "demand" is above average, classified by deprivation score

record the locations of hospices (which provide palliative care to those with terminal illnesses) and the number of bed spaces each has available. They also estimate the travel time between each small area and each hospice and use this, together with data on bed supply, to derive an accessibility index for each area. Relating this to data on the number of adults with cancer allows them to derive a sub-set of the areas that are *both* relatively inaccessible to the service *and* have relatively high levels of "demand" for that service (see Figure 3.11). That sub-set can be mapped, but also related to the distribution of relatively deprived populations (see Figure 3.12); in this way we can identify small areas that may merit additional investment for in-patient hospice services.

A further example comes from work on access to hospitals in Canada that offer specialist facilities for the treatment of patients suffering an acute stroke (Scott et al., 1998). It is recommended that patients suffering such an episode must be evaluated and treated within three hours of the stroke onset; assuming that the immediate hospital care requires one hour then two hours remain for transport from place of onset to hospital. The Canadian research suggested that this corresponds approximately to a distance of 105 kilometers (65 miles) – an assumption that could be modified using more detailed digitized road networks – and so this radius can be superimposed on the locations of all 110 hospitals offering the requisite care. Using population data from the census it is possible to identify the proportion of the population living within 105 kilometers of such centers. Results suggest that 85% of the total population is so served, though this access varies with socio-economic status. Those who are better educated are closer than those who had completed less than nine years of education. Similarly, while 70% of families with incomes in excess of CAN$50,000 were within the critical distance, only 54% of those with incomes less than CAN$20,000 had such access. We need, therefore, to disaggregate population "need" to look at the needs of particular social groups. Similar work has been conducted in the United States (Love and Lindquist, 1995), while researchers in Kenya have used digitized transport networks to estimate that fewer than two-thirds of the population is within one hour of health facilities (Noor et al., 2006).

Interpreting the Geography of Health: Qualitative Approaches

Until quite recently, quantitative studies of disease incidence tended to focus almost exclusively on aggregate spatial patterns; a good example was that considered in Chapter 1, where the spatial distribution of asthma in New York City was of interest. More recently, as we have seen, attention has switched to quantitative analyses based on individual events, and the spatial arrangement of point patterns that represent collections of individuals with disease have been studied. The danger, however, is that the analytical methods give little attention to what the points or dots on the map represent. The dots are not inanimate objects; they are real people, and while the ways in which they are arranged spatially may shed some light on disease causation, the approaches we have explored so far give no consideration at all to the feelings, experiences, beliefs, and attitudes individuals have.

There are several ways in which we can engage with these individuals in a *qualitative* way. We might want to ask them a set of questions, interviewing them at length but allowing them to shape the conversation rather than dictating it ourselves. We might want to spend long periods of time with them, participating in their daily lives in order to better understand their experiences of illness or disability. How we do this, and how we analyze the material we collect, is the focus of this section. We have already noted that there is no simple one-to-one relationship between theoretical stance and methodology; qualitative methods might be used by social interactionists, structurationists, postmodernists, and others. They are unlikely to find much use among positivists, however, since the latter set store by measurability and objectivity. Those using qualitative methods do so because they seek an understanding of human beliefs, values, and actions, and they do so from the standpoint of a balanced relationship between the researcher and the researched. As we will see, however, the choice of qualitative rather than quantitative approaches to research does not imply a sacrifice of rigor. There is a variety of methods for collecting qualitative data. Each of these is discussed below, in turn. We also spend a brief time discussing the analysis of qualitative data, as well as processes for assessing rigor in qualitative research.

Interviews

Structured questionnaires seek to elicit information from a set of individuals, yielding data amenable to quantitative analysis, perhaps permitting some generalizations to be made from the sample to a wider population. In a health context they invariably ask for information about age and sex, as well as asking about health status. But they do not give the respondent any opportunity to give voice to their (health) experiences, presenting instead a series of questions that invite only a limited number of closed responses (yes/no; good/ very good/poor). It might be possible to make room on the questionnaire for the respondents' own comments, and this is frequently done (for example, "and how does this make you feel?") but an alternative strategy would be to engage the respondent in a conversation, the purpose of which would be to allow them to talk about their own experiences of health and life.

This conversation, or *in-depth interview*, gives the interviewee more scope to describe and explain their own experiences of health and illness, perhaps raising issues that had not occurred to the interviewer. The aim is not to obtain an "objective" account of health status, nor does it treat the interviewees as "objects." Rather, the approach is avowedly subjective, treating those being interviewed as people whose experiences, beliefs, and feelings are to be respected and valued as legitimate sources of data. The essentially hierarchical nature of the questionnaire schedule, where the researcher is "in charge," controlling the process in a position of some power, thus gives way (ideally) to a more equal relationship in which the conversation is shaped as much by the interviewee as by the interviewer. This form of relationship is given particular importance by those engaged in feminist research (Dyck and McLaren, 2004).

Those being interviewed are chosen as an indicative rather than a representative sample. For example, if we wished to explore the uses people make of health services, and of particular interest to us were urban and rural settings, and gender variations in usage, we might wish to interview a small number of men and women living in towns and in rural areas. An interview schedule would be prepared, perhaps outlining a brief list of key themes to be explored during the conversation. We need to emphasize that the sample is not a "probability" one, in the sense that there is a known probability of an individual being sampled; rather, the aim is to select individuals who may provide a rich body of information concerning the problem being tackled. Thus, the sample sizes used in qualitative research are much smaller than those used in quantitative research, typically in the range of 20–30 interviews for an in-depth study. Small sample size is often a source of criticism of qualitative research. However, the aim is for an intensive understanding (as opposed to an extensive exploration) of issues related to health and illness. Another potential source of criticism in qualitative research is the nature of sample selection. That is, while in quantitative research a random sample is selected typically from an existing sampling frame (e.g. electoral lists, tax assessment lists, telephone or street directories), a qualitative sample is often purposefully selected in order to represent not random variation, but maximum variation in views, So, a qualitative researcher will often use a key informant to identify potential interviewees, or use what is popularly called the "snowball" method of sampling, whereby one interviewee is asked if there are others in the same situation who they think should also be interviewed. Furthermore, it is also unusual for qualitative researchers to define the sample size a priori. That is, qualitative researchers will continue to collect data until *saturation* is reached; that is, until no new themes emerge from the data. Regardless, it is essential to remember when reviewing the work of qualitative researchers that the intent of the investigation is not to develop generalizable laws of human behavior; rather, to develop an intensive understanding of the issues under study, drawing upon the voices and experiences of those impacted.

A classic example of qualitative work in health geography is that by Eyles and Donovan (1986). They present an account of how people living in the anonymized West Midlands town of "Mossley Green" make sense of illness and opportunities for health care. They draw upon 30 in-depth interviews as a way of uncovering what it means to be ill and to need health care in this particular locality. A number of themes emerge from their field work, including: how health and illness are perceived; how lay people think about the "causes" of sickness; and the social context of health and illness. In discussing perceptions of health and illness the notion of "biographical disruption" comes to the fore, as when one interviewee says that "In the morning, I'm in agony, trying to get my breath. Half me life I've suffered from fibrositis of the spine . . . I've got pins in me ankles and thumb and in cold weather, they're terrible" (Eyles and Donovan, 1986: 418). Perceptions of health care are shaped by social structure and expectations, as when reductions in public expenditure are thought to be a problem: "We've two lovely hospitals in the town . . . they have to send you out of the beds because they haven't got enough staff" (Eyles and Donovan, 1986: 420). Possible causes of sickness (such as smoking and diet) are thought to be both under the control of the individual but also shaped by industrial history: "I have to walk through the industrial site and there is a white car that has been parked there and it has gone all yellow from the pollution . . . there's a foundry and you can taste it sometimes" (Eyles and Donovan, 1986: 423). Clearly, while some of the narratives carry few clues as to the spatial setting, others are couched within a very place-specific context.

The methodology employed by Eyles and Donovan is therefore explicitly qualitative, in the sense of "learning first-hand about people, observing and analyzing 'real life' and studying actions and learning about ideas as they occur" (Eyles and Donovan, 1986: 416). They seek to take the "common-sense" ideas of their sample, to appraise and analyze this material, and to derive constructs and categories from the observational data. This is what others have referred to as *grounded theory* (literally, theory grounded in, or derived from, the data); see below, page 82.

In-depth interviews have been used extensively as an important research tool in health geography. Elliott and Gillie (1998) and Groleau and colleagues (2006) used qualitative interviews to add to our understanding of the experiences of health and illness for immigrant groups in Canada. Wilton (1996) used qualitative interviews to explore the diminished life worlds of men living with HIV/AIDS in southern California. He further (Wilton, 2003) used this methodology to understand the relationships between poverty and mental health among a sample of ex-psychiatric patients living in an urban area. Driedger and colleagues (2004) as well as Crooks and Chouinard (2005) used qualitative interviews in their research related to women living with chronic illness, while Wilson (2003) used qualitative interviews with Canadian First Nations peoples to explore the relationships between culture, health and place.

As we saw in an earlier section, one approach to geographical epidemiology is a "scientific" study that collects data on disease incidence using official sources, then maps and analyzes these for evidence of clustering or elevated disease risk. In contrast, there is a growing literature on popular or "lay" epidemiology, one that gives priority to the accounts, perceptions, and fears of those who worry that there is a raised incidence of disease, possibly caused by environmental contamination. We consider here one example, since it makes extensive use of qualitative methods. The Sydney Tar Ponds are located on the east coast of Canada in the Province of Nova Scotia. The legacy of over 100 years of steel production and coking works operations, the Tar Ponds now contain over 700,000 metric tones of sediment contaminated with polyaromatic hydrocarbons and PCBs, both of which are known carcinogens (Haalboom et al., 2006). Haalboom and colleagues were funded by the Canadian government to undertake research on community environmental health concerns. While the primary health concern of the local population was understandably cancer, not far behind was a concern about reproductive health. There were "urban legends" about couples having difficulty conceiving as well as inordinate numbers of stillbirths, miscarriages and low birth weight babies. All of these adverse reproductive outcomes were being blamed on the tar ponds. A "scientific" epidemiologic study documenting the reproductive histories of 500 women living within 2 kilometers of the tar ponds site showed no significantly raised levels of adverse reproductive outcomes (Burra et al., 2006). The community, however, was not convinced. Haalboom undertook 29 in-depth interviews with men and women living within 2 kilometers of the tar ponds site. The thematic analysis of the interview data reveal that despite the results of so-called "scientific" studies, the community remains worried about their health in general, about the possibilities of contracting cancer, and the potential for adverse reproductive outcomes, to the extent that they are making decisions around conception based on their perceptions. While some delayed conception until after health study results were to be published, others decided on adoption, relocation to another part of Canada, or foregoing having children at all. Furthermore, the community had completely lost faith in the government that was supposed to be protecting their health. While the contamination on site was discovered in 1981, at the time of writing, no steps had been taken towards site clean up. The community also felt marginalized; the fact that this community is located in a socio-economically deprived area of Canada indicated to them that they just did not matter: "If the tar ponds were located in any other part of

Canada, it would have been dealt with by now" (Haalboom et al., 2006: 235). This example reveals something of the value of detailed interviews, in which such qualitative methods are seen as the only way of revealing local community understandings of health problems. If we want to get to grips with people's health worries we can only do this by engaging with them at length. Having said this, there are obvious dangers in the way respondents are sought and in verifying the information they reveal, and few public health researchers would want to leave unchallenged some of the "evidence" presented. They would argue that different cancers, or reproductive outcomes, have different possible causes, that we know little about the sources or duration of exposure to environmental contamination or about possible confounders (e.g. lifestyle factors, genetics), and that unless we place a very local study in a broader regional context we have no way of knowing whether the popular concerns are supported by "hard" facts. This tension between different forms of *evidence* is a recurring theme in contemporary health geographies.

There are methodological problems in conducting interviews, and the interview setting is invariably power-laden, in the sense that the interviewee may feel in an inferior position relative to that of the interviewer. This is less the case than with a questionnaire (and certainly less so in feminist research), but there is still a risk of academic "voyeurism" in which the researcher gains valuable data and the respondent does not benefit. As we saw in the Dyck and McLaren (2004) example in Chapter 2, the researchers went to great lengths to empower their interviewees, so much so in fact that the interviewees co-opted the interview experience to gain what they could from the researchers in terms of information about job, education, and other economic opportunities for new immigrants in Vancouver. Indeed, Dyck and colleagues (Kirkham et al., 2002) have considered the methodological implications of making interviewees feel "safe" during the interview experience, particularly when the focus is on race, ethnicity, class or gender. They suggest the methodological operationalization of the notion of "cultural safety:" The overall goal of cultural safety [in interviewing] is to take remedial action against the very conditions that prompted its formulation: deep-seated disadvantage and inadequate understanding of Maori models of health" (Kearns and Dyck, 1996: 372). For example, in an attempt to provide cultural safety, Dyck and colleagues provided a Cantonese translator for an interview with an Asian woman. The interviewee, however, insisted on speaking in English. Dyck suggests in fact that in this case, even something as simple as language could be seen as a source of social power. "Although culture and its material contexts undoubtedly remain critical determinants of how health and illness are interpreted and experienced, as researchers we must be alert to the colonial inscriptions and hierarchical power relations that tend to follow any attempt to construct discrete cultural groups (Kirkham et al., 2002: 229). Research participants, particularly those from disadvantaged groups (the poor, women, ethnocultural groups, indigenous peoples) now assert their needs and their own power into the research process: "no more about us without us" (Wilton, 2004a).

The taping of interviews can be problematic. Apart from the obvious risks of technical failure, the instrument can itself be a barrier to interaction or, for some, it can be perceived as yet another means of surveillance. This was certainly the experience of one of the authors (Elliott) when conducting in-depth interviews in the former Soviet Union. Moreover, the spatial setting of the interview cannot be ignored; in what place should it occur? The home setting may be comfortable for the interviewee, though issues of the safety of the interviewer must be addressed.

Focus groups

A focus group is a collection of a small number of people (usually between 4 and 12) that meets to discuss a topic of mutual interest, with assistance from a facilitator or moderator. Usually, the group members are "key informants," that is, they represent particular positions or interests. The discussions are informal and consist essentially of exchanges of views, opinions, and personal experiences. The atmosphere is intended to be non-intimidating and to permit discussion that does not necessarily result in agreement or consensus. As with a series of interviews, much thought must be given to the location and physical setting in order to create the optimal environment for interaction. Discussions are taped and transcribed (as with in-depth interviews) for subsequent thematic analysis. Assembling an appropriate mix of people is not always straightforward; nonetheless, focus groups do serve to give people a sense of "ownership" in any research, particularly if findings are fed back to them as part of the process of *member checking*.

A good illustration of the method comes from Garvin's (1995) report on the development of a partnership between female academics and community researchers in order to research women's health and health needs. The setting is Dickenson County, in the Appalachian region of south-western Virginia; this is a rural area, with a history of coal-mining but now suffering from high unemployment. The cooperative research project sought to "give a voice to Appalachian women's experiences, draw attention to their concerns and quality of life, reduce excess burdens of illness, facilitate the development of a culturally sensitive research process, and create a planning process that could be diffused to surrounding counties as women learned from and about other women" (Garvin, 1995: 274).

Six focus groups, comprising in total 44 women, were organized according to location or area of interest. Discussion was focused on a number of issues, including: women's health concerns; health-seeking behavior; perceived barriers to accessing health care; and perceived quality of local health services. Following transcription, the academic researchers identified and summarized key themes that were then sent to the community researchers (including nurses, social workers, and health visitors) for comment. Themes were grouped into three sets of factors affecting use of services. These were: predisposing factors or individual attitudes (such as religious beliefs); enabling factors, concerned with local or federal policies that permit or constrain health behaviors (transport provision, for example); and reinforcing factors that reward or modify subsequent behavior (such as the quality of service). In general, users felt that services were inaccessible, inconvenient, and poorly delivered. They were well aware of the need to adopt a "healthy" lifestyle, but argued that the lack of facilities for exercise, the lack of choice in local food stores, and the expense of getting screening and other preventive care all acted as barriers from acting on knowledge. This is a theme to which we return in the next two chapters, but meanwhile this study serves as a good example of how to elicit information on health services from the people they are supposed to serve.

As hinted at above, however, the transcription – and hence interpretation – of focus group data is complex, to say the least. We are reminded of this by Lehoux et al. (2006) in their analysis of the use of focus groups in an attempt to understand "the patient's view." In so doing, they quite rightly assert that there has been increasing interest in the health services' research on the patient's view, facilitated in large part by the wider use and acceptance of qualitative methods. While focus groups are indeed part of this suite of methods, the authors highlight for us *the* unique feature of focus groups: the social interaction that occurs between participants, and between participants and the moderator/facilitator. In order to accommodate the uniqueness of this constructed social space in the analysis of focus group data, they offer up a template that "acknowledges the processes through which participants attribute authority to the claims of others, including the focus group moderator" (Lehoux et al., 2006: 2091).

Other qualitative methods

While the majority of qualitative research within health geography employs either inter-views or focus groups as a primary data collection method, there are indeed others. Many researchers use non-participant or participant observation as a data collection tool. In the case of the former, a researcher embeds him/herself into a research situation, but in an unobtrusive manner that does not affect the interaction being observed. Hester Parr's work (2002a) using Internet chat room discussions as empirical data for understanding the use of health information on the Internet is a good example. Alternatively, participant observa-tion entails full-scale participation in the events of the research nexus. For example, Johnsen et al. (2005) used participant observation to explore the meaning and experience of "soup runs" for the homeless in Britain. Along with in-depth interviews, one member of the research team worked as a volunteer on several different soup run projects in one city over a four-month period, while another member of the team did so in an alternative city. "Such work proved vital in establishing relationships of trust with service users who can otherwise be wary of talking with 'outsiders', in facilitating interviews with people who have chaotic lifestyles, and enabled observation of the complex dynamics shaping service environments. Notes relating to the informal discussions with staff and service users . . . were used to triangulate survey and interview data" (Johnsen et al., 2005: 326).

Documentary analysis has also been used in health geography. For example, Robinson et al. (2005) analyzed the project reports of three provincial initiatives in Canada to examine the utility of linking systems between public health resources and user organiza-tions for health promotion dissemination and capacity building. Even the analysis of pho-tographs has contributed to our understanding of health geographies. For example, in their analysis of community health and safety, Mitchell and others (2006) gave disposable cameras to young children and asked them to take photos of "dangerous things" in their neighborhood. An analysis of these photos led the authors to conclude that, in order to assess neighborhood safety, one had to look at things through the eyes of those affected.

Qualitative data analysis

In contrast to quantitative approaches, in which the analysis of data follows on, and is quite separate, from the collection of data, qualitative data analysis is often bound up closely with data collection. Typically, data collection is used to refine, discard, and then reformulate ideas, quite differently from the formal testing of hypotheses used in much quantitative work.

Following, in part, Grbich (1999) we shall discuss methods of analysis very briefly under three headings. These are: *enumerative*, where documentary material is analyzed, typically using content analysis; second, *investigative* methods, such as discourse analysis; and third, *iterative* methods, where interview material is analyzed. Clearly, the "distance" between the researcher and the subject decreases from enumerative through iterative methods.

Content analysis examines documents such as text or visual materials, specifying elements or units (such as words, phrases, or images) to be analyzed and noting the relative frequency of particular codes. Discourse analysis sees texts as historically produced materials that embody and reflect power relations. A "discourse" is a theme or subject made up of "discursive practices;" for example, the medical records of an individual, and how he or she interacts with a health professional are discursive practices within a wider discourse of, say, patient health. A key issue is to trace the power relations that have shaped the discourse; such relations might be between doctors and nurses, or those professionals engaged in health promotion and the public they seek to influence. Discourse analysis thus draws to some extent on Foucault's post-structuralism, nicely illustrated in an analysis of health service reform in the UK in the early 1990s (Moon and Brown, 1998).

With iterative approaches, the researcher collects data via interviews or observation, transcribes this information (for example, typing the interview from the recorded conversation), and reflects on what it has to say about the topic under investigation. This is a preliminary analysis that annotates transcripts, highlighting in the text some ideas in a process known as "open coding." It requires a detailed, slow, and careful reading of the text, with the analyst's comments in the margins or sections of text highlighted carefully. Emerging concepts and themes then shape further data collection, and data collection halts when saturation is reached. The skill in qualitative data analysis lies, then, in the coding of the text and the subsequent identification of themes. Broad themes may need disaggregating, in a process known as "axial coding." But there is a continual interaction between the analyst and the data, in order to ensure that nothing is missed and that the codes and themes "fit" the data. This process is clearly inductive, in the sense that one moves from the data to an emerging set of concepts in order to develop a theory. Since this is very much grounded in the data, the inductive nature of the process is known as a *grounded theory* approach (Grbich, 1999: 171–80).

MacKian (2000) has introduced some novel forms of visualization into the qualitative analysis of health narratives. In essence, this has led to her devising "maps" as heuristic devices to allow us to understand a respondent's experiences and what has shaped these. Drawing on her research into people with chronic fatigue syndrome (myalgic encephalomyelitis or ME) she demonstrates very convincingly that these visual representations reveal the structure of experience and the relative locations of the factors that relate to this experience. MacKian's respondents in turn confirmed that the maps were valid representations of their worlds. There is also a substantial literature around the use and integration of multiple qualitative methods to address the same or similar research questions. This method of *triangulation* (derived from the similar process when surveying topography; that is, using three points instead of one to find an accurate "answer" to a locational question) attempts to arrive at *completeness* or *comprehensiveness* when asking a research question using qualitative methods. While many qualitative studies refer to the importance of triangulating models, methods, theories, few provide a how-to of steps to accomplish the task. A recent exception is Farmer et al. (2006) who used triangulation of methods and researchers to understand organizational capacity building in public health around chronic disease prevention.

Regardless of the data collection or analysis method used, it is important to reflect on the growing use of computer software packages designed for use in the analysis of qualitative data (Pope et al., 2006). It is important to realize, however, that until relatively recently there was substantial resistance to the use of computer software in the analysis of qualitative data by those who felt that the use of such packages contaminated a "pure" ethnographic approach. Nothing could be further from the truth; the researcher(s) still must spend substantial amounts of time reading and coding interview transcripts or documents or letters; the use of, for example, NVIVO (http://www.qsrinternational.com/) simply allows for a more elegant and efficient management of large data sets. For example, the average interview lasts one to one- and-one half hours. This translates into about 30–50 pages of transcribed text. The average project involves approximately 30 interviews; that literally means at least 900 pages of transcript. That's a large data set to handle without any help!

Rigor in qualitative research

The very nature of qualitative research leaves it open to criticism by quantitative research-ers; small sample sizes, use of conversations as data, and analysis which – until recently with the introduction of computer software packages for qualitative data – was based on simply reading the transcripts of the conversations held. Under such circumstances, it was not unusual to hear qualitative research referred to as "touchy feely" and/or "subjective" as opposed to systematic and rigorous (Crang, 2003). However, over the past couple of decades, researchers have described how we might evaluate the rigor in qualitative research (Giacomini and Cook, 2000) and indeed how to build it into the research process (Lincoln and Guba, 1985; Baxter and Eyles, 1997).

Lincoln and Guba (1985) identified four criteria for establishing rigor in qualitative analysis. *Credibility* is defined as the authentic representation of experience and is tanta-mount to what a quantitative researcher would think of as *validity*. *Transferability* is defined as the fit of the research findings outside of the specific study situation and is tantamount to the notion of *generalizability* in quantitative research. *Dependability* is defined as the minimization of idiosyncrasies in interpretation and is similar to what quantitative researchers would refer to as *reliability* or *replicability*. *Confirmability* refers to the extent to which biases, motivations, interests or perspectives of the inquirer influence interpreta-tions. This is similar to the quantitative notion of *objectivity* in the research endeavor. These authors go on to describe the strategies that the qualitative research must employ to prac-tice each of these criteria. Any reader interested in undertaking qualitative investigation should be guided by this work.

On the basis of a meta-analysis of qualitative (primarily interview based) studies, Baxter and Eyles (1997) further developed a general set of criteria (as opposed to rigid rules) for assessing the rigor in qualitative research undertaken in social/health geography. As the authors show, these criteria translate into a checklist of questions for use in evaluating qualitative interview research (Baxter and Eyles, 1997).

Concluding Remarks

The distinction between quantitative and qualitative methods is long established, and it cannot be denied that those researching the geography of health have tended to use one or the other set of tools, a conclusion mirrored in health research in general. Others, however, take a more pragmatic and eclectic stance, using whatever methods are appropriate to the problem under investigation. Indeed, for some researchers, the mixing of both quantitative and qualitative methods has proved singularly useful, with insights from in-depth interviews adding color and explanatory power to quantitative studies. For many geographers both sets of methods can shed light on most research questions. We shall see evidence of this eclecticism in the rest of this book.

This chapter ends the section of the book that deals with approaches to, and methods for, conducting enquiries into the geography of health. We turn next to a series of substantive issues. These deal first with the social environment, where it is quite clear that the different styles of enquiry compete for attention. In the second substantive section we look further at issues linking the physical environment to health outcomes; here too the different methods of explanation raise their heads, though perhaps not as prominently as earlier in the book.

Further Reading

Some modern approaches to spatial data analysis, including those for handling both point and area data, are considered in Bailey and Gatrell (1995); this gives several health-related examples. For details of SaTSCan, including a download of the software, trial data sets, and an extensive bibliography, see: www.satscan.org. Helpfully, the bibliography is structured by application area, so if literature on clustering of breast cancer (for example) is of interest, this is readily accessed.

On GIS, the outstanding reference work is Cromley and McLafferty (2002) which *inter alia*, introduces many of the spatial analysis techniques considered in this chapter. For another study of how GIS and spatial analysis assist in the study of environmental justice see Fisher et al. (2006).

On the spatial modeling of health data you should look at Haining (2003) and Congdon (2003), but considerable statistical expertise is required to conduct these forms of analysis. For an empirical study of area-based data that uses several of the spatial analysis methods outlined in this chapter, see the study of prostate cancer in Louisiana (Mather et al., 2006).

On multilevel modeling see Jones (1991) for a pedagogic overview, and Jones and Duncan (1995) for a clear exposition and good example. There are now dozens of applications of MLM in health research, as a literature search will confirm.

The book edited by Flowerdew and Martin (2005) has several chapters on research methods and techniques. In particular, the chapters by Valentine on interviewing, Cook on participant observation, and Crang on the analysis of qualitative material, are all valuable sources for the qualitative researcher, and merit a close reading. Miles and Huberman (1994) is a key source on qualitative methods, while Mason (2002) is exceptionally useful.

Part II

Health and the Social Environment

Chapter 4

Inequalities in Health Outcomes

In this chapter we first describe some of the patterns of inequality relating to health outcomes, particularly as they manifest themselves in the developed world. There is, of course, a large number of possible health "outcomes" on which to draw, including mortality (both all-cause and numerous specific causes), life expectancy, and many indicators of morbidity or disability. Inevitably, therefore, the account will be selective, but we hope that the guide to further reading (page 123) will suggest possible sources for evidence not covered here. There is a huge literature on health inequalities, particularly from those interested in the sociology of health and illness. Much of this focuses on inequalities that are expressed in terms of social class, gender, and ethnicity. We touch on all three, but give particular emphasis to the spatial patterning of health inequalities and to explanations that give some prominence to place.

We adopt an approach that is based on geographical scale, beginning with some patterns of variation at the international scale, before examining some of the broad regional differences in selected countries. Although it is difficult to draw a firm distinction between descriptive and explanatory accounts (if for no other reason than that most studies try to explain their descriptive findings!), we then turn in the second half of the chapter to reviewing some of the competing explanations for health inequalities. As we shall see, some authors consider these to lie in varying health behaviors, while others believe that health inequalities can only be explained by structural factors that are deeply embedded in human societies. Others offer different explanations for these inequalities.

It is also worth pointing out that while the term "health inequalities" has become accepted, we really mean "health inequities." The distinction is important. Inevitably, there will be unevenness or variation in health outcome, whether from place to place or along some other dimension. But what really matters, and what "inequity" implies, is the fact that these differentials may be avoidable, and should be capable of being narrowed; their existence is, in a sense, unethical. Fundamental issues of social justice should be borne in mind in a reading of this and the following chapter.

Patterns of Inequality

Health inequalities: international comparisons

We begin by drawing attention to broad contrasts between countries in the developed and developing worlds. There are numerous health indicators that might be used to describe these contrasts. Mortality data are an obvious source, though the variable quality of death certification makes this problematic. Morbidity data are even more poorly reported in some parts of the world (Phillips, 1990). Some comparative work uses data on Disability Adjusted Life Years (DALYs); DALYs are defined as the sum of years of life that are lost due to premature mortality and years lived with a disabling condition (adjusted for severity); thus they take account of morbidity and disability as well as mortality.

Research conducted at the Harvard School of Public Health in association with the World Health Organization (Murray and Lopez, 1996; Lopez et al., 2006), known as the Global Burden of Disease Study, has used both mortality data and DALYs in order to quantify and compare ill-health across the world. Using data for 2001, in both high-income and low-income countries, ischaemic heart disease (see Box 4.2 below) and cerebrovascular disease (stroke) were the leading causes of death, accounting for 20% of all deaths world-wide. But while lung cancer ranks third as a cause of death in high-income countries it does not figure prominently in poorer regions; here, it is infectious diseases (diarrhoeal disease, tuberculosis, HIV/AIDS, malaria, and lower respiratory infections) that are primary causes of death and disability (see Figure 4.1). Further, of the 10.5 million children aged under five years who died in 2001 (and do please pause to contemplate that number) 99% were in low-income countries and more than 40% in sub-Saharan Africa; under-nutrition and poor access to safe drinking water and sanitation remain the primary causes. It is also worth drawing attention to the high burden carried in Europe and central Asia due to cardiovascular disease (see Figure 4.1).

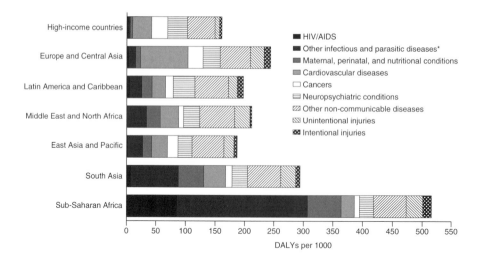

Figure 4.1   The global burden of disease: Disability Adjusted Life Years, 2001

While these statistics offer a reasonably up-to-date snapshot, patterns of mortality and morbidity do change over time (see Lopez et al., 2006 for comparisons between 1991 and 2001, and Ruger and Kim, 2006 for changes over 40 years); in particular, the contribution made by different causes of death is highly variable from country to country, and over time, and this has been encapsulated in the concept of *epidemiological transition* (sometimes known more generally as "health transition"). In essence, this proposes that at an early stage of development causes of death such as parasitic, infectious, and nutritional diseases account for the burden of mortality, but then these give way to non-communicable diseases. Diseases that account for high rates of infant mortality give way to the degenerative diseases, such as heart disease and cancer, in adulthood and old age (Phillips, 1990; 1991). As overall levels of fertility and mortality fall, diseases that are common in the developed world begin to account for a greater proportion of deaths in developing countries.

This is a very simplified and brief account, but we can illustrate the epidemiological transition with reference to countries in south-east Asia (Phillips, 1990; 1991; 1994). In Hong Kong, infectious and parasitic diseases (mostly diphtheria, enteric fever, and dysentery) accounted for nearly a quarter of deaths in 1951, a proportion which had fallen to 3% in 1988 (see Table 4.1). The pattern of cancer deaths was the reverse; in 1988 these accounted for 30% of deaths (and in 2005, just over 31%), compared with 4% in 1951. Rates of heart disease and stroke have also increased (29% in 1988, similar in 2005). This is evidence of very rapid epidemiological transition. Thailand shows a similar pattern; the crude death rate from tuberculosis dropped from 47 to 12 per 100,000 between 1970 and 1983, while the crude death rate for all cancers rose from 13 to 27 per 100,000. Malaysia and Singapore offer a similar picture (Phillips, 1991). Note, however, that many of these figures pre-date the emergence of HIV/AIDS as a global health problem; this, coupled with the re-emergence of diseases such as tuberculosis means that we cannot point (even in the developed world) to an end stage of the epidemiological transition in which simply diseases of "affluence" remain (Smallman-Raynor and Phillips, 1999).

Mathers and Loncar (2006) provide some projections of global mortality and disease burden through to 2030. Their projections are based on regression models of age- and sex-specific mortality, by cause, using as predictors: gross domestic product (GDP); "human capital" (taken here to be years spent at school by adults); and time. The models were estimated using historical data (1950–2002) for 106 countries. After fitting the models, projections of GDP and human capital, as well as smoking impact, were used to derive projections of mortality under baseline, optimistic and pessimistic scenarios. Taking just the baseline scenario we can speculate on the leading causes of death, for both high-income

Table 4.1   Epidemiological transition in Hong Kong

Year	Infectious and respiratory	Cancer	Cardiovascular	Parasitic diseases
	Percentage of all deaths			
1951	23.6	4.2	5.5	27.4
1961	16.2	12.1	7.4	14.8
1975	4.0	24.2	26.6	15.8
1988	3.0	29.8	28.6	17.2

(*Source*: Phillips, 1990: 45)

Table 4.2   Projected ten leading causes of death in 2003, by income group

Rank	High-income countries	Percentage of deaths	Low-income countries	Percentage of deaths
1	Ischaemic heart disease	15.8	Ischaemic heart disease	13.4
2	Cerebrovascular disease	9.0	HIV/AIDS	13.2
3	Respiratory cancers	5.1	Cerebrovascular disease	8.2
4	Diabetes mellitus	4.8	COPD	5.5
5	COPD	4.1	Lower respiratory infections	5.1
6	Lower respiratory infections	3.6	Perinatal conditions	3.9
7	Dementias (incl. Alzheimer's)	3.6	Road traffic accidents	3.7
8	Colo-rectal cancer	3.3	Diarrhoeal disease	2.3
9	Stomach cancer	1.9	Diabetes mellitus	2.1
10	Prostate cancer	1.8	Malaria	1.8

Note: COPD is chronic obstructive pulmonary disease
(*Source*: Mathers and Loncar, 2006: 2023)

and low-income countries (see Table 4.2). Circulatory diseases (ischaemic heart disease and cerebrovascular disease) are likely to account for a quarter of all deaths in high-income countries, and close on a quarter of deaths in poor countries. But in rich countries several cancers, in aggregate, are projected to account for 12% of deaths, while cancer does not appear in the top ten leading causes in poor countries; here, HIV/AIDS is projected to account for over 13% of all deaths, and perinatal causes and road traffic accidents some 4% each. Inevitably, projections are bound by uncertainty; a major new international public health crisis may well lead to a very different outcome in 2030. Nonetheless, the broad contours may be clear from these figures.

What is the response of the major international agencies to these contrasts? The United Nations established in 2000 a series of Millennium Development Goals, a series of eight measurable targets for reducing poverty and improving health and development in industrializing countries (see Box 4.1). All goals are, directly or indirectly, health related, embracing reductions in child and maternal mortality, and reductions in the incidence of infectious disease such as HIV/AIDS, tuberculosis and malaria.

Of course, this broad global canvas masks huge contrasts between countries, even within low-income groups in the same region of the world. In Africa, one of the most appalling sets of socio-economic circumstances is in Zimbabwe, where health status has worsened considerably in recent years. Specifically, life expectancy at birth was, in 1990, 60 years. Ten years later it had dropped to 39.8 years, while the most recent data from the WHO Africa Regional Office (2005) indicate it is currently 37.3 years. Infant mortality (53 per 1000 live births in 1990) had risen to 81 per 1000 in 2005. And, in a five-year period, the maternal mortality rate (per 100,000 live births) had risen from 610 in 1995 to 1100 in 2000 (see Figure 4.2). Contrast this with a neighboring country, Mozambique. Here, while life expectancy was still very poor (42 years in 2005) it had not worsened since 1990; and infant mortality (158 per 1000 in 1990) had improved by 50%. In addition (see Figure 4.2), the

> ## Box 4.1 Millennium development goals
>
> The eight Millennium Development Goals (MDGs) are to:
>
> (1) Eradicate extreme poverty and hunger
> (2) Achieve universal primary education
> (3) Promote gender equality and empower women
> (4) Reduce child mortality
> (5) Improve maternal health
> (6) Reduce incidence of HIV/AIDS, malaria and other diseases
> (7) Ensure environmental sustainability
> (8) Develop a global partnership for development
>
> Within the set of goals and targets there are some 48 indicators. For example, Goal 4 is "reduce child mortality," and the specific target is to reduce by two-thirds, between 1999 and 2015, the under-five years' mortality rate. The indicators within that goal include not only the under-five mortality rate but also the infant mortality rate and the proportion of children aged one year who are immunized against measles. Regular progress charts are produced by the UN, data for 2007 indicating that for some goals poor progress is being made in sub-Saharan Africa and that there is a low probability of meeting targets in many parts of the world by 2015 (the original target date).
> See www.un.org/millenniumgoals/pdf/mdg2007-progress.pdf

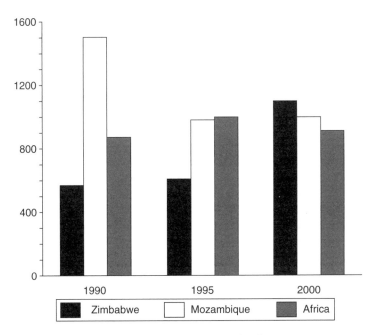

Figure 4.2 Maternal mortality rates per 100,000 live births

maternal mortality rate was, in 2000, on a par with that for Zimbabwe, when it had been 50% higher in 1990 (see Chapter 5, pp. 132–3).

These changes in health status mirror those in the respective economies. In Mozambique, the average annual growth of GDP per capita in the period 1985–95 was 1.9%, but it was 6.1% in the following decade (to 2005). Zimbabwe's average annual growth in GDP per capita was considerably higher in the earlier decade (3%) but was actually negative (−3.4%) over the decade 1995–2005.

There is a clear political context to these differences between two low-income countries. Zimbabwe has been led, since its independence from Britain in 1980, by Robert Mugabe who, in recent years, has embarked on land redistribution (to transfer ownership from white farmers to black). The chaotic nature of this program of land reform has led to severe food shortages; for example, the loss of maize (corn) production has led to the government's "Operation Maguta" – "live well" – that has imposed such production on farmers in regions not well suited to the crop. In addition, his initiatives to destroy shanty towns (such as "Operation Murambatsvina" – literally "drive out the filth") are motivated by a wish to stifle political opposition. In contrast, Mozambique, also a former colony (of Portugal) that achieved independence in 1975, is, after many years of civil unrest, in a relatively settled political environment that encourages foreign investment.

One Europe or many?

As we shall see later, particular attention has been paid, within the health inequalities literature, to varying incidences of circulatory diseases: those involving blood circulation (see Box 4.2). Data from different countries in Europe reveal huge variations, suggesting that a homogeneous late stage in the epidemiological transition is a gross over-simplification. It is evident from Figure 4.3 that there is a very clear gap between eastern Europe and the rest of the continent. For men in particular, Russia has a substantially higher death rate than any other country, followed by the Baltic States of Estonia, Lithuania and Latvia. Former communist countries in central Europe (Poland, Czech Republic) follow close behind, with countries in western Europe showing (broadly similar) lower mortality rates. Taking a more historical perspective, mortality rates since 1970 in Eastern Europe have increased while those in all countries of Western Europe have declined (Watson, 1995). This is gender-biased, however, in that while both men and women in eastern Europe do worse than in the west the differential is far greater among men (see Figure 4.4; (a) shows the graph for men, and (b) that for women). Hungary had, in 1990, the highest death rates among both men and women, but mortality rates among eastern European women had actually improved slightly over the 20-year period. The increased death rates among men were due to a number of causes, including heart disease, respiratory disease, and cancer. But these data refer to the period 1970–90, and pre-date the break-up of the former Soviet Union; what of more recent trends?

Life expectancy at birth, for men in the European Union as a whole, rose from 72 years in 1985 to about 73.5 years in 1994. Life expectancy in the former USSR had always been lower than in the west; and in the Russian Federation the figure was 63 years in 1985. However, life expectancy has declined sharply since then (see Figure 4.5a), and in 1994 was less than 60 years of age; although not quite so dramatic, the same is true for women

Box 4.2 Circulatory diseases

There is not space in this book to discuss in any detail the nature of specific diseases, but since diseases involving blood circulation have been the focus of considerable attention in the health inequalities field they merit brief consideration here.

   The most common form of circulatory disease arises from the deposition of fatty material (atheroma) in the linings of the arteries that supply blood to the heart muscle (myocardium). The resulting narrowing of the arteries and insufficient blood supply is known as *ischaemic heart disease*. In an extreme case, the narrowing of the arteries causes a *myocardial infarction* or coronary thrombosis (commonly known as a "heart attack"). Diets that are high in saturated fats have a high ratio of cholesterol to protein and increase the volume of "low density lipoproteins" in the blood. Since these low density lipoproteins transport fats to muscles they are a risk factor in heart disease. Tobacco products also serve to degenerate coronary arteries and are therefore an additional risk factor. The generic name for diseases of the heart and blood vessels is *cardiovascular disease*.

   Narrowing of the arteries in the brain causes "stroke," a form of *cerebrovascular disease*. Since the brain cannot function without supplies of oxygen, supplies that are provided by the blood flow, strokes are frequently associated with symptoms of paralysis, usually on one side of the body or face. Stroke shares some of the same risk factors as ischaemic heart disease, including diets rich in saturated fats, and smoking.

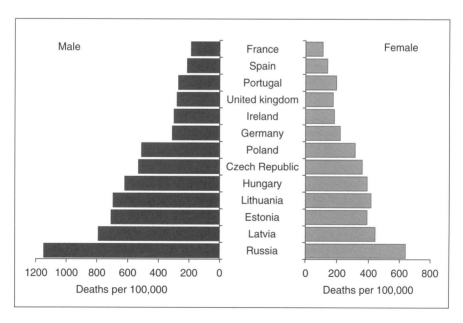

Figure 4.3   Age-adjusted mortality from circulatory diseases in 2004

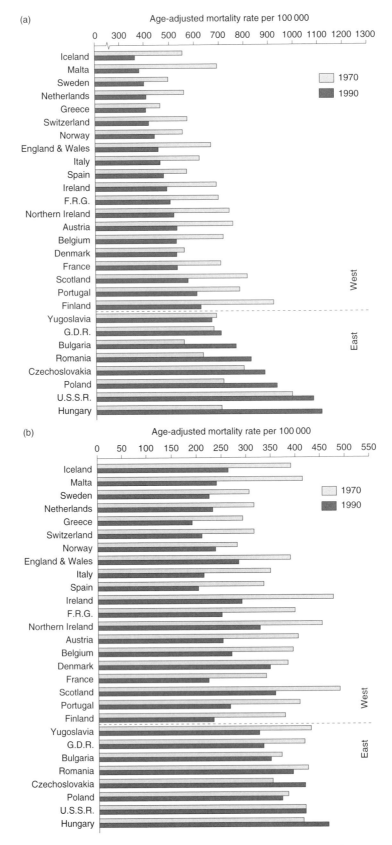

Figure 4.4 Mortality rates in Europe for: (a) men and (b) women aged 25–64 years: 1970 and 1990

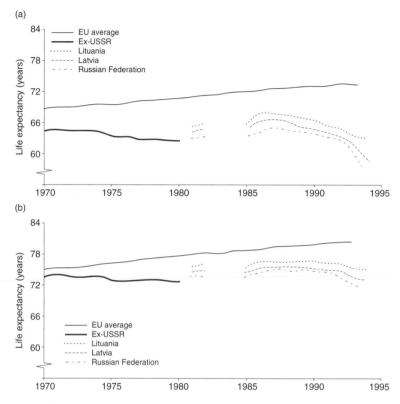

Figure 4.5 Life expectancy at birth for: (a) men and (b) women in European Union and selected countries of the former USSR

(see Figure 4.5b). More recent evidence suggests fluctuations in life expectancy in Russia; improvements in the late 1990s seem to have given way to more recent declines. These figures for Russia as a whole, however, mask considerable spatial variation, as analysis of data for 52 regions reveals (Walberg et al., 1998). More recent data from the European Office of WHO suggest that there are big gaps in life expectancy; for example, in 2002, in the administrative territory of Ingushetia male life expectancy was 75 years but only 55 in the Republic of Tuva.

There are, then, striking differences between countries, but we need to ask whether different social groups within such countries are at equal risk from early death? To what extent are there international differences according to socio-economic status (see Box 4.3)?

We may point up some international contrasts between those people working in manual occupations, compared with those in non-manual classes (Kunst, 1997). From Figure 4.6 we see variations in the probability of a middle-aged man dying between ages 45 and 65, for both the USA and selected European countries. For those in non-manual occupations

Box 4.3   Socio-economic status

In the USA inequality of health outcome is commonly assessed with reference to either race or income. Income is widely used, since the US census collects such data on a decennial basis. In other countries, such as Britain, no income data are generally available. As a result, another dimension, that of social class or socio-economic position, is used instead. This is usually based on occupation. In Britain a six-fold classification is adopted, as the following table (with example occupations) illustrates:

Social class		*Example occupation*
I	Professional	Doctor, lawyer, accountant
II	Managerial	Sales manager, teacher, nurse
IIIN	Skilled non-manual	Clerk, secretary
IIIM	Skilled manual	Bus driver, chef, carpenter
IV	Partly skilled	Farm worker, security guard
V	Unskilled	Laborer, cleaner

Often, classes I and II, and IV and V, are combined, while there is a simple distinction between non-manual (I, II and IIIN) and manual (IIIM, IV and V) occupations. Clearly, this scheme ranks occupations in terms of prestige and social status and, implicitly, in terms of the income each attracts. It is not without problems, however. Married women who do not work outside the home are usually assigned the social class of their husbands. Those out of work, or children, or the elderly, are also imperfectly captured by this scheme.

   As an alternative, we might use household tenure or educational level as measures of status, or car ownership as a possible index of "wealth." There are problems with all of these indicators, but each in its way is strongly associated with mortality and morbidity.

there is relatively little difference, while for those in manual occupations the probabilities of mortality in eastern Europe (again, for Hungary in particular) are up to twice as high as in western countries. This pattern is repeated across other dimensions of social status, such as level of education. Kunst (1997: 149) further shows that in 1990 the ratio of mortality rates among men aged 30–44 in manual, compared with non-manual, occupations, was 2.25 in the Czech Republic and 2.89 in Hungary. Comparable figures for the USA and England and Wales were 1.42 and 1.46, respectively. These differences are maintained for all leading causes of death.

Health inequalities: regional and class divides

We turn now to some of the evidence for health inequalities within particular countries. As in the previous section, some consideration will be given to social class and gender differences, though we introduce ethnicity as a further dimension of health divides. We concentrate on evidence relating to England and Wales, and the USA.

   It is useful to begin by observing that, despite the fact that the literature on health inequalities has mushroomed over the past 30 years, health inequalities have been

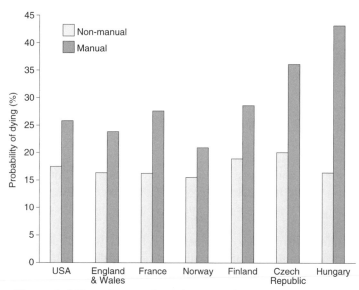

Figure 4.6   The probability of a man dying between the ages of 45 and 65

Table 4.3   Mean age at death of members of families belonging to various social classes in England (1840s)

District	Gentry/professional	Farmers/tradesmen	Laborers/artisans
Bath	55	37	25
Rutland	52	41	38
Bethnal Green	45	26	16
Leeds	44	27	19
Manchester	38	20	17
Liverpool	35	22	15

(*Source*: Macintyre, 1998: 21)

documented since at least the early part of the nineteenth century (Macintyre, 1998). Edwin Chadwick's report into the working conditions of working-class people in Britain showed how life expectancy varied by both social class but also by area (see Table 4.3). The social class gradient is striking, but note, too, the higher life expectancy in the more affluent southern areas of Rutland and Bath; life expectancy there was higher for all social classes.

Much more recently, broad regional differences in health outcome in England and Wales have been revealed. For example, data on life expectancy at birth for males in 2003–5 show that life expectancy in southern regions was on average nearly three years higher than in the north. Although, on average, life expectancy for women was between four and five years more than for men, the same broad regional differences are observed for women too. At a finer geographical scale (English health authorities), data for 2003–5 reveal considerable variation, with the "healthiest" district (Kensington and Chelsea) reporting a life expectancy of 82.2 years for men (86.2 years for women) and the least healthy (Manchester) a life expectancy of 72.5 (men) and 78.3 (women). Put simply, a male born in the most privileged part of London could expect to live almost ten years longer than one born in Manchester, while a female could expect to live about eight years longer. Within these districts there is smaller-scale variation; thus, in parts of Manchester male life expectancy is under 70 years.

A classification of local authorities according to socio-economic characteristics indicates that the differences in life expectancy are magnified, varying from 75.8 (80.4 years for females) in the most prosperous areas to 71.7 (77.5) in port areas. A similar health "divide" exists if we look at how morbidity varies among these broad social areas (see Table 4.4). Rates of heart disease and mental illness are much higher in traditional industrial areas than in the most prosperous ones.

Among many important studies, the work of Shaw and her colleagues (1999) deserves careful reading, since it looks in a very imaginative way at premature mortality (deaths under 65 years) for parliamentary constituencies in Britain. In particular, it contrasts those constituencies with a population of one million persons aged less than 65 years, which have the highest SMRs, with a similar set having the lowest SMRs. There are many findings of interest here; to cite just one, the Scottish constituency of Glasgow Shettleston has an SMR of 234 (1991–5 data), while Buckingham in southern England has an SMR of just 71. Further, this gap has widened since the 1980s (Shaw et al., 1999: 118).

Some of this geographical work highlights broad areal contrasts; we need to begin to understand the extent to which such contrasts simply mirror the composition of the population. Do certain regions have low life expectancy, or high mortality, simply because poorer people, or those belonging to a different ethnic group, live there? These are difficult questions, on which the whole chapter touches, but we can begin to answer them by examining some of the evidence on social class and other gradients.

Table 4.4 Prevalence of treated disease in England and Wales, by sex and ONS area classification (1996)

	Coronary heart disease		Depression or anxiety	
	Men	*Women*	*Men*	*Women*
Coast and country	35.2	19.4	39.8	85.7
Mixed urban and rural	33.1	20.1	34.2	80.9
Growth areas	30.6	16.8	33.4	77.7
Most prosperous	25.8	15.7	27.0	63.2
Services and education	29.8	17.9	26.6	64.5
Resort and retirement	33.6	18.8	38.7	81.0
Mixed economies	36.7	23.6	36.6	83.2
Manufacturing	35.3	23.0	37.7	83.9
Ports and industry	43.0	28.5	41.8	95.4
Coalfields	42.0	26.7	41.2	91.6
Inner London	24.8	14.7	35.4	73.7
England and Wales	34.7	20.8	36.2	81.9

Data are age-standardized rates per 1,000 patients
(*Source*: *Health Statistics Quarterly*, Office for National Statistics (Crown Copyright 1999), reprinted by permission)

Mortality rates in England and Wales show clear evidence of a social class gradient, a gradient which is observed for all causes and which is consistent across most of the major causes of death. For example, mortality rates among women and men in manual occupations are substantially higher than in professional and managerial social classes, as Table 4.5 indicates for deaths from coronary heart disease. For men, the mortality "gap" seems to have widened over the past 25–30 years; in other words, while mortality has improved substantially among non-manual groups, improvements in manual workers have not kept pace. In contrast, the mortality gradient for women has narrowed. Most cancers show higher mortality in lower status groups, though breast cancer shows no evidence of a gradient (and skin cancer mortality – not shown in the table – also tends to be more common among professional people; see Chapter 9).

In Britain, a series of studies of those working for the government in office-based work ("civil servants") has revealed clear social class gradients in health outcomes. These studies (known as the Whitehall studies) began in 1967 and have tracked more than 17,000 individuals. Early work (Whitehall I) focused on mortality and revealed that the lower the grade of staff, the higher is the standardized mortality for heart disease (and, indeed, for virtually all diseases); there is a four-fold difference in rates between the highest and lowest grades of occupation (Marmot et al., 1978). In other words, even within a single occupational group of non-manual workers there are major differences in mortality risk. Later work (Whitehall II) has concentrated on morbidity and has confirmed the social class gradient for heart disease, bronchitis, and self-reported health status; recent interest has focused on the psychosocial dimensions of health (see, for example, Head et al., 2006).

In the USA there is a large variety of health and health-related social indicators that can be used to paint a contemporary picture of health divides. Consider evidence (Miringoff and Miringoff, 1999) on inter-state differences in three such indicators: life expectancy, low birth weight, and births to teenagers. Nation-wide, life expectancy at birth for white males is 73.9 years, but only 66.1 years for black males, a gross disparity. Geographically, life expectancy is higher in the western and mid-western states of Utah, Colorado, Minnesota, and Wisconsin (where some counties show average life expectancy of over 76.5 years). Contrast this with parts of the rural south or inner city areas (Bronx, New York, or Baltimore, for example), where men can expect to live 12 years less. The poorest life expectancy is in six counties of South Dakota (61 years), reflecting the presence of native North

Table 4.5 Age-standardized death rates from coronary heart disease (ages 35–64 years) by sex and social class in England and Wales (1976–99)

	Women		Men	
	1976–81	*1997–99*	*1976–81*	*1997–99*
Professional and managerial	39	22	246	90
Skilled non-manual	56	30	382	117
Skilled manual	85	41	309	141
Partly skilled and unskilled	105	50	363	107
Ratio of total non-manual to total manual	2.23	1.73	1.19	1.50

Source: British Heart Foundation coronary heart disease statistics; www.heartstats.org/temp/Tabsp1.8spweb07.xls

Americans on reservations. Also of interest is to contrast neighboring Washington DC (62.2 years) and Fairfax, Virginia (76.7 years).

There are also considerable variations in the proportion of infants who are of low birth weight (under 2500 grams). At least 8.7% of all births in states in the south (Mississippi, Louisiana, Alabama, South and North Carolina, and Tennessee) are of low birth weight, compared with under 6% in the north-west (Washington and Oregon) and in New Hampshire (4.8%). The same contrasts are observed for births to women aged 15–19 years. Rates (per 1000 such women) are over 70 in states such as Mississippi, Arkansas, and the south-western states of Arizona, New Mexico and Texas, but under 33 per 1000 in the New England states of Massachusetts, Maine, Vermont, and New Hampshire.

As above, we need to know whether these geographical differences reflect social composition; in particular, in the USA to what extent are these broad differences attributable to the varying life chances of black and white people? There is a clear patterning of health outcome according to "race" or ethnicity (the American literature tends to adopt the former term, though the latter is more common elsewhere). For black American adults aged 25–44, the all-cause mortality rate among both men and women is over twice that of white adults (Sorlie et al., 1995). These differences are spatially variable, however. For example, the SMR among black men aged 15–64 living in Harlem, New York City is 411 (1989–91 data), but only 181 among black men in the southern state of Alabama (Geronimus et al., 1996); for black women the same figures are 338 and 189 respectively. Given that the "standard" is 100 for white people then mortality rates are over four times as high for some black populations. Put differently, a 15-year-old black person in Harlem has a 37% chance of surviving to the age of 65; this compares with a 77% chance for blacks as a whole, and 87% for whites. Other evidence (Geronimus et al., 1999) confirms this and indicates that for both black women and men there is a rural "advantage" to living in southern states, compared with cities such as Chicago, New York, and Detroit where both homicide and AIDS have led to worsening mortality between 1980 and 1990 (see Figure 4.7; (a) shows the graph relating to men; (b) that relating to women). Clearly, a focus on race/ethnicity in the absence of "place" will not suffice.

Do these stark contrasts remain when we adjust for income? In other words, do black people whose family income is similar to that of whites have similar health outcomes, or do the differences remain when allowance is made for these influences? The answer seems to be that they do. Regardless of race, those with lower incomes have consistently higher mortality than those who have higher incomes; this is true for both men and women (see Table 4.6). Yet, across any particular category, blacks have higher mortality rates. Using data from the US Longitudinal Study, black male mortality rates (among 25–44-year-olds) are 36% higher than for whites, and 69% higher among black women, even when adjustment has been made for employment status, income, and education (Sorlie et al., 1995).

Similar inequalities according to ethnicity emerge in Britain, though particular care needs to be taken with "non-whites" in this context since it is problematic to assign an "ethnic group" to an individual, while health outcomes vary substantially from one group of non-whites to another. Allowing individuals to categorize their ethnicity according to family country of origin, and looking at data on self-reported health status, evidence suggests (see Figure 4.8) that non-white ethnic minorities as a whole are 20% more likely than whites to report poor or only fair health. This masks important differences, however, and even the crude "south Asian" category cannot distinguish the significantly poorer health

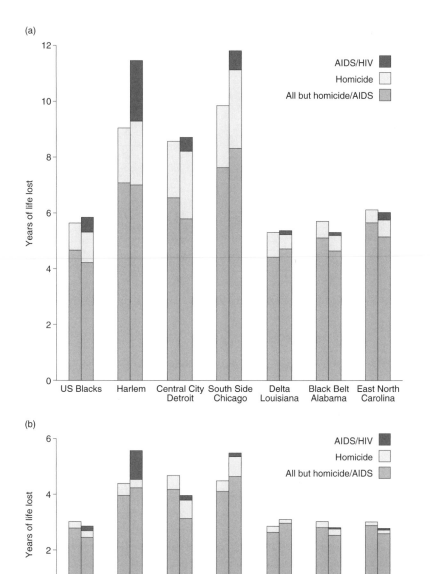

Figure 4.7 Years of life lost in areas of USA in 1980 and 1990, by cause: (a) black men and (b) black women

Table 4.6   Mortality rates in USA (25–64-year-olds) in 1986[a]

	Women		Men	
Annual income ($)	White	Black	White	Black
<9000	6.5	7.6	16.0	19.5
9000–14999	3.4	4.5	10.2	10.8
15000–18999	3.3	3.7	5.7	9.8
19000–24999	3.0	2.8	4.6	4.7
>25000	1.6	2.3	2.4	3.6

[a] Age-adjusted death rates, per 1000
(*Source*: Macintyre, 1998: 23)

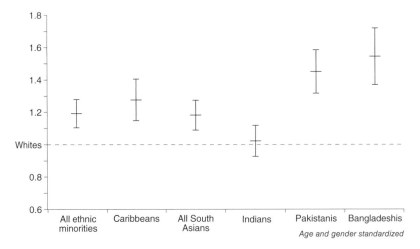

Figure 4.8   Relative risk of non-whites reporting fair or poor health compared with whites: risk for each sub-group, with 95% confidence interval around each

of Pakistanis and Bangladeshis from the much better health of "Indians." The latter is itself a very crude grouping, masking the very different cultural, religious, and geographical origins of people; for example, Muslims tend to report worse health than Hindus or Sikhs (Nazroo, 1998).

Explaining Inequalities in Health Outcomes

Inevitably, having outlined some of the huge range of descriptive findings relating to health inequalities it has not been easy to avoid hinting at some of the possible explanations. We have seen, for example, that there are socio-economic gradients, and that health outcomes vary by sex and ethnicity. We need now to consider some of the types of explanation put forward to shed light on such inequalities.

We can picture health status in terms of a series of layers of influence, with a set of fixed factors (age, genetics, and sex) at the core, surrounded by a series of factors that are potentially modifiable. The first layer comprises lifestyle factors representing behavior(s) that may or may not be conducive to good health. The second represents social networks and community influences, while the third comprises living and working conditions, as well as access to services and facilities in the local area. Finally, there is a set of wider structural determinants representing the influence of macro-economic and broad-scale social conditions on individual health (Wilkinson and Marmot, 2003).

This schema is an advance on the traditional distinction between behavioral (lifestyle) and material conditions; in other words, a simple "human agency" and "structure" dichotomy. Yet it, too, is an oversimplification, since the different layers impact on each other, as when our individual "choices" about health behaviors (whether or not to smoke, for example) are in part affected by wider-scale structural influences (such as the freedom of cigarette manufacturers to advertise). Interestingly, from a geographical perspective there is something of a correspondence between these layers and spatial scale, since we move from individual influences, to local or neighborhood effects, through to broader national scale determinants. With these observations in mind, we now examine some of the evidence relating to each "layer."

The programming hypothesis and the lifecourse

As we shall see in the next section, the conventional causal model in much chronic disease epidemiology emphasizes risk factors or behaviors, such as smoking, diet, and lack of physical activity, as determinants of health status. This has been challenged by a number of workers and research groups, who claim that chronic diseases (the major causes of mortality and morbidity in much of the western world) such as diabetes and cardiovascular disease are biologically "programmed" in early life (in the womb or in infancy). David Barker and his colleagues have produced a considerable body of evidence in support of this theory, particularly with respect to cardiovascular disease (Barker, 1992; 1994; Barker and Bagby, 2005).

From a geographical point of view, there is compelling evidence in support of Barker's hypothesis. For example, those areas in England and Wales where people were suffering poor health in the early part of the twentieth century (as measured by infant mortality between 1921 and 1925) are the same areas in which adult heart disease rates are high between 1968 and 1978 (see Figure 4.9); there is a close correlation, both for men ($r = 0.69$) and for women ($r = 0.73$). If infant mortality is further divided into neonatal and post-neonatal mortality (deaths within the first 28 days, and between one and 12 months, respectively), the association with coronary heart disease (CHD) is stronger with the former. Barker argues that in the areas of high neonatal mortality infants in general were at risk; while the weakest died, others survived into adulthood, but their health was compromised by the poor health and physique of their mothers as well as poor nutrition. These visual and statistical correlations are, however, only suggestive; we cannot be certain that the relationship is with early, as opposed to more recent, poor social conditions. When we

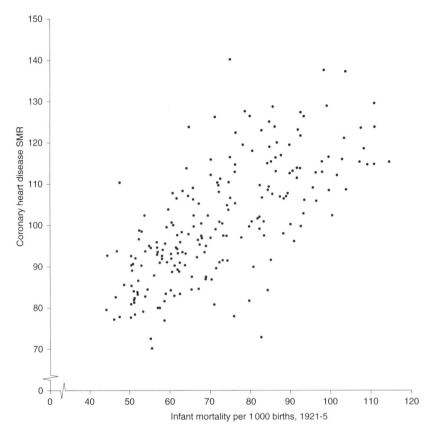

Figure 4.9   Relationship between adult male mortality due to heart disease (1968–78) and infant mortality (1921–5), districts in England and Wales

provide controls for social conditions in adulthood we find that the relationship between infant mortality and adult mortality is weakened, suggesting that geographical variation in mortality is explained by adverse social circumstances in adulthood as much as by early life experiences (Ben-Shlomo and Davey Smith, 1991; Heyman et al., 2006). What we need as evidence are disaggregated data for individuals rather than data aggregated over areal units. Barker himself (Barker and Bagby, 2005) provides a summary of the evidence of biological mechanisms using individual data (see also Nuyt, 2008). In terms of individual data there is a growing body of evidence to confirm the programming hypothesis. For example, Barker and his team have uncovered historical data collected by midwives working in Hertfordshire, England, between 1911 and 1930; such data include birth weight and weight at one year for 37,000 live births (Barker and Bagby, 2005). Using the National Health Service Central Register it was possible to trace the subsequent mortality of these children. Analysis of the relationship between birth weight and CHD mortality is revealing (see Table 4.7); hazard ratios for coronary heart disease fell with increasing birth weight. This association between low birth weight and CHD has now been replicated in many studies, in both men and women, in the USA, Europe and India.

Table 4.7 Hazard ratios (95% confidence intervals) for death from CHD according to birth weight at year 1 in 10,636 Hertfordshire men

	Death from coronary heart disease	
Weight (in pounds)	*Before 65 years*	*All ages*
<=5.5 (n = 486)	1.50 (0.98 to 2.31)	1.37 (1.00 to 1.86)
5.6 to 6.5 (n = 1385)	1.27 (0.89 to 1.83)	1.29 (1.01 to 1.66)
6.6 to 7.5 (n = 3162)	1.17 (0.84 to 1.63)	1.14 (0.91 to 1.44)
7.6 to 8.5 (n = 3308)	1.07 (0.77 to 1.49)	1.12 (0.89 to 1.40)
8.6 to 9.5 (n = 1564)	0.96 (0.66 to 1.39)	0.97 (0.75 to 1.25)
>=10 (n = 731)	1.00	1.00
p for trend	0.001	0.005
1 year of age		
<=18 (n = 715)	2.22 (1.33 to 3.73)	1.89 (1.34 to 2.66)
18.1 to 20 (n = 1806)	1.80 (1.11 to 2.93)	1.58 (1.15 to 2.16)
20.1 to 22 (n = 3404)	1.96 (1.23 to 3.12)	1.66 (1.23 to 2.25)
22.1 to 24 (n = 2824)	1.52 (0.95 to 2.45)	1.36 (1.00 to 1.85)
24.1 to 26 (n = 1391)	1.36 (0.82 to 2.26)	1.29 (0.93 to 1.78)
>=27 (n = 496)	1.00	1.00
p for trend	<0.001	<0.001

(*Source*: Barker and Bagby, 2005)

On the basis of clinical and other empirical evidence Barker and Bagby (2005) posit three processes through which this relationship is operating. First, someone whose fetal growth was restricted will have fewer cells or functional units in key organs such as the kidneys. A second process through which slow fetal growth may be linked to chronic disease in later life involves hormones and metabolism; that is, an undernourished fetus may establish a thrifty gene for coping with low energy (i.e. food) intake in later life. Third, Barker puts the programming hypothesis into a larger context by suggesting that the link between low birth weight and chronic disease in later life may be because people who were small at birth may be more vulnerable to adverse environmental influences in their later life.

If we accept Barker's findings, the implications for public health are quite profound; namely, to put as many resources as we can into improving the nutrition of expectant mothers in order to prevent or inhibit chronic disease in adults. However, some researchers are skeptical about the programming hypothesis. Some dispute the link between early biological markers (such as birth weight) and adult disease, arguing that this may simply reflect being born into poor families; the social may matter as much, if not more, than the biological. Others prefer to argue that risks of developing chronic disease cannot be attributed solely to either early life or adult experiences. Rather, such factors (which may be both biological and social) are likely to operate cumulatively throughout life; this is the so-called *life-course* approach to explaining variations in disease incidence (Kuh and Ben-Shlomo, 1997). As Frank and colleagues (2006) point out, coronary heart disease – the major cause of death in the developed world, and the primary determinant of hospitalization, after childbirth – is multifactorial, having more than 20 known risk factors of biological significance, as well as dozens of other factors acting over the entire life course. What evidence favors this mode of explanation?

The life-course perspective clusters around three generic models. The first – the *latency model* – suggests relationships between an exposure at one point in time in the life course and the probability of a health outcome years or decades later, regardless of any intervening experience. For example, exposure to asbestos in the workplace can elevate the risk of various cancers decades after exposure has stopped. The *cumulative model* refers to multiple exposures over the life course that combine to impact health. For example, an individual may live in chronic poverty, live in an unhealthy neighborhood and be a regular smoker. The *pathway* model suggests a series of pathways between childhood and adulthood. These pathways link the parents' economic circumstances to low birth weight and to the child's subsequent socio-economic, physical and psychosocial environments; low birth weight may well shape adult disease, but so, too, do socio-economic conditions throughout life. For example, the divorce of someone's parents in early childhood may reduce that child's readiness for school, which may in turn affect school performance; this could, in turn, affect employment opportunities and hence economic status and therefore health throughout life (Hertzman and Power, 2006). Powerful support for these ideas comes from longitudinal studies that trace a cohort of people through from birth to adulthood. For example, a National Child Development study was set up in Britain in 1958 to examine a birth cohort; these individuals are now in their forties and their health at this age can now be studied. Power and her colleagues (1999) constructed a lifetime socio-economic score, based on the father's occupation when an individual was born and at age 16, together with the individual's own social class at age 23 and 33. There is a striking relationship between the proportion of people reporting poor health at age 33 and this lifetime score.

It is complicated to assess whether the current, or much earlier, stage in the life course impacts upon health. As a further example of this, note how birth weight (taken to represent "early environment") is confounded with obesity (a socially patterned adult risk factor) in terms of its impact on heart disease incidence (Barker and Bagby, 2005). The incidence does indeed increase for those of lower birth weight, but this relationship is mediated by obesity; the gradient is much higher for more obese people and much shallower for those who are not. This, and much other, work indicates that research into social circumstances, and health behaviors, throughout the life course will continue to be productive. Hertzman and Frank (2006), in fact, articulate a long-term research agenda that will help us to explore these truly transdisciplinary questions. The next section addresses the contribution of lifestyle behaviors.

Behavioral (lifestyle) factors

Here, we consider the argument that people experience poor health because they are more likely to participate in health-damaging behaviors, and that since these behaviors are socially patterned (with those on low incomes or of otherwise lower socio-economic status more likely to engage in unhealthy behaviors) such behaviors are likely to be a key determinant of health inequalities. Typically, the literature deals with four such aspects of lifestyle – diet, alcohol consumption, smoking, and lack of physical activity. We first look at the UK experience (see Table 4.8), then move on to review the data from the USA and Canada. We pay particular attention to the "new" global epidemic: obesity.

Table 4.8 Health behaviors in England

Social class	Professional and managerial	Skilled non-manual	Skilled manual	Partly skilled and unskilled
Men				
Eats wholemeal bread	21	15	15	14
Eats fruit, vegetables, and salad daily	66	51	59	52
Drinks skimmed or semi-skimmed milk	74	75	65	58
Mean weekly units of alcohol	16.8	15.1	18.9	29.0
Total current cig. smokers	22	30	36	39
Smoking banned at work	35	32	15	18
Smoking not allowed in house	35	32	24	24
Percentage ever using any illegal drug	39	45	33	41
Frequency of exercise (five days/week or more)	32	35	47	55
Uses sun cream	72	65	55	51
Two or more sexual partners in last year	18	35	25	31
Women				
Eats wholemeal bread	30	25	29	23
Eats fruit, vegetables, and salad daily	81	71	66	68
Drinks skimmed or semi-skimmed milk	84	76	74	64
Mean weekly units of alcohol	7.5	8.0	6.7	6.8
Total current cig. smokers	20	28	35	34
Smoking banned at work	47	43	31	33
Smoking not allowed in house	41	30	25	24
Percentage ever using any illegal drug	29	24	21	25
Frequency of exercise (five days/week or more)	31	25	39	37
Uses sun cream	84	79	66	64
Two or more sexual partners in last year	13	17	9	17

Figures are percentages unless otherwise specified; social class based on last or current job

(*Source*: Office for National Statistics (Crown Copyright 1997), *Health in England 1996*, reprinted by permission)

It is well-established that excess consumption of fats, salt, and unrefined sugars contributes to heart disease. There is evidence from British data that food consumption patterns vary with socio-economic status, those of lower social status groups being less likely to consume fresh fruit and vegetables, foods that have a high dietary fiber content, or milk with a lower fat content. Consumption of alcohol varies across social class for men, but not for women, though the data do not indicate what types of drinks are consumed.

Smoking is a well-established risk factor for much chronic disease, notably lung cancer and heart disease. The class gradients for smoking are clear, for both men and women. Interestingly, this social class difference did not exist 40 years ago. In Britain, smoking behavior depends very much on social circumstances; nearly 70% of single parents on low incomes, living in social (public) housing, working in manual occupations, and with few educational qualifications, are regular smokers (Townsend, 1995). The context in which smoking takes places also varies; those in higher social classes tend to tolerate it less in the home, while legislation in many parts of the developed world has banned it in many indoor environments.

Smoking is also implicated as a cause of sudden infant death syndrome ("cot death"). But smoking tends to affect perinatal mortality (stillbirths and deaths in the first week of life) only among women of lower social status, so although smoking may lead to low birth weight (and therefore has a potential impact on perinatal mortality as well as adult health – see the previous section) this does not seem to be true for middle-class women, whose risks of having a poor pregnancy outcome are presumably mitigated by other factors we are considering in this chapter.

Similar patterns exist in the USA and Canada, with individuals of lower socio-economic status exhibiting much higher rates of health damaging behaviors. For example, data from the American's Changing Lives longitudinal survey (Lantz et al., 1998) show statistically significant differences in levels of smoking, alcohol consumption, body mass index, and physical activity both by education level as well as income (Lantz et al., 1998: Table 2). A national survey of 35,000 Canadians indicates a similar pattern, with education playing a greater role in the explanatory models than income (Potvin et al., 2000). Similar findings emerge in many other countries in the world, including France and Northern Ireland (Yarnell et al., 2005) and Sweden (Molarius, 2003).

This body of work on the socio-economic differentials in the uptake of risky health behaviors extends to what is currently referred to by WHO as a global epidemic: obesity (see Caballero, 2007). Essentially, the obesity problem is an easy one to understand; that is, it represents an imbalance between calorie intake and use of those calories for physical activity. For the last couple of decades, we have seen in North America and much of the developed world a situation where we simply take in too many calories and do not expend enough of them, therefore resulting in large proportions of the population being overweight or obese. Many factors contribute to this situation: the rise of sedentary leisure activities (television, video games), automobile dependent cities, sedentary jobs, as well as increasing access to high fat fast foods (see Box 4.4). With rates of overweight and obesity affecting one-third of the adult population in both Canada and the USA, evidence suggests that we live in a generation where children – also victims of the obesity epidemic – may no longer outlive their parents. Policy responses have involved the reintegration of physical activity programs in schools as well as the banning of junk food. Data also indicate socio-economic differentials in the risk factors that contribute to overweight and obesity, with lower levels of physical activity among members of lower socio-economic status groups.

Box 4.4   Obesity and food environments

Considerable attention is currently being paid to the so-called "obesity epidemic" in the developed world. We considered some studies in Chapter 3. As with many health problems, debates center on the extent to which such problems are, or should be, under the control of the individual or the household, or are shaped by "context," whether at a local, national, or even international scale. To what extent does national food policy, the strategies adopted by food producers and retailers, and (from a geographical viewpoint) the locations of food outlets, have a bearing on geographies of obesity?

This has been considered by several writers, but a good and brief overview is in Cummins and Macintyre (2006). They suggest that proximity to poor quality food environments (such as shops and stores selling relatively little fruit and vegetables, or fast food outlets) "amplifies individual risk factors for obesity such as low income, absence of transport, and poor cooking skills or knowledge" (Cummins and Macintyre, 2006: 100). In the USA, in general, low income and minority neighborhoods have poorer access to healthier foods, while fast food outlets are often located in poorer areas. However, the authors claim that this finding does not seem to extend outside the USA. Despite this, some governments are taking "context" seriously. For example, the British Government in early 2008 announced a new Healthy Living strategy, which looks at food labeling, cookery lessons in the classroom, healthier school meals, as well as urban planning initiatives to restrict the locations of fastfood outlets near schools, and policies to encourage walking and cycling.

Whether unhealthy eating is due to poor "choices" or other factors is hotly debated, as Box 4.4 suggests. Lifestyle or behavioral explanations of health inequalities have attracted plenty of criticism, not least because they have an element of "victim-blaming." The explanation is very individualistic and it is too easy to blame people for their "feckless" or "irrational" behavior. For a young mother seeking to cope with small children in circumstances of severe poverty, smoking may be quite understandable, stress-coping behavior. Others would wish to see strategies tackling root causes of poverty, or to challenge business interests that seek, via subtle or not-so-subtle means, to capture new smokers, particularly among the young. However, behavioral explanations of inequalities have certainly appealed to governments reluctant to engage in public spending. Consequently, public health strategies have long advocated an emphasis on setting targets for risk factors (reductions in smoking, improvements in diet, increased levels of exercise, and so on) as the chief way of acquiring health gains. A classic example is the COMMIT study, undertaken in North America in the 1990s. Over $50 million was spent, over five years, on community-based interventions, to try to get individuals to quit smoking. The results were very modest; despite the scope and intensity of the intervention, only 3% of heavy smokers across 12 intervention communities reported having quit smoking, while none of the light or moderate smokers reported having quit. Further, the overall quit rates in the intervention (3.5%) and comparison (3.2%) communities were virtually the same (COMMIT Research Group, 1995).

To what extent, then, can health behaviors "explain" health variations? Does the social class gradient in smoking, for example, explain the social class gradients in heart disease?

On the face of it, the answer to this question should be "yes," since smoking and dietary fat consumption are major risk factors for heart disease and are elevated in those from lower socio-economic groups. The Whitehall I study, however, showed that cholesterol levels were greater in the higher grade occupations, so it does not appear that cholesterol variations can account for higher rates of heart disease in lower grade occupations. More generally, the "classic" risk factors such as smoking and diet explain only about one-third of the variation in heart disease rates, whether among British civil servants or using international data.

Disentangling behavioral from other factors (particularly the "material" circumstances considered later) is not easy. This is because although some health-damaging behaviors are more common among groups of lower social status, such groups are also those with poorer access to material resources, such as adequate incomes and hence decent housing and food or secure employment. The general conclusion concerning lifestyle-based accounts of health inequalities is that they are only a very partial explanation of mortality and morbidity; structural factors and material conditions shape behavior, influence choices and have direct impacts upon health.

Social and community influences

There is a growing body of opinion that social isolation and lack of engagement in local community life contribute to poor health and even early death. In a sense, this idea is not new, since the French sociologist Durkheim showed that suicide was more common among socially isolated individuals. More recent epidemiological work suggests that individuals' social networks – the social connections and relations they have with friends and family – are a predictor of all-cause, and cause-specific, mortality or of poor health. A prospective study of over 37,000 American men aged 42–77 years in 1988 (Kawachi et al., 1996) showed that those who were socially isolated (defined as unmarried, having fewer than six relatives or friends, and playing no part in community groups) had an elevated risk of mortality from heart disease, even after allowing for other risk factors. Kiecolt-Glaser and colleagues (2002) studied isolated, elderly, spousal care-givers of Alzheimer's patients by giving them a small puncture wound and observing the subsequent healing process. They found dramatically reduced rates of wound healing compared to a control group.

The suggestion from various studies is that social networks provide social support, whether of a tangible (perhaps financial) nature, or in the form of advice and information (perhaps health-related), or more general emotional support in times of stress. In other work this notion of social support has been extended to the wider concept of *social capital*, which embraces a variety of aspects of the social environment. These include, for example: feelings of trust and safety; the quality of connections among people living in neighborhoods; the number and quality of connections with family and friends; and the extent to which people are members of community groups, societies, clubs, and other organizations. Notwithstanding the difficulties in adequately capturing the concept of social capital, evidence from a number of studies suggests that it has some value in explaining health variations. For example, work in the USA by Kawachi and others (1997) measured social capital in 39 states, using data on the membership of voluntary groups and levels of social trust. Low levels of social trust, and of voluntary group membership, are strongly correlated with mortality from most causes, an effect which is only modestly attenuated when an adjustment is made for poverty. In other words, regardless of mean income, states that have poorer stocks of social capital have poorer health. But this is a study of aggregate (and large!) spatial units. Mohan and colleagues (2005) undertook an extensive small-area analysis of the effects of social capital (measured using indicators such as levels of voluntary activity, political involvement, interactions with neighbors, and participation in organized social activities) on health (using the 1985 English Health and Lifestyles (HALS) survey). They conclude, in fact, that there is "little support, at this spatial scale, for the proposition that area measures of social capital exert a beneficial effect on health outcomes" (Mohan et al., 2005: 1267). This begs the question: to what extent might this construct be applied at the individual level?

A study of local-scale health inequalities in parts of north-west England (Gatrell et al., 2000; 2001) found that while loneliness, diet, smoking, and income were significant predictors of self-reported health, some measures of social capital, such as connections with family and friends, and participation in the local community, also contributed explanatory power. Quantitative findings were supported by in-depth interviews. For example, a woman from a deprived area spoke of "more crime in this area that wouldn't have occurred even five to six years ago . . . car crime, thieving of cars, people abandoning cars and setting fire to them." She indicated that "I find it hard to relate to some people from this area" (Gatrell et al., 2000: 165). Another example comes from Cohen and others (2006) who find that collective efficacy (defined as the willingness of community members to look out for each other and intervene when trouble arises) is directly related to obesity among adolescents in Los Angeles County.

Despite the growing empirical support for social capital as a partial explanation of health divides, it has not gone unchallenged. Some of the variables that comprise indices of social capital are somewhat suspect. For example, simple membership of social groups (the Ku Klux Klan, or exclusive golf clubs, perhaps?) does not necessarily produce social capital. In addition, from a political standpoint it is quite convenient for liberal governments to ask communities to improve social capital, since it diverts attention from more fundamental issues of poverty. George Bush's *healthy marriage initiative* (Struening, 2007) is a very good example. Navarro (2002) critically assesses the theoretical foundations for social capital as a determinant of health, while others (e.g. Muntaner and Lynch, 1999; Lynch et al., 2000) assert that this is the "sociological equivalent of 'blaming the victim', where communities, rather than individuals, are held accountable for 'not coming together' or 'being disorganized'" (Muntaner and Lynch, 1999: 72). Further, from an empirical point of view evidence suggests that material circumstances (see below) are a more important determinant of health divides than is social capital (Cooper et al., 1999; Gatrell et al., 2000; 2001; Carlson 2004).

One aspect of "community" that has attracted considerable interest in recent years, primarily in the USA, is that of racial residential segregation. Some evidence suggests that the risk of potentially adverse health outcomes (particularly those relating to pregnancy, such as low birth weight or pre-term delivery) among African-American women is increased when those women live in areas within which they are a minority group (Pickett et al., 2005). The converse is that when they are not in a minority they enjoy richer social networks that lead to improved birth outcomes. Others (Bell et al., 2006) suggest that – depending on how it is conceptualized and measured – segregation can be sometimes protective and sometimes health-damaging. There is a fascinating emerging literature on this topic that merits a careful reading and further analysis.

Working conditions and local environments

Here, we consider the environments within which people work and live, given that both have major impacts on health outcomes.

Many of the occupations in which those of low social status are employed can be hazardous, and the risks from accidents because of exposure to dangerous machinery or chemicals, or poor workplace practices, will certainly go a long way towards explaining some of the social class gradients in mortality and morbidity. Possible associations between occupation and cancer, for example, have been the subject of much research (Clapp et al., 2006). In many cases, workers have had to fight for recognition of their occupational illnesses. A classic example is Smith's (1981) historical analysis of black lung disease in West Virginia in the 1960s. Local coal miners fought not only for the definition of the disease (against physicians who were hired and paid for by the mining companies) but for compensation for the families they left behind when they died prematurely (see also Lilienfeld, 1991 on the asbestos industry). Other aspects of the workplace environment are possibly health-damaging. Those employed in jobs that are routine and mundane, and where there is little scope for autonomous decision-making ("job decision latitude") are more at risk of ill-health than those who have variety in, and control over, their daily work (Alterman et al., 1994; Elovainio et al., 2004). This may go some way to explaining the health gradients among the Whitehall workers discussed above, all of whom were engaged in non-manual administrative work but some of whom had more varied workloads, as well as higher social status and higher incomes.

Domestic working environments can also be sources of hazard and stress. For example, a woman who works either part-time or full-time in paid employment often carries the burden of most or all of the domestic tasks. Anyone "managing" the home will find this easier if resources such as an operational washing machine, a telephone, and a car are readily available. Some women working as paid, rather than unpaid, domestic laborers can find themselves exposed to harassment and intimidation by their employers, quite apart from the expectation that they should work long hours for little pay. Clearly, at a micro-scale, the nature of the home environment, whether physical or social, can shape the health of those working, as well as those living, there.

Outside the workplace, in what ways does the nature and quality of the local environment impact upon health? In the previous chapter we talked about the community of Sydney, Nova Scotia, Canada. This community was exposed to environmental pollution through the air, water and soil for over 100 years as a result of major steel and coking works. That legacy continues in the form of the Sydney Tar Ponds. But this environmental exposure has to be examined within the wider historical and cultural context: a relatively poor community, stigmatized by high rates of risky lifestyle behaviors, and geographically isolated from the central core of Canadian urban life, with limited access to sophisticated health services. We need to ask more searching questions about the characteristics of places or localities; what they are really like, and what their histories tell us about likely health outcomes. Such histories need to consider not merely the physical environment (such as varying air quality), but also local cultures and social environments.

Some of the most significant work along these lines has been done by medical sociologists working in Glasgow, where extensive comparisons have been made of contrasting parts of the city (Macintyre et al., 1993; Sooman and Macintyre, 1995). The Glasgow team asks whether there are features of local areas that are health damaging or health promoting. Under this heading are included such things as access to leisure and recreation facilities, public transport, a clean and safe environment, reasonably priced and healthy food, and good health care provision. Even the simplest tallies of resources and facilities suggest that access is worse in the more deprived south-western districts of Mosspark and Pollok (see Table 4.9); there are higher rates of crime against the person in these areas, and residents there perceive a higher level of threat than those in the north-west of the city. A similar in-depth examination of local variations in health determinants has been undertaken in Hamilton, Ontario, Canada, an industrial city located at the western end of Lake Ontario, one of the five Great Lakes. It is located in what is euphemistically referred to as the Golden Horseshoe of Canada due to its prime location relative to markets – via the Great Lakes – in Canada, the USA and beyond. The primary industry in Hamilton is steel. The result is a legacy of air quality issues, a relatively blue collar culture and the stigma of being an industrial city. The researchers involved in this study focused on understanding the determinants of health in Hamilton at the local level. They divided the city into four homogeneous and relatively distinct neighborhoods (Luginaah et al., 2001), across which they found significant differences in socio-demographic characteristics, social support/social capital, and self-reported health outcomes. They found a relationship between social exclusion and self-rated health (Wilson et al., in press), social capital and health (Veenstra et al., 2005), an important role for both the physical and social neighborhood environments, both actual (Keller-Olaman et al., 2005) and perceived (Wilson et al., 2004), as well as access to health care (Law et al., 2005).

Table 4.9 Local social environment and health outcomes in Glasgow, Scotland

Area indicator	West End (NW)	Garscadden (NW)	Mosspark (SW)	Pollok (SW)
Access to services				
Pharmacy[a]	97.9	96.0	87.2	92.0
Post office[a]	98.4	97.7	89.4	96.4
Grocery store[a]	99.5	99.4	97.9	98.5
Reported problems				
Discarded needles[b]	4.3	8.5	10.8	18.0
Poor public transport[b]	13.4	16.7	20.5	18.8
Assaults and muggings[b]	28.2	34.5	47.9	56.1
Health outcomes				
Health for age[c]	35.1	22.6	29.8	13.2
HADS anxiety[d]	16.5	17.7	29.8	25.6

Notes:
[a] Percentage of respondents reporting amenities within half a mile
[b] Percentage of respondents reporting selected problems
[c] Percentage of respondents reporting health as excellent
[d] Percentage of respondents reporting anxiety
(*Source*: Sooman and Macintyre, 1995)

Much of this neighborhood and health work is based on the premise that *the neighborhood* has an influence on health, over and above its status as a collection of individuals living within a particular space. For example, Boardman (2004) finds that the *stability* of the *neighborhood* acts as a mediating factor between stress and health in poor black neighborhoods in the USA. The role of *neighborhood* as an independent variable impacting upon health continues to be investigated. The challenge of course is to be able to disentangle this determinant from a broad range of others, as well as various scales of influence.

Material deprivation and health

As we have seen in earlier sections, there is some evidence to link poor health with a lack of services, amenities, and resources in local neighborhoods, as well as a poor physical environment. We have also suggested that a lack of participation in society and a limited set of social relations are implicated in poor health. The first of these factors leads to *material deprivation*, while the second can be characterized as *social deprivation* or *social exclusion*. In this section we concentrate on material deprivation, recognizing that it also embraces lack of income and wealth, poor quality housing, and unemployment (whether real or threatened).

A concern with health and deprivation is not new. The French writer Villermé (Macintyre, 1998) presented evidence for Parisian districts in the early nineteenth century, the results revealing a striking relationship (see Figure 4.10) between death rates and poverty (as measured by the proportion of properties exempt from taxation because of poverty).

Over the past three decades a large number of studies have examined how material deprivation has varied by area. This is done by using a set of indicators (usually taken from population censuses) for either quite large or (often) quite small, areal units. The indicators (e.g. unemployment, income inequality, education, home ownership, crowding,

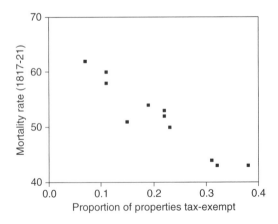

Figure 4.10 Relationship between mortality rates and tax-exempt properties in districts of Paris, 1817–21

car ownership) are combined, either in a simple additive way, or perhaps using more complex multivariate statistical methods (such as principal components or factor analysis). This then provides a deprivation "score" for an area, to which aggregated health data may be related. Such indices have a number of potential applications (Davey Smith et al., 2001). First, they can be used when data describing an individual's socio-economic status cannot be, or has not been, collected directly. Second, they can be useful in informing the distribution of health care service resources. Third, they allow researchers to control for possible socio-economic confounding when undertaking ecological studies of local environmental conditions on health. Finally, they are very useful when examining the effects of characteristics of *place* on health. Such measures are now widely available in many developed countries, for example Britain (Senior, 1991; Gordon, 2003), New Zealand (Salmond et al., 1998; Blakely, 2002), the USA (Singh, 2003; Eibner and Sturn, 2006) Spain (Benach and Yasui, 1999), Sweden (Sundquist et al., 2004) and Canada (Frohlich and Mustard, 1996; Langlois and Kitchen, 2001). Indices have been developed not only to assess material deprivation (such as the Townsend index of deprivation); Congdon's anomie index is a measure of social fragmentation, based on Durkheim's theoretical concept of social integration (Davey Smith et al., 2001). Studies have shown that these different indices are indeed related to health, but the strength of association varies by health outcome (see Table 4.10; see also Adamson et al., 2006).

To what extent does area deprivation explain variations in health outcome? We can answer this question with reference to a handful of studies drawn from a potentially vast set. These studies relate to a variety of health outcomes.

Examining data for the USA, Singh (2003) has shown that area deprivation gradients in US mortality increased substantially between 1969 and 1998 and, further, that higher mortality rates were seen in more deprived areas. In their study of 633 parliamentary districts in Great Britain for 1981–1992, Davey Smith and colleagues (2001) found strong correlations between the Townsend index of deprivation and all-cause mortality. Data on life expectancy reveal the same picture. Woods and colleagues (2005) examined life expectancy in England and Wales and found that life expectancy at birth varies with deprivation quintile and is highest in the most affluent groups. Using these data, these investigators found a clear north-south gradient (see Figure 4.11).

Table 4.10 Correlations between the Townsend and social fragmentation indices and standardized mortality ratios, Great Britain, 1981–92

	Townsend Index		Social Fragmentation Index	
Mortality	*Women*	*Men*	*Women*	*Men*
All cause	0.82	0.87	0.35	0.46
CHD	0.66	0.67	0.12	0.13
Stroke	0.58	0.67	0.12	0.24
Lung cancer	0.81	0.84	0.51	0.44
Stomach cancer	0.60	0.60	0.15	0.15
Suicide	0.38	0.58	0.58	0.71
Cirrhosis	0.63	0.70	0.56	0.67

(*Source*: Davey Smith et al., 2001)

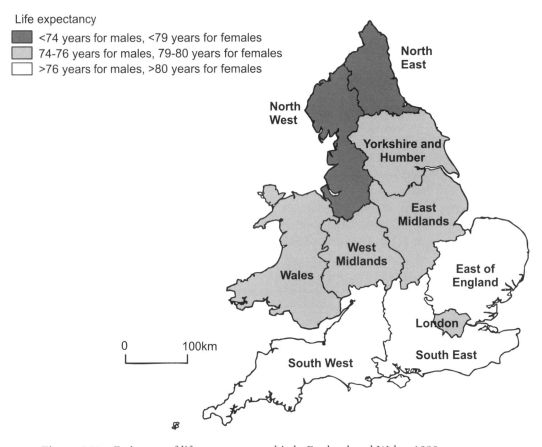

Life expectancy
- <74 years for males, <79 years for females
- 74-76 years for males, 79-80 years for females
- >76 years for males, >80 years for females

North East

North West

Yorkshire and Humber

East Midlands

West Midlands

Wales

East of England

London

South West

South East

0   100km

Figure 4.11 Estimates of life expectancy at birth, England and Wales, 1998

Elsewhere in Britain, Senior and his colleagues examined the relationship between mortality and deprivation in Wales, focusing particularly on how this relationship changed between the early 1980s and early 1990s (Senior et al., 1998). The areal units are electoral wards (the smallest units of local government in England and Wales, typically comprising populations of under 5000). The map of all-cause SMRs (1981–3) for those aged under 65 years, ignoring small areas where there are fewer than 20 deaths, reveals high levels of premature mortality in the industrial south and north-east of Wales, in some of the poor inner-city and suburban districts of the major urban areas (Cardiff, Swansea, Newport, and Wrexham), and in some small rural towns and seaside resorts. There is a highly significant, linear relationship between mortality and deprivation (Townsend score), one that is repeated for the 1991–3 data. However, the gradient of the slope relating mortality to deprivation is higher in the 1990s, leading to the conclusion that there has been a widening of mortality differentials over the decade.

In a study of suicide and attempted suicide (para-suicide) in part of south-west England, Gunnell et al. (1995) looked at 24 localities (groupings of electoral wards). Taking data on over 6000 cases of para-suicide between 1990 and 1994, and nearly 1000 suicides recorded between 1982 and 1991, they show a clear linear relation between standardized ratios and deprivation. Inner-city parts of the main urban area, Bristol, show SMRs well over twice the district average and five times higher than those in more rural areas. Despite gender and age differences (para-suicide is commoner among young women, suicide among men), and despite the absence of data on individual histories and circumstances, the conclusion is that deprivation – and particularly unemployment – is a key factor explaining variation in suicide and para-suicide. At a national scale, Whitley and colleagues (1999) have shown that social "fragmentation" (as measured by levels of mobility, single person households, and those renting in the private sector) is at least as important as deprivation in accounting for spatial variation in suicide. Davey Smith and colleagues obtained similar results for the 633 parliamentary districts in Great Britain. This indicates that both material and social factors are important in predicting suicide.

Morbidity data are linked to deprivation too. For instance, dental decay among children has been shown in a number of studies to be associated with socio-economic deprivation. Conventionally, this is measured by a "DMF" index, representing the number of teeth which are decayed (D), missing (M), or filled (F). Jones et al. (1997) examined the mean DMF score for a number of small areas (electoral wards) in three districts of northern England, and plot the relation between mean score and deprivation (Jarman score). There is a clear linear relationship (see Figure 4.12). However, the three districts are characterized by different types of water quality. One district, Salford/Trafford, has no fluoride in the water supply; another, Newcastle/North Tyneside, has had the water artificially fluoridated, while the third, Hartlepool, has naturally high levels of fluoride. As the graphs show, the relation between dental health status and deprivation varies between the three areas. Children living in the less deprived areas show much less variation in DMF score than those in very deprived

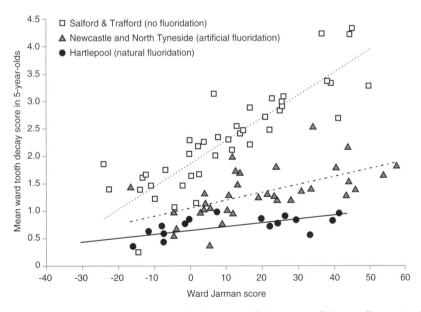

Figure 4.12 Relationship between tooth decay and "deprivation" by small area in three districts in England

areas, suggesting that the latter would benefit in particular from fluoridation. In Britain at least, fluoridation is a politically charged topic and there are strong lobbies arguing that artificially fluoridating water infringes personal liberties (see pp. 211–12 below).

All these studies are conducted at the area, not the individual level. Data for individuals – on their cause of death, dental health, birth weight, or another measure – are aggregated to form a count for an areal unit; such studies are therefore referred to as "aggregate" or "ecological." In some cases, however, individual data may be used to look at area deprivation effects. For example, Adamson et al. (2006) have examined inequality in disability among older adults, using data from the West of Scotland Twenty-07 study. This study is a unique longitudinal examination of a cohort aged approximately 15 years, 35 years and 55 years when first contacted in 1987. Both socio-economic and psychosocial variables were measured as potential determinants of disability in old age. Results suggest that while both sets of variables are important, there was stronger evidence for material conditions as a predictor of disability; those in the most deprived material group were 2.5 times more likely to report severe disability. Salmond et al. (1999) have looked at the prevalence of asthma in New Zealand adults and examined the link to material deprivation. Asthma among those living in the most deprived areas is 50% higher than in the least deprived areas, a difference that is only slightly attenuated when allowance is made for individual age, gender and ethnicity. But there are other ways in which we can separate individual-level effects on health from the effects of living in a deprived area. We saw in the previous chapter that techniques such as multi-level modeling can be used to sift out the contextual from the compositional, and to see whether there are "area effects" on health outcomes, over and above those operating at an individual level. We can illustrate this with specific reference to health inequalities by examining recent US studies on low birth weight.

Roberts (1997) analyzes data for over 130,000 births in Chicago in 1990, and rates of low birth weight show considerable spatial variation (see Figure 4.13), from less than 1% to 20% in some community areas. Given detailed individual data, Roberts introduces controls for known risk factors such as age, parity (number of previous children), and level of pre-natal care. This is important, since otherwise we would not know the extent to which high rates of low birth weight simply reflected a preponderance of young mothers, for example (known to be more likely to give birth to low birth weight babies) in particular areas. At the individual level, maternal race/ethnicity accounts for a considerable amount of variation, with African-American mothers more likely to have a low birth weight infant. At the community level, an index of economic hardship (combining measures of unemployment and poverty) is associated with high numbers of low birth weight babies. Roberts suggests that "women in high-poverty, high-unemployment communities have fewer material resources and therefore run higher risks for malnutrition, lower quality health services, and stress" (Roberts, 1997: 600). He also indicates that social support networks are likely to be more fractured in such areas.

In a parallel study of Baltimore, O'Campo and her colleagues (1997) use multi-level modeling to bring out the area or neighborhood effects. They are thus able more explicitly to answer the question: "Are neighborhood-level variables directly related to an increase or decrease in risk of low birth weight?" As with Roberts' study, individual-level variables such as later initiation of pre-natal (ante-natal) care contribute to the risk of low birth weight. But at the neighborhood level those living in census tracts where per capita income was under $8,000 were at significantly higher risk than women living in census tracts with per capita

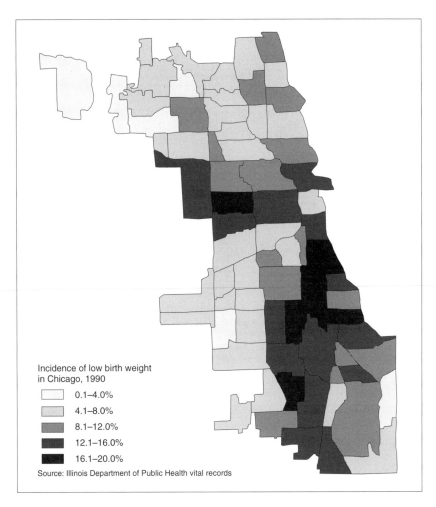

Incidence of low birth weight
in Chicago, 1990

☐ 0.1–4.0%
▨ 4.1–8.0%
▨ 8.1–12.0%
▨ 12.1–16.0%
▨ 16.1–20.0%

Source: Illinois Department of Public Health vital records

Figure 4.13   Incidence of low birth weight in Chicago, by community area, 1990

income exceeding this figure. Interestingly, there is evidence that the effect of individual-level variables varies between neighborhoods, and according to neighborhood characteristics, as shown in Figure 4.14. For example, while women with low levels of education are more likely to have low birth weight infants, the effect of this is stronger in high-crime rather than low-crime neighborhoods. On the other hand, in wealthier neighborhoods, where unemployment rates are lower, the benefits of early ante-natal care seem to be greater than in poorer neighborhoods. In any event, this kind of study (see also Bell et al., 2006) indicates that explanations, and also health policies, need to target areas as well as individuals.

Much of the work we have reviewed to this point provides good *descriptions*, but is less good at offering *explanations* of health inequalities. What is it about "material deprivation" that contributes to poor health? In terms of markers of poor housing (such as

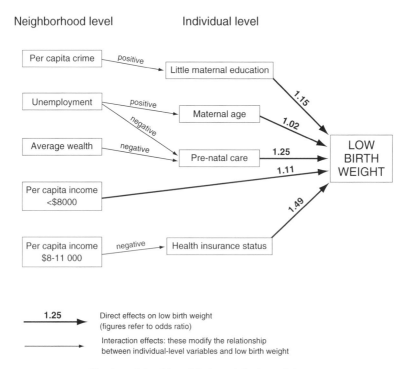

Figure 4.14   Factors affecting risk of low birth weight in Baltimore, 1985–9

overcrowding), high-density occupation may encourage more rapid spread of infections. Damp housing, too, will produce fungal spores that lead to poor respiratory health, while lead-based paint or water pipes are further potential causes of morbidity. Low incomes have major indirect consequences for health, limiting dietary choices; it is well known that a good diet is important, but the absence of this from the household may not reflect lack of knowledge so much as poor access to a range of healthy foods, or a restricted choice. Unemployment, too, will constrain household budgets and have direct health effects, such as psychosocial stress and depression.

Despite increasing evidence for the operation of "place" effects (O'Campo, 2003) there is continuing debate as to whether area deprivation matters less than individual deprivation. The general weight of evidence suggests that there are "place" effects on health, though what matters most is the level of poverty in the individual household (Shaw et al., 1999; Macintyre et al., 2002).

For Wilkinson (1996), the causes of health inequalities lie in the deeper underlying structural inequalities within societies. This theoretical position may be used to account for health variations at a number of spatial scales. For example, among developed countries there is only a weak correlation between life expectancy and average incomes. However, there is a much more striking association between life expectancy and the distribution of income within countries; those countries where wealth is more evenly distributed are those with higher life expectancy (Wilkinson, 1996: 72–6). Research both within the USA and Britain confirms this. All-cause (and cause-specific) mortality in US states correlates inversely with the percentage of household income received by the least well-off 50% of the population (Kaplan, 1996). In Britain, Ben-Shlomo et al. (1996) have shown that mortality is associated with variations in deprivation within local authorities. Both these studies controlled for average income and deprivation, respectively. At a yet finer spatial scale, Boyle and colleagues (1999) have shown that the variation in deprivation within quite small localities in England and Wales (electoral wards and those adjacent to them) predicts self-reported long-term illness. This adds a further spatial dimension to the literature on associations between ill-health and deprivation. Watson (1995) adopts an explanatory framework similar to Wilkinson's, rejecting the idea that behavioral factors, the quality of health care, or levels of environmental pollution, play a significant part in worsening health outcomes in eastern Europe. She, too, stresses the importance of "relative deprivation"; the raising of the Iron Curtain meant that those living in countries behind it had raised expectations of improved living standards that began to approach those in the west. For her, the health crisis is "the outcome of socialism's (ultimately unsuccessful) struggle to modernize in an increasingly global context" (Watson, 1995: 928).

A study that also links community and social factors to wider socio-political processes is that by Barnett and colleagues (2005) in which they examine the effects of changing levels of inequality upon ethnic variations in smoking rates in New Zealand, 1981 to 1996. This was a time of extensive and rapid structural change in New Zealand's economic and social welfare systems, which impacted on marginalized groups – particularly the Maori – in a more profound way than others. While rates of smoking during this period declined for both dominant groups (the Maori as well as the Pakeha – persons of European descent), the gap in smoking rates *between* the two groups increased dramatically indicating that "levels of social inequality between Maori and Pakeha have an independent effect on Maori smoking rates and that communities which experienced increased social inequality during both the 1980s and 1990s were more likely to have higher Maori smoking rates" (Barnett et al., 2005).

Over two decades of concerted research on the links between income inequality and health have sparked many debates as well as varying findings. Wilkinson and Pickett (2006) undertook a meta analysis of 168 separate analyses in 155 papers that had reported research findings on the association between income distribution and population health. They found that 70% of the studies "suggest that health is less good in societies where income differences are bigger" (Wilkinson and Pickett, 2006: 1768).

Wilkinson's work, and the further research it has spawned, argues for changes to broader social structures, and ultimately a redistribution of material resources. For example, Wilkinson and Marmot (2003) identify what they consider to be the weight of evidence for ten broad social determinants of health: the social gradient, stress, the events of the early life years, social exclusion, social support, work, unemployment, addiction, food and transport. They identify the policy implications associated with each, for example, "Through policies on taxes, benefits, employment, education, economic management and many other areas of activity, no government can avoid having a major impact on the distribution of income. The indisputable evidence of the effects of such policies on rates of death and disease imposes a public duty to eliminate absolute poverty and reduce material inequalities" (Wilkinson and Marmot, 2003: 17). In short, risk factors such as those representing health behaviors are seen as merely surface causes of ill-health; the deeper causes are those lying in racism, discrimination, hostility, unemployment, and, more broadly, a lack of access to material wealth (Nazroo, 1998; Wilkinson and Marmot, 2003).

A historical perspective on this is given in work on the incidence and spread of tuberculosis (TB) in South Africa (Packard, 1989). Some contemporary explanations of the growing burden of TB among urban black Africans in the early nineteenth century tended to blame the victims for their poor adjustment to urban living conditions, their ignorance of basic hygiene, and even their adoption of European clothing. Others recognized the lack of proper housing conditions and diet but chastised those infected, as if they chose these conditions rather than having them forced upon them as a powerless workforce. The response to the disease burden was to enforce sanitary segregation. A series of health acts forced the removal of thousands of Africans and "coloreds" to quickly constructed ghettos on the edges of large cities such as Cape Town, Johannesburg, and Port Elizabeth. This had little impact on the TB problem; however, "slum clearance did not mean slum removal in the sense of eradication but simply the physical transfer of slum conditions beyond the city limits" (Packard, 1989: 53). An inadequate supply of decent housing, poor wages, and high rents encouraged overcrowding and increased the likelihood of developing TB.

In later years, the rise of the Nationalist Party after the Second World War saw that the "natural home" of black Africans should be the homelands. This displaced the disease from the main urban centers into "rural dumping grounds" and continued the structural policy of "dealing with TB through the application of exclusionary social controls" (Packard, 1989: 252). Although there have been fluctuations, disease notifications among black Africans and colored groups have been, on average, at least six times higher than among whites. Poor quality housing, poor sanitation, and the long journeys to work on overcrowded trains and buses, all provided continued fertile conditions for the spread of TB infection. Mustering a convincing structuralist argument, Packard argues forcefully that state legislation, a basic neglect of housing, and especially the apartheid policy of creating independent homelands, meant that the black TB problem was methodically removed from the white view.

Concluding Remarks

The explanation for health inequalities is the subject of intense ongoing debate. While earlier work pointed to the importance of *either* structuralist *or* behavioral (lifestyle) explanations, more recent work suggests that other factors operate. Of these, one's social position throughout the life course merits attention, as do psychosocial factors. The idea of social capital has generated considerable research interest in recent years and does seem an additional determinant of health status. Yet many of the factors interact in complex ways. Indeed, the determinants of most common diseases involve a complex set of interactions across the life course between environmental and biological/genetic factors, embedded within a social context.

We have seen evidence in this chapter of the differing approaches to explanation considered in the first two chapters. Much of the classical epidemiological literature is distinctly positivist in conception, trying to disentangle the relative effects of particular variables whilst providing controls or adjusting for the effects of others. A much more modest literature (e.g. Popay et al., 2003) seeks to hear the individual voices of those on whom health divides have an impact, and calls for a more social interactionist perspective. Other very persuasive work adopts an implicitly structuralist perspective, arguing that health divides are rooted in deep social and economic structures; here, it is poverty and income inequality that cause health inequalities.

But are poor health outcomes also a function of variable provision of health services? As far as child dental health is concerned, for example, is this simply a function of behavior (parents not getting children to brush their teeth, or letting them eat too many sweets), a function of wider structural factors (such as lack of fluoridation), or a function of poor provision: an inadequate supply of dentists, or the cost of accessing such care? Are high rates of heart disease in part dependent upon access to services for prevention and treatment, as well as a lack of income, or social support, or poor health behaviors? We therefore need to consider one further possible explanation for inequalities in health outcomes, namely, the varying provision or organization of health services, and the differential use of such services. To what extent do place-to-place variations in morbidity and mortality reflect differences in the provision and organization of care? We therefore examine, in the next chapter, inequity in the provision and use of health services.

Further Reading

A good overview of health inequalities from a geographical perspective is provided by Boyle et al. (2004). Another key collection includes Heymann et al. (2006); see, for example, the chapters in that collection by Dunn et al. as well as Ross et al. and Mustard et al.

At a global level we have referred to the Global Burden of Disease Study. Others have undertaken similar global comparisons using the World Development Indicators database from the World Bank. Such studies (Ruger and Kim, 2006) indicate that health inequalities – whether for adults or children – between the richest and poorest countries continue to grow. An important collection is that edited by Kawachi and Wamala (2007).

Comprehensive and up-to-date overviews of health inequalities, of both a theoretical and empirical nature, may be found in several of the essays collected together in Bartley et al. (1998). In particular, the chapter by Curtis and Jones is an excellent starting point for the geographer, while some useful cross-country comparisons are made in the chapter by Shaw and others that covers some of the same material reviewed here. The book by Shaw and her colleagues (1999) is by far the best overview of work on health inequalities in Britain. See also the collection edited by Krieger (2005).

Kunst's (1997) monograph has a wealth of comparative material for west Europe and the USA. Summaries of the rich vein of work on programming may be found in Barker (1992; 1994), Barker and Bagby (2005) and Nuyt (2008). For a comprehensive overview of the life-course perspective see the edited collection of Heymann et al., 2006, especially the chapters by Clyde Hertzman. Some recent empirical analyses of the role of social capital have begun to invoke the theoretical work of French sociologist, Pierre Bourdieu (see, for example, Gatrell et al., 2004 and Veenstra, 2007). There is a strong and growing literature around neighborhoods and health. A very useful point of departure is the collection edited by Kawachi and Berkman (2003).

There is a substantial literature on the association between inequalities and health, and social capital and health, only some of which we have considered. For a carefully written and skeptical discussion of these associations you should read Kunitz (2007).

Much of the literature on health inequalities finds its way into a relatively small number of journals. Of these, *Social Science & Medicine, American Journal of Public Health, Health & Place,* and the *Journal of Epidemiology and Community Health* have been very prominent.

Chapter 5

Inequalities in the Provision and Utilization of Health Services

We turn now from inequalities in health outcomes – variations in mortality and morbidity – to look instead at inequalities in terms of the supply and usage of health services. In examining supply, we are concerned with where services are located and how they are configured, in short, how they are *provided*. We shall look at this in a variety of spatial scales, from international contrasts in provision through to variations within countries and local provision. But we also discuss how people use such services and whether such use is shaped by location, including the costs of overcoming travel distance to such facilities; service *utilization* is therefore a second key theme running through this chapter. We cannot escape broader issues, however, including how the "need" for services is determined and how decisions about resource allocation are made.

In examining patterns of inequality of health outcome in the last chapter we drew upon evidence from a variety of countries. We do so again in the present chapter. We do not present here a comprehensive picture of the way health care services are organized and administered in different countries; this would be a daunting task, not least because new governments have a habit of reforming health care systems soon after they take office! The essential distinction is between those systems that give precedence to the "market," versus those that make provision available, at least in principle, on the basis of need rather than willingness or ability to pay (see Box 5.1 below). The former expects the health care user to pay at the point of use, or to buy into private health insurance that meets the costs of treatment. In the latter, health services are funded via income from taxes or compulsory insurance.

Principles of Health Service Delivery

Levels of health care provision

As outlined briefly in Chapter 1, it is conventional to separate out different levels of health care provision, usually into three categories. First, *primary* health care is provided

Box 5.1   Models of medical care

Essentially, the developed world offers two types of organizational models of health care service provision: "collectivist" (e.g. the UK, Canada) and "anti-collectivist" (e.g. the USA).

In Britain, the ideology is fundamentally collectivist, with a National Health Service (NHS) established in 1948 to provide health care on the basis of need rather than ability to pay. It is funded both by general taxation and taxes on those in employment. People access the health care system usually on the basis of their registration with a general practitioner (GP) who makes referrals, where necessary, to secondary health care (hospitals). The state provides funding to GPs largely on the basis of the numbers of patients registered with them. Primary Care Trusts (PCTs) are given budgets by the government to spend on hospital and community health services and are charged with the responsibility of monitoring the health and health needs of the population within their areas. The number and boundaries of PCTs (which have replaced former Health Authorities) have undergone numerous changes in recent years. Hospital services are now organized into "Acute Trusts" that provide the health care commissioned by PCTs. Specialist Trusts offer emergency (ambulance) care or mental health services.

Similar to the UK, Canada has a collectivist type of organization of health care service delivery. However, there are two fundamental differences. Although we speak of a universal health care system in Canada, health care is a provincial responsibility; there are ten parallel health care systems, one for each of the provinces (but note that the northern territories are a federal responsibility, and hence their health service delivery system is as well). The Canada Health Act (1984) brings these parallel systems together by way of five unalterable principles: universality, comprehensiveness, accessibility, portability (which means one can move from one province to another and not be denied access to care for financial reasons), and public administration, The second major difference is that secondary types of services are not covered under the universal system in Canada; rather, these are typically covered – except for the very poor – by employer-based or private insurance schemes.

In the USA the ideology is "anti-collectivist," with funding of health care coming from private health care insurance schemes (arranged by the individual, but generally tied to employment) and from two state schemes (Medicare and Medicaid, introduced in 1965). Medicare provides insurance cover for people aged 65 years and over, while Medicaid provides cover for people on low incomes. Nonetheless, high proportions of the population are uninsured (see Table 5.6 below). Unlike in Britain or Canada, individuals are not required to consult a general practitioner for referral to secondary care; they can consult specialists directly. The unevenness in health care provision, and the burgeoning costs of such provision, as well as insurance schemes, led to the creation of health maintenance organizations (HMOs), a form of what is called "managed care." Here, people pay monthly premiums for a comprehensive package of health care to be provided by a contracted group of health practitioners and organizations.

The Clinton administration sought in the 1990s to introduce further state regulation into the provision of health services. It endeavored both to control rising costs, and also to widen health care coverage. Political opposition, as well as pressure from some vested insurance and medical interests, saw the failure of these attempts. However, the 2008 US election has put universal health care back on the political and policy agendas.

in the home, or in a clinic or health center. Such care could include basic medical attention (from a general practitioner, nurse practitioner, physician assistant, or a family physician, depending upon where in the world you live), with a range of services offered, such as prescribing of medication, childhood immunization, screening for some diseases, and/or nutrition or smoking cessation counseling; it therefore covers many aspects of preventive medicine and health promotion. Primary care also includes other health services, such as dental and ophthalmic care and facilities for the dispensing of medication (such as community pharmacies). All such services need to be offered locally, since they are accessed frequently.

Secondary care may be thought of as that offered in hospital settings, where patients may be admitted for treatment that cannot take place in a health center or clinic. The route for accessing such care may, as in the UK and Canada, be via the primary health care system, so that a doctor or physician in the community refers a patient to hospital for further investigation, diagnosis, and treatment, including more complex surgery beyond the minor procedures that might be performed in the primary care setting. In essence, the secondary sector offers more specialized care and will often be less concerned with prevention and more with cure. Patients requiring further specialist care may be referred to a *tertiary* center that will have facilities not available in smaller hospitals. For example, cancer patients may be referred to tertiary centers for investigation, surgery, or radiation or chemotherapy treatments, while those in need of specialist heart surgery would also have these procedures performed in such centers. The boundaries between these three levels are often blurred; nonetheless, the distinction is a convenient one that will be used below. Of course, health care is offered in other settings, such as nursing homes and, for those with terminal illnesses, hospices. Health care can certainly take place outside the clinic and the hospital.

Geographies of rationing

The majority of developed nations are currently experiencing a *health conundrum*; that is, as life expectancies increase, so too do the costs of the health care system (Evans et al., 2000). As a result, the provision of health services is inevitably tied up with issues of resource allocation, priority-setting, or "rationing." We might all like to have a well-staffed primary care clinic in our immediate neighborhood, open at all hours, but the finances available to any health care delivery system rule this out. Government constraints on health care expenditure mean that we cannot provide all the care we might wish to and therefore decisions have to be made about the nature and range of services to provide. These are questions of a geographical, as well as an economic, political, and managerial nature, since such services will have to be provided *somewhere*. Even if such services are mobile, so that provision need not be delivered at a fixed site, issues arise concerning where to locate the supply of vehicles; this arises most obviously in emergency care, where emergency response vehicles such as ambulances must be located somewhere before they are called upon.

The allocation of scarce resources and the rationing of services might be thought of as having little or nothing to do with geography. However, there is massive geographical disparity in the availability and provision of treatment and therapy. Even in a country like Canada that provides universal access to health care free of financial barriers, there exist underserviced groups as result of the extreme costs of provision to remote areas of the country. For example, Newbold's (1998) analysis of the Aboriginal Peoples Survey indicates chronic under-servicing of Canada's Aboriginal Peoples, particularly those living on reserve. In Canadian provinces with large and dispersed populations, such as Ontario, we find further inequalities in access – in this case between urban and rural areas. For example, when young doctors graduate from medical school and wish to set up practice, they tend to be attracted to major urban centers rather than remote rural areas. In response, the government of Ontario has implemented several versions of incentive schemes to encourage a redistribution of family physicians, but to little effect (Kralj, 2001). Where access is not free of financial barriers, rural and remote areas still feel the impact of geographical access to health care services (see for example, Arcury and colleagues' (2005) investigation of health care utilization in the Appalachian area of North Carolina, USA).

Efficiency and equity

When making decisions about the provision of health services there are different criteria that can be adopted. First, we can aim to provide services that are *efficient*; by this we mean how to provide services that maximize health benefit while minimizing cost. Second, we can aim to provide services that are *effective*; are the treatments we offer having real benefit, or are we wasting resources on providing services that offer little health gain? Third, we can aim to provide services in an *equitable* way; are services being provided uniformly to the populations they are designed to serve? These are complex issues, and we do not plan here to go into detail. However, from a geographical perspective there are important implications. For example, as a general rule there is a conflict between efficiency and equity in all planning and resource allocation; there are tensions between the wish from "providers" (or funders or managers) to manage resources efficiently and the wish of users to have those resources provided conveniently and equitably. In a locational sense this means a tension between providing one large site or facility (such as a hospital that offers a comprehensive range of services) and providing several smaller facilities that are closer to the population that needs them. The first solution offers economies of scale for the provider, while the other minimizes travel costs for the user. It is not quite so simple, however, since patients might well benefit more from traveling further to see specialists who are located in centers of excellence. There are, therefore, important debates to be had about the geography of provision, particularly of secondary and tertiary services, since while the spatial concentration of services might seem "unjust" on the surface, the costs of travel might be greatly outweighed by the quality of service provided (see Box 5.2).

Equity in the provision of services must not be confused with equality. Geographically, equality suggests that there should be an even distribution of services per head of population. But what matters much more is equality in relation to need, and this is what we understand by equity. To what extent are services being provided to those who need them? Some have gone as far as to argue that an "inverse care law" operates, where those in most need of health care are least likely to find it available (Tudor Hart, 1971). This clearly goes against one of the founding principles of collectivist health care systems. Much of the present chapter is an attempt to consider the evidence for such an "inverse care" thesis. But before doing so we need to unpack the notion of "need" and its assessment.

Box 5.2 Does size matter?

As the text suggests, an important question is the extent to which patients benefit from being treated in a larger hospital, one with a high volume of activity (number of procedures and operations), as opposed to a smaller one that is perhaps located closer to their place of residence. For example, do hospitals in which specialists treat larger numbers of women with breast cancer, or which conduct more coronary artery bypass grafts, see better patient outcomes such as improved survival? This issue is explored in a number of chapters in the collection edited by Ferguson and his colleagues (1997). Some of the issues are considered here.

One advantage of larger, more concentrated units is that these are more likely to be centers of clinical expertise and excellence, since the carrying out of larger numbers of procedures permits skills to be developed and refined. Research evidence suggests that outcomes (such as survival) improve as volume of activity increases. But to what extent is this due to the fact that low-volume hospitals might be dealing with more serious illness, older people, or more emergency cases? Here, outcomes may be poorer not because of less expert staff but because they are dealing with more severe cases, the prospects for which would be poor anywhere. Few studies adequately provide controls for severity or case-mix.

A further advantage of larger centers is that they can reap economies of scale (size) and scope (using the same facilities for a variety of purposes). In this case, health care costs are lowered, since the costs of staffing and running one large unit of, say, 100 beds will be less than managing 10 units each with 10 beds. Work shows that economies of scale are exploited in hospitals with between 100 and 200 beds, and that hospitals with many more beds may suffer diseconomies. In England, over half the hospitals have more than 300 beds and this suggests that the scope for further concentration is limited.

Set against these benefits are the increased travel costs, and possibly lower rates of utilization, since, as we shall see later, rates of use decline with distance. Because of these factors, and the evidence that economies of scale have already been reaped, Ferguson and his colleagues conclude that there are no compelling grounds for seeking yet further concentration of hospital services, or merging these onto a smaller number of sites.

Of course, in the early twenty-first century, one cannot underestimate the role of "access" to health care "services" via the Internet. The amount of health information available to the average person – outside the formal health care system – has grown exponentially in the past decade. Whether or not the information available is useful, or indeed can end up doing more harm than good, is debatable (Berland et al., 2001). But the point here is that the size and scale – and location – of such provision is irrelevant.

The need for health care

The need for health care takes on a very different perspective according to whether one is looking at population requirements in the developed, or developing, world. In the latter, the basic needs for life, such as access to safe drinking water or an adequate supply of nutritious food, are far from universally met, although they are taken for granted in the developed world. In developed countries, "need" takes on a different dimension. But while millions have basic needs met, others die because their needs are not met. Good examples are where the expertise needed to treat cancer patients is not provided, or where there are insufficient resources devoted to heart surgery.

The gap between need and demand (sometimes referred to as "expressed need") for health care is often a fine one. For example, some people will demand surgery to change their sex, surgery which they argue is needed in order to live a fulfilling life; others would argue that this is not a health "need" and would prefer to see scarce resources devoted to other forms of health care. Some people living with the problems of multiple sclerosis demand the legalization of cannabis, a drug that they argue alleviates their symptoms; for them it is a need. Difficulties, though, arise when health professionals believe a treatment is ineffective, and this conflicts with the contrasting beliefs of lay people.

One of the clearest examples of expressed need in the Canadian context is waiting lists. Canadians' dreams of universal health care become a nightmare for politicians when faced with questions associated with long waiting times for essential treatments (Hadorn, 2000), resulting in the emergence of what some would consider private health care services in Canada. In the USA, private insurance companies are attempting to define need through a rationalization of coverage, going through a process of de-listing what were formerly insured services. The best known example of this is the state of Oregon in the USA (Oberlander et al., 2001). A lengthy consultative process resulted in a de-listing of over 200 conditions (including, for example, sexual dysfunction and impulse disorders, no longer covered by any health insurance agencies in the state). Despite reports of mixed success, the Oregon plan has attracted the interest of several Canadian provinces as well as the UK NHS.

Inequalities in the Provision of Health Services

This section examines the extent to which there is inequity in the provision or availability of health care. It does so with reference to both the developing and the developed worlds, looking at broad variations as well as more local variations in provision.

Health care provision in developing countries

Recall from Chapter 4 that it was possible to make comparisons between groups of countries (see Figure 4.1, page 88) according to the common metric of Disability Adjusted Life Years. How is this global burden mirrored, and perhaps ameliorated, by health expenditure (see Figure 5.1)? The answer is that there is an inverse relation, with those countries in "burdened" regions accounting for a much smaller fraction of global health expenditure than is the case in established market economies. "Low and middle income countries account for only about 18% of world income and 11% of global health spending . . . yet 84% of the world's population live in these countries, and they bear 93% of the world's disease burden" (WHO, 2000: 7). Further, the range of risk factors for health varies dramatically between the developed and the developing world. In their investigation of risk factors and the global burden of disease, Ezzati and colleagues (2002: 1347) found that "in the poorest regions of the world, childhood and maternal underweight, unsafe sex, unsafe

Health expenditure by region
(% health expenditure world wide, 1990)

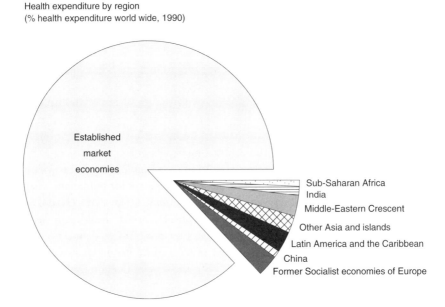

Figure 5.1   Health expenditure by world region, 1990

water, sanitation and hygiene, indoor smoke from solid fuels, and various micronutrient deficiencies" were major contributors to loss of healthy life.

There are three fundamental objectives of any health system (WHO, 2000):

(1)   Improving the health of the population they serve;
(2)   Responding to people's expectations; and
(3)   Providing financial protection against the costs of ill-health.

Over the past 25 years, global health systems have undergone essentially three (overlapping) stages of reform. These have been prompted not only by the apparent failures of health systems, but also by a desire to meet the expectations and demands of a population. The twentieth century saw the evolution of national insurance schemes to provide health care services to populations in, for example, the UK, Canada, and Australia. While initiated in developed countries, these systems soon migrated to low and middle income countries. But after a couple of decades, these universal systems experienced substantial stress in terms of financial and human resources and it was discovered that they had not solved the problem that they had set out to solve: universal access for all, free of financial barriers. Indeed, these systems – to this day – are accessed more by the better-off members of society, and efforts to reach the poor have failed. For example, where a European model of health care was implemented in African countries under British rule, it was primarily intended for colonial administrators and expatriates, not indigenous Africans (WHO, 2000). In most developing countries, where the populations were concentrated in rural areas, major urban hospitals received about two-thirds of government funding for health care.

A second generation of reforms thus saw the emphasis on creating more cost-efficient, equitable and accessible health care systems, hence a radical shift in emphasis from secondary to primary care. This evolved through the Alma Ata meeting held in what is now Kazakhstan, in 1978, with the stated goal of "Health for All by the Year 2000." The model was based on empirical evidence from several low-income countries, where life expectancy had been increased by as much as 20 years with a relatively small investment to provide a minimum level of health services for all, food and education, and adequate access to safe water and sanitation. The "Health for All" declaration led to some positive outcomes. For example, there was a substantial effort in many countries to hire and train community health workers (over 100,000 in India) who could deliver basic, cost effective services in modest, rural facilities in areas that previously had no access to care (WHO, 2000: 14). But the "Health for All" declaration was not the panacea it was hoped for. Inadequate resources and facilities, lack of training and proper equipment were barriers that prevented community health workers from doing their jobs effectively.

The final evolution of health system development described by WHO (2000) is a hybrid model referred to as the *new universalism*. This model retains the strong fundamental principles of primary care but focuses on the high-quality delivery of essential care for everyone, defined primarily on the basis of cost-effectiveness, rather than all possible care for the entire population or only the simplest and most basic care for the very poor. This model also emphasizes public, or publicly guaranteed and regulated, finance, but not necessarily public *delivery* of services. For example, Bobadilla and colleagues (1994) assessed the impacts of a national package of health services applied to the populations of sub-Saharan Africa, Latin America and the Caribbean. The content of the packages included basic public health (e.g. school health program, HIV/AIDS prevention program) and clinical (e.g. family planning, prenatal and delivery care) services. In so doing, they estimate that the package could eliminate 21–38% of the burden of premature mortality and disability in children under 15 years and 10–18% of the burden in adults, at a cost of about US$12 per person per year in low income countries.

The very selectivity of much primary health care in the developing world has been called into question, not least from a structuralist perspective which argues that only the effects, and not the deeper causes, are being treated. Lives may well be saved, but the children will still be susceptible to the respiratory and water-borne infections that contribute so much to the global burden of disease. Better housing, nutrition, and safe water are required, otherwise the "cured" children simply become re-infected. Comprehensive strategies are therefore still required to counteract the appalling lack of access to basic needs of safe drinking water and adequate sanitation (see Table 5.1). These strategies are meant to evolve out of the setting of the Millennium Development Goals (MDGs; see Box 4.1 and pp. 90–2 above). Most of the eight goals have implications for health in developing countries and while progress is being made, it remains slow and characterized by substantial regional variation (United Nations, 2005). Goal number 1 is to eradicate extreme poverty and hunger; we know the tremendous impacts this can have on health status. A related target established for this goal was to halve, by 2015, the proportion of the people in the world whose income is less than $1 per day. By 2002, this rate had fallen – globally – from 28 to 19%. However, the major gains here were made in Asia; sub-Saharan Africa remains essentially unchanged. And there have only been very slight decreases in the proportion of people who suffer from chronic hunger. The goal of achieving universal primary education is seeing some gains, but there remain significant gender gaps. More children are

Table 5.1 Access to safe water and adequate sanitation in selected countries (2004)

Country	Percent houses in urban areas connected to a water supply	Percent houses in urban areas connected to a sewerage system
Afghanistan	15	6
Benin	25	0
Cambodia	36	23
Central African Rep.	9	0
Ethiopia	32	2
Gambia	39	35
Haiti	24	0
Madagascar	16	7
Malawi	29	1
Niger	35	0
Sierra Leone	30	n.d.
Somalia	3	n.d.
Sudan	46	1
Zambia	41	29

n.d. = no data
(*Source*: WHO, 2008)

being vaccinated, child mortality rates are decreasing, and maternal mortality rates are decreasing, but not in sub-Saharan Africa or Southern Asia (United Nations, 2006). Of all the MDGs, the least progress has been made around increasing access to adequate sanitation and safe drinking water: "With half of developing country populations still lacking basic sanitation, the world is unlikely to reach its target" (United Nations, 2006: 18).

Some authors argue that existing programs are not always targeted at those most in need, and that the ethos is wrong, taking the form of passive hand-outs rather than integrated, participatory, health care. Wisner (1988) argues that some interventions prioritize a biomedical model rather than a more appropriate social model of health care. He further suggests that poverty is too often seen as "natural," with population growth a cause of poverty rather than a symptom. However, addressing this "cause" via a focus on female education and family spacing, for example, merely stresses the "ignorance" of the women involved and bypasses the broader, and more difficult, socio-structural causes. In addition, a selective approach to maternal and child health ignores other health care needs of women in developing countries. The health risks to women of working both inside and outside the home deserve fuller attention than they have received to date (Lewis and Kieffer, 1994).

Turshen (1999) has pointed to the massive structural adjustments in health care since the mid-1970s, which she attributes to the declining role of the World Health Organization and the growing hegemony of the World Bank. The World Bank is the major donor to developing countries and sets fiscal targets, the response to which in Third World countries has been to cut social rather than military expenditure and to enlarge the role of the private sector in health care delivery. She further argues that the reduction in health expenditure has had severe impacts on women (see Table 5.2). For example, the introduction of user charges in Congo led to a steep drop in numbers of ante-natal visits and visits

Table 5.2 Maternal mortality rates and health care provision in parts of sub-Saharan Africa

Country	Estimated maternal mortality rate		Estimated percentage of births attended	
	(1990)[a]	*(2005)*	*(1990–6)*	*(2006)*
Angola	1500	1400	15	47
Mozambique	1500	520	25	48
Uganda	1200	550	38	39
Zambia	940	830	51	43
Tanzania	770	n.d.	53	46
Kenya	650	560	45	42
Zimbabwe	570	880	69	73
Industrialized countries	31	9	99	99

[a] Rate per 100,000 live births

(*Source*: Turshen, 1999: 20, and WHO Department of Reproductive Health & Research, 2008)

to child care clinics (Turshen, 1999: 19). Note from Table 5.2 that, with the exception of Zimbabwe, maternal mortality rates have, over the last 15–20 years, fallen sharply in some countries.

Among other examples, Turshen points to Zimbabwe as a country whose health care delivery has worsened because of structural readjustment. Zimbabwe was an enthusiastic adopter of the Alma Ata declaration, and despite the geographic concentration of health care resources in the two main cities (Harare and Bulawayo) it was able to deliver adequate access to health services to 80% of the rural population; the number of rural health centers rose from 500 in 1980 to more than 1000 in 1990. The proportion of the health budget devoted to preventive care rose from 8% in 1980 to 14% in 1983. In 1992, however, the imposition of structural adjustment by the World Bank saw the abolition of 1200 health and nursing posts and a loss of doctors to more lucrative positions overseas. Fees were introduced – in effect, a way of rationing health care – which had drastic and immediate impacts on health outcomes. Exemptions for the poorest people resulted in only limited improvement. For Turshen (1999: 40), the public sector is in disarray and the growing private sector caters only for those willing and able to pay. However, the extent to which Zimbabwe's plight is due to structural readjustment rather than the policies promulgated by its leaders (see pp. 90–2 above) is debatable.

The role played by multi-national drug companies in providing health care in the developing world is well documented; so, too, is their lack of willingness to provide essential drugs to the poor in those countries (Henry and Lexchin, 2002). "Access to essential drugs is a basic human right often denied to people in poor countries … such rights will not soon become a reality" (Pécoul et al., 1999: 367). Pécoul and colleagues (1999) suggest four main factors affecting access to quality drugs in developing countries. First, the drugs are just too expensive; for example, we have long known this to be the case for essential antiretroviral drugs for use in patients who are HIV positive (Creese et al., 2002). Further analysis of drug costs in low and middle income countries by Mendis and colleagues (2007) supports this view; for example, they find that the cost of one month of asthma treatment is equivalent to 9.2 days' wages in Malawi. Further, the cost of name brand drugs in some countries is ten times higher than their generic equivalent. The second major limiting factor is related to the production and distribution of counterfeit and substandard products. For example, an extensive vaccination campaign was organized in Niger in 1995 to deal with a major meningitis epidemic. Pasteur Merieux and SmithKline Beecham donated 88,000 vaccines, to be administered by a team from Médecins Sans Frontières. However, the vaccines were an odd color and consistency; it was discovered that they were made by a local producer with counterfeit labels. Over 60,000 people received the ineffective vaccine. Third, drugs necessary for the treatment of certain tropical diseases are simply disappearing from the market as large multinational pharmaceutical companies do not see them as profitable. For example, Human African trypanosomiasis (HAT), more commonly known as sleeping sickness, is caused by a single-celled parasite transferred to humans by infected tsetse flies. The parasite attacks the central nervous system and causes serious neurological disorders. In some parts of sub-Saharan Africa, prevalence rates of 20–50% have been reported. In these instances, HAT represents a greater cause of morbidity and mortality than HIV/AIDS. Without treatment, this disease is fatal. There are several treatments that could be used; most of these are not in production due to lack of commercial interest (Pécoul et al., 1999). Essentially, the pharmaceutical industry is the most profitable business venture in the world, with an average of 16% profit and some as high as 40% (in comparison, financial companies bring in just under 12%, and beverage companies around 10%; Henry and Lexchin 2002). Finally, Pécoul and colleagues discuss at length the need for, and barriers to, sufficient research and development for cost effective drugs for use in developing countries, along with the impacts of trade agreements that control access to patents and the availability of generics. Regardless, even if essential drugs were made available at low cost in developing countries, there are not enough practitioners to prescribe them appropriately (Kessel, 1999).

We want, finally, to say something about regional variation in service provision. Some evidence from India points to marked disparities (Akhtar and Izhar, 1994). Per capita expenditure on health and family welfare in 1986–7 shows that this is low in the densely populated northern states of Uttar Pradesh and Bihar (which have a history of poor resources, and a poor agricultural and industrial base) but much higher in states such as Nagaland, in the north-east. The same picture emerges for staffing levels, with Uttar Pradesh having a ratio of one doctor per 15,600 persons while Chandigarh (in Northern India) has one for every 820. Similarly, there are problems of differential access in urban and rural areas, with 35% of the rural population having access to some health care within 2 km, while 84% of those living in urban areas have similar access. In part, this reflects an unwillingness of doctors to work in rural areas (Akhtar and Izhar, 1994).

The "primacy" of settlement structure in many developing countries, whereby population is concentrated in the largest city, means that such cities receive resources over and above what they might expect. For example, the Manila urban region in the Philippines has about 25% of the total population, yet 43% of the hospital beds. Since political power is concentrated in capital cities the health care needs of those in more remote rural areas are neglected. In Ghana in the 1980s, spending on specialist tertiary care amounted to 40% of the health budget, yet this benefited only 1% of the population; primary health care spending was only about 15% of the budget. According to Turshen (1999), Sierra Leone, Malawi, Togo, Tanzania, and Liberia still spend over 80% of their health budget on hospitals.

Urban bias in the provision of health care is shown in other contexts. For example, in South Africa the provision of X-ray services is highly uneven spatially (Walters et al., 1998). While some districts are relatively well resourced, particularly in the Johannesburg region, others are not. One-quarter of all districts in the country – representing over 7.5 million people – have no X-ray facilities. This spatial inequality parallels that for health services as a whole in South Africa. In Andean Bolivia, Perry and Gesler (2000) have made basic use of GIS to examine the provision of primary health care facilities in a remote mountainous part of the country. Much of the region is only accessible on foot. The authors use global positioning systems to record the locations of small settlements, and then construct circular buffer zones around existing health centers, as well as more complex shapes that represent one-hour walking times to and from such centers. The most inaccessible region, Charazani, has numerous settlements located far from the nearest health center, and the authors show how additional modest investment would ensure that nearly half the population in this remote area would be within one hour's walk; currently, one-fifth of the population have to travel for more than a day to get access to basic primary care.

Health care provision in the developed world

Here, we consider some aspects of service provision within the developed world. In particular, we look at selected aspects of the provision of primary care facilities in Britain and then examine some issues concerned with the unevenness of provision of secondary care in the USA. This is a very selective picture, but some key themes are highlighted.

In Britain, after the creation of the National Health Service in 1948 the distribution of general practitioners came to match more closely the distribution of population, since restrictions were introduced as to where such GPs could practice. However, the need for primary care is not simply a function of population distribution. Needs are greatest where the burden of disease and illness is high. As a result, some authors suggest that we need to take account of morbidity data (such as that on limiting long-term illness, available from the census) in shaping the provision of GPs. For example, a more discriminating form of needs indicator shows that some rural areas are relatively over-provided with GPs compared with deprived urban areas (Benzeval and Judge, 1996).

As Table 5.3 suggests, there have been major changes in recent years in the structure of general practice in England and Wales. The proportion of those working alone has shrunk considerably, while the number working in larger practices (six or more GP partners) has increased from under 10% to nearly a quarter in the last two decades. Single-handed GPs tend to be older than others and are more likely to be male. Geographically, they are concentrated in urban areas, particularly in inner city parts of London and Birmingham (see Table 5.4). The number of patients registered with a practice ("list size") varies from one part of the country to another (see Table 5.4). In the London area and outskirts (such as Essex) there are over 1800 patients, on average, on each GP's list, while doctors in the south-west have less than 1500.

While some work has been done on the provision of general practice care in Britain, other authors have looked at different aspects of the provision of primary care. It is important to consider "access" not merely in terms of geographical proximity, but also in terms of the quality of provision. Rogers and her colleagues (1998) have adopted an ethnographic

Table 5.3   Size of general practices in England and Wales

Percentage of practices	1991	2005
Single-handed	32	22
Six or more GPs	9	23

(*Source*: General and Personal Medical Services Statistics, Department of Health, 2005)

Table 5.4   Average list size and practice size in England and Wales (2005)

Health region	Average list size per GP	Single-handed practices (per cent)
England	1666	22.8
Wales	1674	19.2
English SHAs		
North East London	1882	33.8
Essex	1830	31.9
Birmingham & Black Country	1785	38.1
Cumbria & Lancashire	1687	30.1
Norfolk, Suffolk & Cambridgeshire	1544	9.7
South West Peninsula	1445	10.0
Somerset & Dorset	1438	5.7

Note: SHA is Strategic Health Authority
(*Source*: General and Personal Medical Services Statistics, Department of Health, 2005)

approach to equity of access to, and provision of, community pharmacy services in the UK, couching this within a structurationist framework in which the local physical environment and internal spatial arrangement of the pharmacy shapes the social interaction that takes place there. Simply mapping the distribution tells us nothing about the quality of care (advice) offered in different locations. Using a variety of pharmacy settings in north-west England, the authors recorded and transcribed interactions between staff and patients over the period of a week. Particular interest lay in the comparison of inner-city, small town, and rural localities. The inner-city pharmacy was characterized as a "fortress," encased in iron bars and razor wire, in which health-giving advice was negligible. The main business was dispensing prescriptions for methadone, though it did so in a caring and valuable way. The rural pharmacy was a small and old-fashioned "haven," free from the surveillant eye of security cameras, and operating with a low turnover. Here, advice-giving was much more frequent and the conversations were informal and wide-ranging. One patient asks for advice about his hand and is advised by the pharmacist: "I would go to the doctor's with that," to which the patient responds: "I would if I could walk straight in." The pharmacist substitutes for the general practitioner. As the authors note, "inequalities may not simply relate to the use of services but to the more subtle and less tangible aspects related to the type of service provided" (Rogers et al., 1998: 372). Those with the greatest need seem to be receiving an inferior service, another manifestation of the inverse-care law.

Table 5.5 Ratio of physicians to total population for contiguous states in USA (2004)

	State	*Rate (per 100,000 residents)*
Less than 200 per 100,000	Idaho	169
	Oklahoma	171
	Mississippi	181
	Nevada	186
	Iowa	187
	Wyoming	188
More than 300 per 100,000	New Jersey	306
	Rhode Island	351
	Vermont	362
	Connecticut	363
	New York	389
	Maryland	411
	Massachusetts	450
	District of Columbia	798

(*Source*: United States Census Bureau, Statistical Abstract 2007, Table 154)

There have been several studies by geographers looking at the spatial distribution of health services in the USA. The provision of general physicians per head of population reveals relatively good provision in north-eastern states, and much poorer provision in states with dispersed populations; see Table 5.5 for those states with the highest and lowest rates. But the spatial organization of health care systems cannot be divorced from the political and economic context. This "structuralist" perspective on health care has been promoted by Whiteis (1997; 1998), who sees the contemporary American health care system as an "ascendancy of individualistic, intervention-oriented, corporate-sponsored medicine" (Whiteis, 1997: 230). He points to disinvestment in health care in major urban areas, as corporate interests withdraw health and medical services from unprofitable poor communities and switch their investment to more profitable areas. Thus, health maintenance organizations (HMOs), health care providers that put hospitals, doctors (physicians), and other health services under a single group, are becoming increasingly commercialized and oligopolistic. "Liquid assets at corporate HMOs have risen by annual rates of over 15% since the early 1990s; a popular strategy has been to funnel these excess profits into the acquisition of other health plans, thus further consolidating corporate hegemony over expanded market areas" (Whiteis, 1998: 801).

This "corporatization" has spatial consequences, as instanced by McLafferty's (1982) study of hospital closure in New York City. McLafferty looked at 39 hospitals that closed between 1970 and 1981, with particular reference to community hospitals. She showed that those communities having a high proportion of non-white residents were more likely to have hospitals closing, a finding supported by other work (Whiteis, 1998: 800). Closures occurred in areas that had already been designated as areas short of medical personnel; these result in longer travel times, especially where the supply of public transport is poor. In addition, corporate health care providers are reluctant to tolerate patients with HIV and AIDS, or others whose care is likely to be expensive. A further impact of closure programs is the loss of employment opportunities to those formerly working in the hospitals;

thus, the income base of the local community is further denuded. However, evidence from Winnipeg, Canada (Roos et al., 2006), suggests that hospital downsizing has no measurable effect on the health status of the population as a whole or on vulnerable groups such as the elderly.

Contrasts in the provision of health care between urban and rural areas in the USA are often extreme. Knapp and Hardwick (2000) show that rural areas are poorly served by primary health care professionals (physicians and dentists). Nationally, the ratio of primary care physicians to population is 95 per 100,000, but this drops to only 4.2 per 100,000 in the most rural areas. Similarly, there are 76 dentists per 100,000 people in the country as a whole, but only 29 per 100,000 in the extreme rural areas. Graham et al. (1995) point to the relative under-provision in rural areas of services for people living with HIV or AIDS. In rural areas, rates of infection are rising, particularly among black American women. But many rural hospitals lack the services and facilities for comprehensive AIDS care. There is poor access to basic primary care, consultation, diagnosis, early treatment, home care, and other support services. This differential is reflected in survival rates; for example women in metropolitan Atlanta, Georgia, had significantly better survival than those elsewhere in the state. In part, this reflects the lack of health insurance, but it also reflects the closure of hospitals in rural areas. As Graham and colleagues (1995: 444) report, people living with AIDS "in rural areas near urban centers may obtain adequate care given transportation, but in more remote areas, especially in frontier regions, distance can pose an insurmountable barrier."

Disparities in the availability of health insurance in the USA are extreme (Carrasquillo et al., 1999; Miringoff and Miringoff, 1999). Overall, the percentage of the population without health care coverage rose from 11% in 1970 to 16.1% in 1997, though it dropped – marginally – to 15.7% in 2004. Despite the existence of Medicaid, designed to provide health insurance for poorer people, a quarter of those on incomes less than $25,000 have no insurance. While 15% of the white population is uninsured, 20% of the black population, and one-third of Hispanics, lack health coverage (see Table 5.6). These inequalities work themselves out in space too; of those living in southern states, well over 15% were without insurance in 2004, compared with under 12% in more northern states (see Figure 5.2).

Finally, we should not think of provision solely in terms of a surgery or hospital or other fixed location; health care can be provided in other ways. For example, "linkworkers" can be employed as bridges between lay people and health professionals, as is commonly done among ethnic minority communities. Mobile services can be used to provide health care, as in the screening of breast cancer, for instance. Volunteers and family members provide health care to particular groups, such as the elderly. Even here, distance constrains the ability of some caregivers to devote time to elderly relatives, as Canadian work illustrates (Joseph and Hallman, 1998). Those living less than 30 minutes from their elderly relatives are likely to devote on average 4.27 hours to care per week, compared with only three hours for those traveling for more than two hours (see Table 5.7). However, this is gendered inasmuch as only male involvement in care-giving declines significantly with increasing travel time.

Increasingly, too, health care is being provided remotely, via what is known as *telemedicine* or *telehealthcare*. Here, the health professional offers a consultation, or diagnosis, at a distance, using modern telecommunications technology to talk to the patient, or perhaps

Table 5.6 The (health) uninsured in the USA

Percentage of the population without health insurance (2004)	
Persons aged 18–24 years	31.4
Persons aged 25–34 years	25.9
Persons aged 35–44 years	18.7
Persons aged 45–64 years	14.3
Hispanic population	32.7
Black population	19.7
White population	14.8
Income <$25,000	24.3
Income $25,000–$49,999	20.0
Income $50,000–$74,999	13.3
Income >$75,000	8.4

(*Source*: US Census Statistical Abstract, 2007)

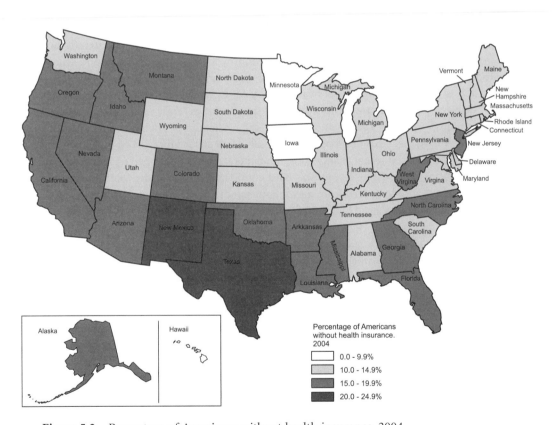

Figure 5.2 Percentage of Americans without health insurance, 2004

Table 5.7 Mean number of hours of care devoted to elderly in Canada, by travel time and gender

Travel time	All respondents	Males	Females
1–30 minutes	4.27	3.55	4.66
31–120 minutes	3.84	2.75	4.63
More than 120 minutes	3.06	2.31	3.74
	p = 0.018	p = 0.029	p = 0.353

Figures are hours per week
(*Source*: Joseph and Hallman, 1998: 635)

to view a scanned image of some part of the body, rather than requiring the patient to travel a considerable distance. The applications, both real and potential, of this technology are many and various, but are particularly appropriate for isolated communities where a primary health care professional can get an expert opinion from a major health care center (Cutchin, 2002; Marcin et al., 2004). The technology offers some way of circumventing the problem (see Box 5.2, page 128) of wishing to provide specialist care in large centers without disadvantaging those in peripheral regions. Many specialists are now using tele-medicine in this way, whether in cardiology, dermatology, obstetrics, or psychiatry, though there is as yet little empirical work on how this improves access to health care and critical reflections on access to, and uses of, these technologies are merited (Curtis, 2004: 131–3).

Last, it is important to note that health care needs to be seen in terms that are broader than "delivery" by health care professionals in formal settings (Parr, 2003). The experience and meaning of care goes well beyond that provided by doctors/physicians, even in primary care. Increasingly, the needs of those who are disabled, or sick, vulnerable or frail, are being met in the home and in recent years a substantial literature has developed on care delivered within the domestic setting. Researchers have considered care provided both to children and older adults with long-term care needs. For example, Yantzi and her colleagues (2006) looked at the challenges faced by mothers of children needing 24-hour care, living in both urban and rural Ontario. Her interviews suggested three main types of challenges: the physical ones of planning any outing and navigating unfavorable environments; the social ones of lacking robust networks of family and friends to provide additional support; and the service ones of a failure to provide adequate respite care. And, of course, relative loca-tion makes a difference; out-of-home respite care is more accessible in the urban area, but in rural areas may be over an hour's drive, if available at all. Yantzi's work makes sober reading. So many of us take for granted the mundane activities of sleeping, eating, getting dressed and going shopping, but for Yantzi's respondents all these activities require negotia-tions and meticulous planning. Although she herself does not focus directly on the conse-quences for the mothers' own well-being, others have examined the impacts of care provision on the health of the carers themselves.

For example, Christine Milligan has been at the forefront of research on care provision for older adults, particularly the "old old" (typically, those aged over 85 years, a growing proportion of the population in high income countries). In a series of books and papers – for example, Milligan (2001; 2006) – she explores both the roles of the voluntary sector and care homes in providing care to such groups, with particular reference to rural settings. As she notes, declining investment in state or public sector provision of care in the developed world has led to the burden falling on family members (and particularly women) and a shift in setting from the formal to the informal. Even in care-homes considerable care is actually provided by the family member (Milligan, 2006), with consequences for the emotional well-being of the carers who face unanticipated changes in their lives and identities. Whether such consequences are positive or negative depend on whether the places of care are inclusive or exclusive, welcoming or discouraging of the involvement of the carer.

In essence, we are seeing in recent years a trend towards a mixed economy of health care and a blurring of the boundaries between formal and informal care (Milligan, 2006: 320). The mixing is of care in institutional settings, in the community, and in home spaces, and the actors involved include family members, volunteers and a range of health professionals. We can expect these features to continue to characterize care in the years ahead.

Utilization of Services

We concentrate in this section on the uptake of health services in the developed world, organizing the material in terms of the level of care sought. We look first at the use of primary care services, before turning attention to the use of secondary and tertiary services. But first some introductory remarks are in order.

In Britain, access to secondary health care (hospitals) is, with the exception of accident and emergency (A&E) services, through the general practitioner. People "consult" their GP, who may or may not "refer" them on to a hospital specialist. To what extent are there place-to-place differences in rates of consultation and referral? Does relative location (distance from clinic or hospital) predict attendance? Is attendance more, or less, likely from those living in relatively deprived areas? If deprivation is a reasonable measure of "need," we should expect higher rates of consultation from more deprived areas, other things being equal. The relationship between utilization, distance, and deprivation or need is complex. Demonstrating that utilization is higher nearer clinics or hospitals may mean that proximity encourages attendance; but it might simply mean that those attending have greater need, because such hospitals may be located in relatively deprived areas. Without adequate data we cannot separate out distance effects from those of deprivation. Distance as a constraint on health-seeking behavior has been studied by geographers and others for many years (see Joseph and Phillips, 1984, for a synthesis of some of this literature).

Use of primary health care services

For both primary and secondary care there is a frequently demonstrated relationship between service use and area deprivation. As with the study of health outcomes (Chapter 4), an important research question is the extent to which this is mirrored at an individual level. Multi-level modeling can be used to shed some light on this (Carr-Hill et al., 1996). In a survey of consultation rates for 60 practices in England and Wales, data were collected on the socio-economic characteristics of both those consulting and their area of residence. Distance from area of residence (enumeration district) to approximate location of practice was also measured, and an indication of whether the patient lived in a rural area or not was also used. At the individual level, higher rates of consultation were associated with those who were permanently sick, unemployed, living in rented accommodation, of south Asian origin, and living in urban areas, although the magnitude of these effects varied with age and gender. Area characteristics, such as levels of housing tenure and car ownership, had limited additional explanatory power, suggesting that service utilization depends largely on individual factors. Those living closer to the clinics were more likely to consult more frequently, both in urban and rural areas. Results presented elsewhere by Carr-Hill and his colleagues (1997) show that patients living within 2 km (1¼ m) of a surgery are more likely to consult than those beyond that distance, regardless of age and sex (see Table 5.8). There is thus clear evidence of a "distance decay" relationship between utilization and distance from the health center.

There is other evidence that people living further away from surgeries and clinics consult less than those living closer, even when socio-economic status is allowed for; in other words, it is not because of a lack of need that those living further away consult less frequently. Work in Norfolk, England, by Haynes and Bentham (1982) showed that those living in remoter villages consulted less frequently than others. In rural Vermont, USA, Nemet and Bailey (2000) confirm the inverse association between distance and utilization. Elderly people having to travel more than 10 miles to see a physician are likely to do so much less frequently than those who live nearer their physician. But a more significant predictor of utilization is whether or not the physician is located within a broader "activity space," that is, the wider set of places that people visit regularly. While the authors' claim that this is linked to "sense of place" (Nemet and Bailey 2000: 1200) is exaggerated there is certainly scope for going beyond simplistic measures of distance in the study of health care use. Evidence from the developing world (for example, Muller et al., 1998) also points to the way in which attendance at primary health care clinics declines markedly with distance.

Table 5.8 Consultation rates in general practice (UK) by distance from surgery

Age group	Distance (km)	Male	Female
Under 5	<2	5.1	4.8
	2–5	4.6	4.5
	>5	4.4	4.1
5–15	<2	2.2	2.5
	2–5	2.0	2.2
	>5	1.9	2.3
15–64	<2	2.5	4.9
	2–5	2.4	4.4
	>5	2.2	4.1
65+	<2	5.2	5.7
	2–5	4.8	5.1
	>5	4.7	5.2

(*Source*: Carr-Hill et al., 1997: 39)

What explains distance decay effects? It may simply be the friction of distance – the fact that distance, whether in terms of time or cost, deters people from consulting. Alternatively, patients may trade off the "costs" of a consultation against the possible benefits. And when we do consult, what characterizes the consultation? Communication between doctor and patient may be ineffective, either because patients do not understand what they are being told, or because the doctor does not spend sufficient time with the patient. Merely looking at "consultation" or the length of time spent on this is less useful than knowing how effective (from both points of view) such health service use is. Research indicates that middle-class patients spend more time with GPs than do those from working-class backgrounds and that they also find it easier to communicate with their GPs (Benzeval et al., 1995).

Other work takes a more sophisticated view of "access," seeing this more in terms of possible time-space constraints on service use. Survey research in north-west England (Senior et al., 1993) sought to use a variety of variables to predict immunization uptake. These included accessibility and mobility (travel time to clinic, availability of car), child care commitments, social class, educational attainment of the parent(s), whether or not the parent was single, and whether the child was ill when called for immunization. Statistical analysis showed that educational attainment of both the mother and father, child care commitments, and illness of the child were all significant predictors. In-depth interviews with parents supported these findings and referred to the practical difficulties of attending the clinic: "I'm on my own with three kids and pregnant. I just don't go. I would have to walk or go to the Precinct and get two buses. I have no money for taxis." For another woman, "My husband was at work, and it's a lot of messing taking the other children with me."

Studies in other countries have backed up these findings. For example, Wellstood and her colleagues (2006) assess the role of both individual and "system" (service) characteristics in seeking to understand experiences of health service use – and problems of access – in two socially contrasting neighborhoods (one low income, one high income) in Hamilton, Ontario. Semi-structured interviews were conducted with a small number of people in each area. Of the 41 interviewed, 32 mentioned difficulties in accessing family doctors' services; waiting times were one theme, but another was the problem of getting to and from the surgery, especially for those without their own transport. For those at work, the surgery hours proved problematic, as other studies have shown. All these "barriers" were common across both types of neighborhood. Parallel quantitative survey work in the same city (Law et al., 2005) indicates that there were only modest differences between neighborhoods in GP usage, once individual factors had been controlled for.

Other dimensions of primary care are the use of screening and ante-natal services. How does the uptake of screening for cancer vary from place to place? Two studies of screening for cancer, both based within quite small regions of Britain, are illustrative.

Bentham et al. (1995) looked at screening for cervical cancer in Norfolk. The present UK system requires all eligible women aged 20–64 to be screened every 3–5 years; screening is often done at the general practice surgery (physician clinic). The following variables are significant predictors of uptake: the presence of a female GP; practice size (larger practices have higher uptake rates); and deprivation (the higher the Townsend score, the worse the uptake). However, distance from the surgery has no significant effect; women are not dissuaded from attending by the costs of overcoming distance. Gatrell et al. (1998) considered practice variation in uptake of screening for breast cancer in South Lancashire,

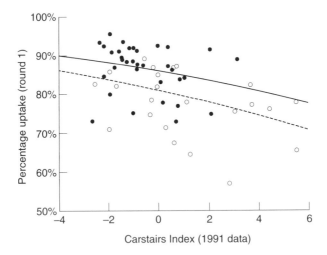

Figure 5.3   Uptake of screening of breast cancer in South Lancashire, UK, 1989–92

in north-west England. Although the screening is not done in the practice setting it is possible to determine an uptake rate by knowing which women are registered with particular practices. Uptake is predicted quite well by the deprivation characteristics of the practice catchment area; practices in more deprived areas have lower rates of screening for breast cancer (see Figure 5.3). The presence of a female practitioner (solid circles in Figure 5.3) is also a contributory factor. This suggests that, although not directly involved in the screening, a female doctor perhaps plays a role in giving encouragement to the woman to attend for screening. Whether or not a lack of screening translates into higher morbidity or mortality is an important question that we consider later.

Let us look finally at pre-natal (ante-natal) care. Larson et al. (1997) studied late pre-natal care in the USA between 1985 and 1987. The data they studied relate to over 11 million births, of whom just under 10,000 died before their first birthday; the overall mortality rate was 8.94 per 1000 live births. Particular interest centers on rurality: specifically, whether residence in a non-metropolitan county has a significant impact on late (or missing) ante-natal care, after adjustment for maternal age, race, education, and other factors. Results indicate that 22% of all births are in non-metropolitan areas and that in such areas there is a significant proportion that has had late or no pre-natal care (see Table 5.9). This is true of all groups except African-Americans. Regardless of area of residence, the proportions of non-whites without adequate care are well over twice those of the white population. Geographically, non-metropolitan residents in most states are at significantly greater risk than urban residents of delaying, or not receiving, ante-natal care. Larson and his colleagues attribute this to the costs of overcoming distance, to declining participation in obstetrics by family physicians, and to higher proportions of uninsured women in rural areas. It should be noted that the proportions of women who access care late, or not at all, have dropped sharply since Larson's study was published, as the final column in Table 5.9 suggests.

Table 5.9 Inadequate pre-natal care in the USA, by place of residence and race (1985–7 and 2003)

	Percentage of total births	Percentage with late or no prenatal care (1985–7)	Percentage with late or no prenatal care (2003)
Whites			3.5
Metropolitan	76.7	4.90	
Non-metropolitan	23.3	5.11[a]	
African-Americans			6.0
Metropolitan	84.1	10.72	
Non-metropolitan	15.9	10.72	
American Indians			7.6
Metropolitan	37.5	11.54	
Non-metropolitan	62.5	13.83[a]	
Other			
Metropolitan	92.5	6.44	
Non-metropolitan	7.5	6.87[a]	

[a] significant difference (p < .01)
(*Source*: Larson et al., 1997: 1750, and *United States Census Bureau, Statistical Abstract 2007*, Table 84)

Use of secondary and tertiary health care services

Several studies from different countries have examined the way in which the use of hospital services varies with distance and deprivation. We might expect that distance would be more of a barrier where the health problem is less serious. Conversely, if patients see some prospects of great benefit from their treatment, they are more likely to make use of services. In this case distance may be less of a constraint, although this assumes an ability to pay the costs of overcoming such distance.

Slack and his colleagues (1997) examine rates of hospitalization in the East Midland region of England in an attempt to look at the role of deprivation in access to services. Access is modeled in a more sophisticated way than simply measuring straight line distance between area of residence and hospital. The authors look at the proportion of the population within particular travel times of the hospitals, according to both private and public transport. Results indicate that both deprivation and travel time are significant influences on hospitalization rates; small areas that are more deprived, and closer to the hospitals, are more likely to have higher rates of hospitalization.

Rates of referral for chronic kidney failure (requiring either dialysis or transplantation) in south-west Wales varied with distance to the hospital of treatment (Boyle et al., 1996), although only for patients aged 60 years and over. This holds even when socio-economic status and ethnicity are controlled for. The explanation for this distance decay effect is uncertain; it may be that general practitioners looking after older patients feel that the difficulties involved in traveling long distances outweigh the benefits of treatment. But in order to ensure equity of access it may be necessary to have further outreach clinics in more local hospitals. In contrast, research in Scotland looking at referrals of patients with testicular cancer to specialist cancer centers (Clarke et al., 1995) found that these were fairly uniform regardless of place; those living in rural areas were as likely to be referred for treatment as those in urban areas.

As with primary care, studying the use of hospital services and access to such services from a geographical perspective requires more than an examination of the effect of distance on use. Let us consider some studies that have explored this broader theme.

Roos et al. (2006) present evidence of a gradient in medical care utilization according to level of education. Using data from the city of Winnipeg, Manitoba and assigning residents to one of five neighborhood groups (according to average level of educational attainment), they reveal a clear gradient of hospital admissions and rates of surgery (see Figure 5.4), a pattern that is repeated for primary care. Those living in the neighborhoods with the lowest average education spent 40% more days in hospital, and had 30% more admissions, than those living in areas with the highest educational levels. The pattern is consistent over time.

One area of tertiary care that has generated some interest is the use of investigative and surgical procedures for treating heart disease, such as angina. There are several such procedures now in use, of which surgery known as coronary artery bypass grafting (CABG) is prominent. Ben-Shlomo and Chaturvedi (1995) have looked at the use of this in north-east London and how it relates to deprivation. Uptake ought broadly to match "need" (here taken to be represented by age-standardized mortality), but results indicate that there is some inequity in relation to need, especially for men. Similar work has been carried out in two health authorities in north-west England, some of which has used qualitative methods to shed light on patterns such as these (Chapple and Gatrell, 1998). This work involved interviews of both the general practitioners looking after patients, and the cardiologists and surgeons to whom they were referred for investigation and surgery. One important finding was the difficulty faced by some patients of South Asian origin, for some of whom there were barriers to adequate care. Those small areas with high concentrations

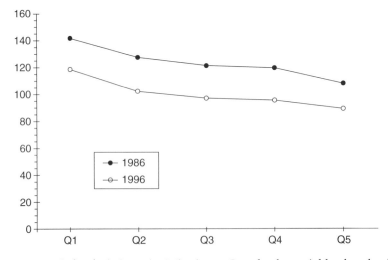

Figure 5.4   Hospital admissions in Winnipeg, Canada, by neighborhood education quintile

of this group appeared to have very low rates of treatment. As one GP noted, "we perceive that they are not as aggressively treated as the indigenous population. I don't know if that is perceived racism or whether it is actually happening" (Chapple and Gatrell, 1998: 156). In England as a whole, other authors (Dong et al., 1998) have shown that women are significantly less likely than men to have cardiac surgery; even allowing for socio-economic factors, smoking behavior, and other illness, the probability of a man having such surgery is nearly three times higher than for a woman. Worryingly, the possibility that this reflects "a more generalized process of doctor discrimination is an uncomfortable notion but is something that requires further investigation" (Dong et al., 1998: 1778).

In Maryland, USA, access to CABG surgery was significantly poorer in patients traveling more than 80 miles (130 km) (Gittelsohn and Powe, 1995). Other work in the USA has looked at variation in the use of beta-blockers (drugs used in the treatment of heart problems, particularly in reducing the risk of death following a heart attack). Wang and Stafford (1998) show that these drugs are under-used, notably among patients aged 75 years and over, among non-white patients and among those without private insurance. Clearly, then, there are similar health divides operating here, too. Regionally, patients in the north-east are much more likely to be offered beta-blockers than those elsewhere in the country.

Von Reichert and his co-authors (1995) conducted a study of access to obstetric care (maternity services) in rural Montana, with the specific goal of comparing the experiences of white and native Americans. Their interest is in determining whether there are differences between the two groups in terms of travel to access care. However, both groups suffer from rural isolation, since there are relatively few physicians at primary care level and few rural hospitals. The more densely populated counties have hospitals offering a high level of obstetric care, while most counties have smaller hospitals that do not provide specialist care and some have no hospital services at all. The Indian Health Service provides medical care in the form of birthing facilities and clinic services on reservations.

During the study period (1980–9) there were over 130,000 births, of which nearly 90% were to white mothers and 10% to native Americans. The authors found that while only 19% of whites traveled outside their county of residence, 37% of native Americans did so. In addition, the latter traveled further to access a lower level of obstetric care. There are no intrinsic differences between the two groups; both prefer to use local services if the quality of provision is high, while both groups will travel if there is a lack of such facilities. In other words, rural residents are at a disadvantage in terms of accessing health care locally. However, the unequal provision of services acts as an additional spatial constraint for the native Americans; since this group is spatially concentrated in sparsely populated areas, and since service provision is also spatially concentrated, they must travel further (see Figure 5.5). Clearly, innovative health care delivery must be sought, perhaps via mobile clinics and further use of non-medical staff; yet such improvements cannot mask the underlying structural inequality, manifested in part by the ghettoization of the native Americans on reservations.

The use of health services is therefore a function of both their supply and the demand for them. A good example of this comes from work on the spatial variation in abortion rates in the USA (Gober, 1994). Abortion rates have varied considerably from state to state, and over time. Until 1973 there was considerable variation, with most terminations performed in New York and California. Legislation by the Supreme Court in 1973 evened

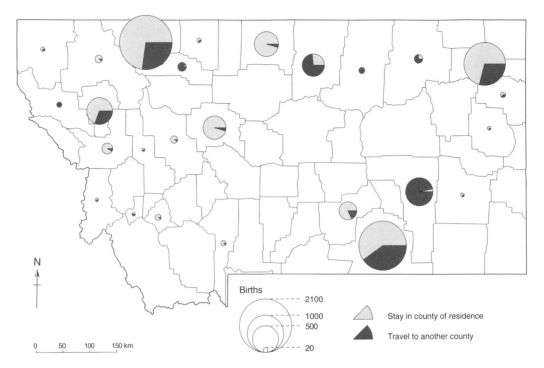

Figure 5.5 Travel for birthing of native American women, 1980–9

these variations for a few years, although individual states were given greater freedom in the late 1980s to assert legislative control. In 1988 estimated rates varied from over 40 per 1000 women aged 15–44 years in California, Nevada, Washington, DC, and New York, to under 10 in North and South Dakota. Gober models this variation using both supply and demand factors. Among the latter, the proportion of the state's population residing in metropolitan areas is a key determinant; the proportion that is black also has a direct and positive relationship with the abortion rate. Other supply-side factors operate indirectly: for example, per capita income influences state funding for abortions, which in turn predicts the rate. Both supply and demand factors therefore need to be taken into account if rates are to be modeled accurately. Of course, this is an aggregate study which cannot take into account some local social movements that have used violent methods to oppose the operation of abortion clinics in some metropolitan areas. Such opposition may well shift utilization from one area to another and cannot be ignored in a full understanding of service use. More recently, Gober and Rosenberg (2001) have contrasted the stances taken by state governments in the USA, and provincial governments in Canada, showing that while some support the right to choose (by bearing the costs of the terminations), others adopt a range of strategies to restrict access. There are clear geographies to a woman's right to choose.

Do Provision and Utilization Affect Outcome?

An important question is whether or not access to, or use of, services has any impact on outcome. Here, we first ask whether, if distance does indeed have an effect on health service use, as appears to be the case, outcomes are adversely affected. We can look at the evidence in terms of both emergency and non-emergency care. Second, we ask whether the use of health care has any impact on health status; here, we draw on a Canadian study as an example.

Let us look first at outcomes following road traffic accidents (RTAs), involving those traveling in vehicles rather than pedestrians. There is some evidence from the USA that fatalities from RTAs are greater in those rural areas that are remote from a hospital or where emergency response times are long. Research in Michigan found that even after adjusting for driver's age and gender, and the characteristics of the crash, the relative risk of fatality was 1.51 in remote rural areas. Bentham (1986) looked at mortality due to RTAs among men aged 15–24 in all local authority areas in Britain; those areas without an Accident and Emergency Department had significantly higher mortality than those which possessed one. This is, however, a very simple measure of provision. More recent work in Norfolk (Jones and Bentham, 1995) suggests that the likelihood of a death from RTA is associated with age (those aged over 60 years have a higher risk) and road speed (death is more likely on roads where speed limits exceed 60 mph.), but, importantly, the ambulance response time does not affect mortality risk. There is an obvious explanation of why this contrasts with American findings, and this concerns the relative travel times to and from the scene of the accident. In more remote parts of the USA these times might exceed two hours, whereas average ambulance travel times in England (to the accident scene and thence to hospital) are unlikely to exceed 25 minutes.

Turning to non-emergency care, we might ask whether survival following surgery for cancer varies with location (see Box 5.3). Research on colo-rectal cancer in southern England by Kim et al. (2000) showed that there were marked variations in survival following surgery. To a large extent, variation in survival depends upon stage at diagnosis; if the cancer has already spread extensively before surgery the prospects are much worse than if it is not advanced. It is also influenced greatly by age – older people have poorer prospects – and by whether or not the surgery is an emergency operation. Kim and her colleagues (2000) demonstrate that, once these major factors are controlled for, clear geographical differences in survival remain. Survival in some districts is 75% worse than in others. The reasons are unclear, but may have to do with the volume and quality of surgery being performed (see Box 5.2 above, page 128). However, Kravdal's work in Norway (2006) suggests that small hospitals do not, in general, have poorer outcomes, and that travel time to hospital is unimportant. Regardless of geographical factors, simply ranking areas, or hospitals, on the basis of survival rates or other "performance indicators," is fruitless unless it provides controls for case-mix (the type of patient and disease being treated).

Breast screening is of particular interest given the high incidence of breast cancer in developed countries, since poor uptake of screening is likely to be reflected in higher disease rates in future years. American research suggests that women in lower-income groups were less likely to have had a mammogram within the past year; only 9.2% of women on incomes less than $10,000 in 1987, compared with 20.9% whose income exceeded $35,000

Box 5.3 Survival analysis

In survival analysis we seek to describe and explain the risk of occurrence of an outcome (usually death) as a function of time. For example, we might measure survival time from the date of diagnosis to date of death, or the period from date of an operation to date of death; but the outcome might be any other designated type. Here, we assume that death is the outcome. Of course, in any given study a fraction of the population under investigation will not have died at the time of study; such data are referred to as "censored" data.

Survival analysis begins with the construction of a survival curve, which shows the proportion of people surviving beyond a given point in time. An imaginary example is shown in Figure 5.6. Initially, all patients will be survivors, but as time progresses a proportion of them fail to survive; as a result, a survival curve is a typically step-like function. "Flatter" curves represent better overall survival than those which drop more rapidly. Of more interest than one such curve, however, is to compare curves corresponding to groups of patients or particular variables that are hypothesized to affect survival. In a non-geographical context this would typically be to compare patients who either have, or do not have, some form of treatment or therapy. But in terms of the focus of attention in this book, we could contrast patients living in different geographical areas, whether deprived or more affluent. For example, Kim et al. (2000) sought to examine whether survival following surgery for colo-rectal cancer was affected by deprivation; do patients in more deprived areas have worse survival rates? Survival curves (see Figure 5.7) indicate that those in the most deprived 25% of electoral wards in a part of southern England have poorer prospects than in less deprived areas. It is important to allow for the effects of other variables, however, in the statistical analysis. For example, it might be that those in the most deprived areas are less likely to come forward to diagnosis than others; they might also be older people. Consequently, survival analysis seeks to fit models (usually *Cox proportional hazards* models) to estimate the separate contributions that different predictor variables make to the analysis.

Survival analysis can be performed using proprietary statistical packages. For basic introductions see McNeil (1996) and Altman (1991).

(Wells and Horm, 1992). This translates into delayed diagnosis, particularly among black and Hispanic women (Richardson et al., 1992), as well as into poorer survival rates – as research from a wide range of countries suggests (Schrijvers and Mackenbach, 1994).

Roos et al. (2006) ask the question: what evidence is there to assess the potential of health care delivery to both improve health status and to reduce health inequalities? As Wellstood and colleagues (2006) note, the received wisdom is that health care provision makes a marginal difference to population health in the developed world. We noted above (see Figure 5.4, page 146) that there is a clear gradient in utilization of secondary care by education status, in Winnipeg, Canada. We might now ask whether the receipt of care in a hospital setting leads to health gain. Evidence (see Figure 5.8) suggests that the greatest improvements in health occurred in those groups receiving the least care; and these are the least "deprived," in terms of educational status. Of course, there are caveats: this is an ecological study of an association, the causal nature of which can be questioned; and we need to control for many other possible factors impacting on health improvement. As the

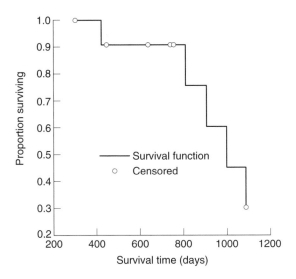

Figure 5.6 A survival curve

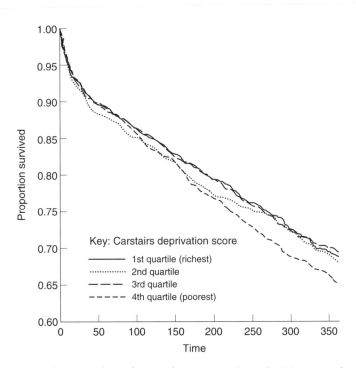

Figure 5.7 Survival curves for colo-rectal cancer patients in Wessex region, southern England, by category of deprivation

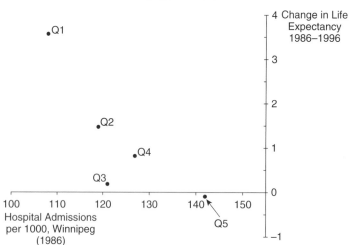

Figure 5.8 Relationship between changing life expectancy and hospital admissions in Winnipeg (1986–96)

authors note, "[I]t may simply be that those most in need of care face multiple symptom or illness complexes against which 'reactive' medical care is relatively impotent" (Roos et al., 2006: 124). Nonetheless, the general issue of what difference investment in services really makes to population health is a key one in policy terms and begs much further research.

We should not imagine that the links between service provision and health outcomes are only to be assessed in quantitative terms. A good example of how this may be examined qualitatively is a study of women's and community health centers in Adelaide, South Australia (Warin et al., 2000). These centers offer an alternative model of health care to the "fee-for-service" (and more medically focused) primary care that is more commonly provided in Australia. The health practitioners in such centers are concerned with a "social" model of health, and with health promotion in the community. Interviews with users reveal that this model of provision is very much welcomed and that it improves mental health and well-being. As one user put it: "they still have that time to talk to you as a person whereas in general practice … I feel it's like a factory production line, an assembly line" (quoted in Warin et al., 2000: 1873).

Concluding Remarks

Inevitably, there is geographical variation in the provision and use of health services. The real question is whether this "variation" also represents "inequity;" is there a poor relation between unevenness in provision or uptake of services, and the need for health care? This chapter has sought to examine some of the evidence, which suggests that there is indeed in general a mismatch between "need" and provision. We have also looked at the relationship between service provision and health outcomes.

In the developing world, issues of service provision revolve less around poor access *relative* to others in such countries, and more to whether or not there is adequate *absolute* provision. In other words, less attention has been devoted to whether low-income or deprived groups suffer relative to more wealthy ones. Of more importance is whether there is an adequate overall level of health care delivery (especially primary care). Even so, there is evidence of geographical disparities, with too great a concentration of resources in urban areas.

In the developed world, we have seen that geographical distance constrains access to, and uptake of, health care facilities. But many other factors are important, and "access" to health care in the USA, for example, depends in many cases on the affordability of health insurance. Distance decay effects are commonly observed in the utilization of services, but this is also shaped by whether or not the area of residence is "deprived." More research is needed on whether poor access to services has negative consequences for health outcomes; the evidence on this remains thin.

Further Reading

An excellent overview of health service consumption is given by Curtis (2004, Chapter 5). This makes some reference to care in industrializing countries, though for a fuller account of service provision and use in such countries see the collection of papers edited by Phillips and Verhasselt (1994).

Joseph and Phillips (1984) cover in more detail some of the ideas, if not the more recent studies, addressed in this chapter. The collection of papers in Ferguson et al. (1997) is well worth reading. There have been several recent texts that the reader will find very useful. The first two focus more on the USA/developed countries more generally: Kawachi and Kennedy (2002) and Jacobs and Morone (2005). Bennett et al. (2007) and Holtz (2008) take a much more global focus on the inequalities in health care delivery in developing countries.

Chapter 6

People on the Move: Migration and Health

Migration is the permanent or semi-permanent change of residence of an individual or group of individuals. We consider here the links between migration and both health and health care, though, as we shall see, it is the associations between migration and health outcomes that have been given priority in research to date.

We examine the health consequences of human movements that involve a change in residential location. Other forms of human movement dwarf migration in terms of the volume of social interaction and may have potentially dramatic consequences for health and health care. International migrants form only about 1% of those traveling as tourists or on business. The widening of travel offers the potential for exposure to diseases not encountered at home. For example, there have been outbreaks of malaria around airports in mid-latitude countries (such as Geneva, in 1989), where infected *Anopheles* mosquitoes have found their way onto airplanes, survived the flight and subsequently escaped to infect local residents (Haggett, 1994: 102). More generally, passengers may be infectious but symptomless while traveling, but the speed of travel means that the trip is over and passengers long dispersed throughout a country before any disease symptoms develop. Another example of the link between tourism and health would be the impact of sex tourism on the health of both local and visitor populations, typically involving those from the developed world visiting developing countries. Time-space "convergence" (the shrinking of distances) or "globalization" thus has potentially major public health implications.

Here, we focus mainly on permanent, or quasi-permanent, migration. The examples are mostly contemporary, although, as we saw in Chapter 2, migration was linked intimately with colonial history and the spread of disease in the New World and elsewhere. Migration can take place over short distances, perhaps from one part of a town or city to another, but can also be inter-regional, that is, movement from one part of a country to another; and, of course, it can be international.

Migration interacts with health and health care in several ways. First, it may have either short- or long-term consequences for health outcomes. We consider the short-term impacts of both voluntary and forced population movement on individuals and groups, including those left behind. We also look at the possible impacts of migration on the spread of disease, and ask whether the incidence of disease in geographical areas is elevated as a direct consequence of population movement. We consider whether those migrating acquire the health status of those they join rather than those left behind. We then ask whether our understanding of the spatial distribution of disease is incomplete unless we take into account the impact of migration.

Second, we ask if the health status of migrants differs from those of non-migrants. To what extent are migrants a non-random sample of the population as a whole, in terms of their health status? Is their health status generally better, or worse, than the people who do not move? We shall consider here this issue of health "selection," that is, whether migrants are inherently healthier than those left behind.

Third, we look at the impact that migration has on the use, or lack of use, of health services. Do people migrate in order to get better access to health and social care? To what extent does the movement of health care professionals impact upon health care provision? How do those providing health care manage the health of migrant groups?

All these are questions that this chapter seeks to answer. Each of these three broad relationships is treated in turn.

Impact of Migration on Health

Migration and stress

There is evidence that migration, especially to a new country, leads to stress and depression, as a result of alienation, the need to come to terms with a new culture, as well as the poor reception that immigrants may get from the host population (Allotey and Zwi, 2007). This is clearly gendered, in that migrant women may have fewer opportunities for social integration, as well as working in unskilled occupations and on low incomes. One outcome may be the adoption of unhealthy behaviors; for example, Carballo and Siem (1996) refer to high rates of drug abuse among young Puerto Ricans moving to New York.

As one example of the health impacts of the migration experience, consider the movement of Fijian women to British Columbia in Canada. Elliott and Gillie (1998) report on a qualitatively based study set within a social interactionist framework, in which 20 women were interviewed in depth about their health and health status, but also about their migration experience. Some reported problems with back pain as a result of the move, while others referred to problems in adapting to a very different climate, constraints on their mobility, and a struggle to maintain cultural traditions. The search for new work resulted in extreme fatigue for some, while easier access to alcohol for her male partner resulted in the physical and emotional abuse of one woman. More generally, there are problems of loneliness and homesickness, especially for older women. The pressures to perform a variety of roles (wife, carer, mother, homekeeper, wage-earner) are exacerbated by the separation from friends and family and the need to fulfill such roles in an unfamiliar setting.

Age-adjusted rates of self-reported long-term illness seem to be much higher among foreign born than indigenous people in Sweden (Sundquist and Johansson, 1997), even after controlling for social and material circumstances, smoking and exercise. The odds of needing treatment for severe long-term illness are over twice as high for both men and women born in southern European countries compared with Swedish-born people. The same picture of worse mental health among migrants to Sweden emerges from research on suicides (Johansson et al., 1997), where women from other countries (notably Russia and Hungary) living in Sweden are at significantly higher risk of suicide than women born in Sweden (see Table 6.1). The risks are much higher than for women still living in eastern Europe. Johansson suggests this is linked to disrupted social and cultural networks and poor material circumstances, as well as possible discrimination and xenophobia. The social and cultural "distance" from Sweden may simply be too great to overcome, and the consequences are severe.

Immigrants often find themselves exposed to higher risks in the workplace, whether from agricultural and industrial accidents (where those employed illegally will often be working in hazardous conditions) or in domestic environments (where young women in particular may be at risk from physical or sexual abuse). For example, in a comparison of occupation-related accidents among the native-born and immigrants in western Europe, Bollini and Siem (1995) show that rates among the latter are two or three times greater than among the former (see Table 6.2).

Ugalde (1997) refers to research on agricultural labor migrants to Spain that shows high rates of poisoning from exposure to pesticides. In addition, agricultural firms were ignoring safety regulations and failing to protect their workforce from agricultural hazards. Accidents at work were not reported, in order to mask the fact that labor conditions were "irregular." Ugalde quotes some African immigrants: "There are many who cannot sleep, they think that they are not going to be given the papers, and if the police stop you . . . too many problems and you cannot sleep." "If you have papers you are fine, otherwise you are worried . . . nervous" (Ugalde, 1997: 94). Further, given that employment and incomes are unstable among immigrant laborers, that families are separated, and that racial discrimination occurs within the working environment and beyond, the burden of mental

Table 6.1 Relative risk of suicide (and 95% confidence intervals) for Swedish-born and other women (1985–9)

Country of birth	Relative risk in Sweden	Relative risk in country of birth
Sweden	1.0	1.0
Finland	1.68 (1.43–1.98) +	0.92 (0.80–1.05) [positive and negative sorts]
Poland	1.63 (1.04–2.56) +	0.48 (0.43–0.53) –
Russia	3.71 (2.05–6.71) +	0.74 (0.69–0.81) –
Germany	1.44 (0.96–2.16)	0.77 (0.71–0.84) –
Hungary	3.39 (2.04–5.63) +	1.84 (1.68–2.03) +
Norway	0.99 (0.63–1.50)	0.57 (0.48–0.68) –

Relative risks are adjusted for age: + denotes significantly elevated, – denotes significantly lower than reference category (Sweden)

(*Source*: Johansson et al., 1997)

Table 6.2   Accidents among native-born and immigrants (Western Europe)

Country	Native-born	Immigrants
Netherlands[a]	32	92
Germany[b]	79	216
Switzerland[c]	158	230
France[d]	11.4	21.5

[a] Rates of occupational accidents per 1000 insured workers
[b] Rates of occupational accidents per 1000 workers, all industries
[c] Rates of accidents per 1000 workers in building trade
[d] Rates of accidents per 1000 workers in building trade, involving permanent disability or death
(*Source*: Bollini and Siem, 1995)

illness among such immigrants is not hard to understand. This suggests that in order to understand health outcomes among migrant communities we need to look at the organization of labor in agriculture and industry and ultimately at wider underlying inequalities in society and the economy.

Migration also has both short- and long-term impacts on the health and welfare of those in the areas from whence the migrants have come. For example, over the short term those migrating may transfer some of their earnings to the families they leave behind and this may mitigate the loss of economically active people from particular areas. Kanaiaupuni and Donato (1999) show how those Mexican communities which have lost population to the USA are areas where infant mortality is high, but that the remittance of income back home ("migradollars") goes some way to offsetting the disruptive economic, if not social, effects of migration.

Other work indicates that, perhaps over the longer term, the migration of more skilled people simply leaves behind a pool of poorer people and serves to reinforce regional disparities in economic and social welfare. Thus, Thomas and Thomas (1999) suggest that in one county in North Carolina (referred to anonymously as "Step County") a rural ghetto emerged over many years as better-educated black people moved out to the central town ("Sparksburg") or out of the county altogether. This reduced the stock of social capital and reinforced residential segregation. The polarization of black and white groups, coupled with the low ratio of males to females (because of the out-migration of many men), may lie behind currently high rates of infection by sexually transmitted diseases. Similar work in the rural Johnson County, Tennessee, hit by closure of manufacturing plants (Glenn et al., 1998) also suggests that over only four years (1990–3) the health of non-migrants worsened, largely as the result of reduced household incomes; at the same time, younger, more affluent people moved into the area as the result of an economic "rebound." This regeneration, and the influx of healthier, wealthier people, masks the problems of the "stayers" who are not participants in the improved economy. Migration then, whether of out-migrants or in-migrants, alters economic and social landscapes and may have long-lasting health consequences for those who have not moved.

The health of refugees

The forced dislocation of people from their homes has drastic health consequences, the nature of which dwarfs many of the less serious health outcomes discussed in this chapter. Important though they are, the stresses of the migration event, and the possible subsequent adoption of new disease profiles, hardly bears comparison with the mortality and morbidity wreaked by civil war, political persecution and the "ethnic cleansing" of the late-twentieth century. Forced migration does not only embrace refugees but also includes, for example, those compelled to leave because of major engineering projects, such as dams (Boyle et al., 1998; Roy, 1999). Allotey and Zwi (2007: 162) point out that over a million people were displaced by the Three Gorges Dam across the Yangtze River in China, with an accompanying resurgence of infections. However, for the purpose of this present section we shall consider refugees to be those forced, whether by invasion or domestic conflict, to leave their usual place of residence in order to seek refuge in another country. It remains to be seen to what extent population movement is exacerbated by climate change, as sea levels rise and flooding (and water and food shortages) forces people from their homes; global environmental change and its health consequences are considered in Chapter 9.

Data from the United Nations High Commission on Refugees (UNHCR) may be used to map geographical variation in the size of the refugee population (for example, in Africa; see Figure 6.1), though this does of course fluctuate from year to year. Kalipeni and Oppong (1998) have reviewed the health impacts of the displacement of refugees in Africa, considering these effects under several headings. First, these movements disrupt livelihoods, the production of food, and the operation of health services. This results in malnutrition. For example, during the conflict in Somalia in 1992 nearly 75% of children under five years of age died, almost all from malnutrition linked to the war. Second, the overcrowding of refugee camps causes further food shortages, as well as poor sanitation; as a consequence, diseases such as cholera, dysentery, hepatitis, and measles are likely to break out. Third, sexual violence is common: UNICEF reported the widespread rape of young girls during the Rwandan genocide of 1994 (see Box 6.1), while sex may be exchanged for food among the most desperate, an exchange which, like the shortage of uncontaminated blood, does little to halt the spread of HIV across the continent. Fourth, the trauma of watching people, often family members, killed and mutilated has devastating consequences for long-term mental health. UNICEF reported a survey in Angola in 1995 which revealed that two-thirds of children had seen people murdered, and over 90% had seen dead bodies. In addition to these health impacts, the chaotic mixing of populations exacerbates the diffusion of newly emerging infections.

Banatvala et al. (1998) studied levels of mortality and morbidity among Hutu refugees from Rwanda who had been repatriated from the Democratic Republic of Congo (formerly Zaire) in late 1996 (see Box 6.1). The genocide in 1994 had resulted in serious depletion in the numbers of health workers, and a consequent over-stretching of health care delivery. Banatvala and his colleagues paint a graphic picture of the burden of disease and ill-health, including attending to bullet wounds and limb injuries caused by machetes. In earlier work a group of doctors working in Goma, the main town in Zaire into which the refugees moved, estimated a crude mortality rate of 20–35 per 10,000 per day, due mostly to outbreaks of severe diarrhoeal disease (Goma Epidemiology Group, 1995). This contrasts with a crude mortality rate of only 0.6 per 10,000 per day in pre-war Rwanda.

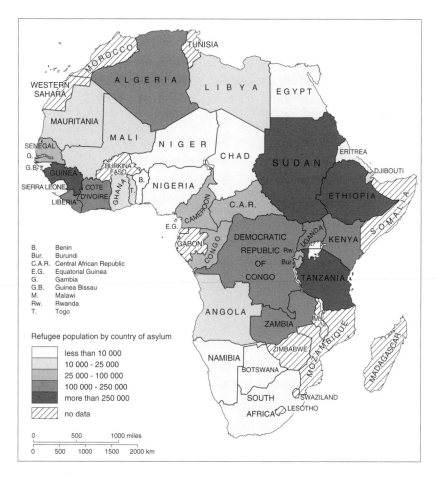

Figure 6.1   Refugee population in Africa, 1998

The impact of migration on the spread of disease

Immigration had consequences for the spread of tuberculosis (TB) in South Africa during the late-nineteenth century (Packard, 1989: 38–43). Two groups in particular contributed to the spread of disease. One was from England, a group already suffering from "consumption" (the archaic term for TB) and seeking a cure in a warmer climate; often, they infected fellow passengers on board ship, in overcrowded cabins. Upon arrival, some were employed as private tutors and acted as sources of infection among middle-class white families, while other infection arose as a result of local African servants taking home left-over food from consumptive employers. Death rates from TB were socially patterned (see Chapter 4); in one district, Aliwal North, mortality among whites was 20 per 1000 in 1897, while for black Africans it was approximately 62 per 1000. A second source of "imported" cases was from eastern Europe, where Polish and Russian Jews were escaping persecution. A high proportion was already infected with TB, and the lack of restriction on movement in some parts of South Africa meant that racially mixed slums grew up that were ideal for disease transmission.

Box 6.1   Conflict in Rwanda

The history of ethnic unrest in Rwanda is a long one but centers on the juxtaposition of a majority Hutu population and minority Tutsi population. The latter held sway during the colonial period, but after independence in 1962 the Hutus took power. In 1994 the Tutsi-dominated Rwandese Patriotic Front overturned the government, forcing over a million Hutus into exile in North Kivu region, Zaire (now the Democratic Republic of Congo), during July. By December 1995, over 2.9 million refugees had fled their homes. One million had gone to Zaire, half a million to Tanzania, and 1.2 million were internally displaced.

Many refugees entered Zaire via the town of Goma, thousands dying there and others moving on to refugee camps further north. With poor water supply and worse sanitation there were outbreaks of cholera that killed an estimated 50,000 refugees by the end of August. In addition, between 18% and 23% of children aged under five years were acutely malnourished, a figure that compares with "only" 5–8% in non-refugee populations of Africa. The distribution of both food and medical supplies was hindered by local Rwandan military and political leaders. As the Goma Epidemiology Group (1995: 343) points out, "the world was simply not prepared for an emergency of this magnitude."

Other associations can be found between migration and TB in South Africa in the late-nineteenth and early-twentieth century, reflecting the role of migrant labor in spreading the disease into rural areas (Packard, 1989, Chapter 4). Several hundred thousand African mine workers moved to and from their workplaces and rural homes each year. "This general pattern, in combination with the desire of sick migrant workers to return home and the mining industry's practice of repatriating all diseased or injured miners, guaranteed that the epidemic of urban-based TB quickly spread to the rural hinterlands, infecting the rural households from which industry drew its labor" (Packard, 1989: 92). While reliable data are not widely available, reports and case studies from local medical officers are highly suggestive. One concluded that TB accounted for over 30% of notified deaths between 1904 and 1906 in some parts of the Transkei and Siskei "homelands." Disease transmission was aided by overcrowded housing; in some areas there were taxes on the building of huts, thus discouraging a reduction in residential densities.

Tuberculosis remains a major public health problem and continues to be associated with migration. Carballo and Siem (1996) report on studies that show high rates among Haitians moving to the USA, Asian migrants in Australia, and those seeking asylum in Switzerland. But they argue that the high rates are not so much a function of their movement from high-risk areas, as of the poor, overcrowded living conditions in which newly arrived people find themselves. Evidence for this comes from several European countries, including Moroccan migrants to France and Cape Verde Islanders moving to Portugal (Carballo et al., 1998). Poor-quality housing clearly has other health impacts, such as increasing the risk of accidents in the home.

Rates of HIV infection, too, are influenced by migration. For men moving to work in construction, mineral extraction, and forest exploitation, their contact with sex workers is likely to facilitate the spread of the virus. A good illustration of this comes from migrant laborers in South Africa, where there were nearly three million living away from home in 1986, about 400,000 of whom came from nearby countries such as Lesotho, Mozambique, and Malawi. Jochelson et al. (1991) conducted in-depth interviews with a small number of those employed in mining, and this qualitative material reveals mechanisms for coping with separation from families: "After washing my workclothes I go out of the hostel because it's lonely there and there is nothing to while away this loneliness. I go to the likotaseng [domestic worker quarters] to look for women" (quoted in Jochelson et al., 1991: 164). The authors argue that the migrant labor system institutionalizes a geographically based network of relationships for spreading sexually transmitted diseases. A mixed hierarchical-contagious pattern of disease spread is posited: "once HIV enters the heterosexual mining community it will spread into the immediate urban area, to surrounding areas, from urban to rural areas, within the rural areas, and across national boundaries" (Jochelson et al., 1991: 169). But it is the economic system of labor migration that is the deeper cause of disease spread; the behavior of individuals cannot be separated from the vulnerability of family relationships, the low wages paid to the laboring men, and separation from the women left behind. All of these factors are shaped by the underlying socio-economic structures.

More recently, Kazmi and Pandit (2001) have examined the impact of refugee movements on malaria incidence in the North West Frontier Province (NWFP) in Pakistan. The Province is adjacent to Afghanistan, sharing a border of over 1000 km. The Soviet invasion of Afghanistan in 1979 led to a major influx of refugees across the border, followed by further refugee movements as civil war and political instability increased. By the late 1980s there were over 2.1 million refugees in NWFP. Some refugees arrived already having been infected with malaria in Afghanistan. Others, with immunity lowered by malnutrition, developed malaria shortly after arriving in NWFP. Overcrowding in refugee camps, coupled with the creation of new breeding pools for the mosquitoes, has had a lasting effect on the ecology of the disease.

As noted in the introduction, population movement need not involve a permanent change of residence, and in some parts of the world more seasonal movements can have major consequences for health. Prothero (1965; 1977) has illustrated this convincingly in his classic work on malaria during the 1950s and 1960s. For example, Somali pastoralists spend the dry season (November–March) in the north, with access to well water. But the land cannot support cattle throughout the year and they move south to the Haud, a plateau whose hollows fill with water during the wet season and provide ideal breeding grounds for the mosquito vectors of malaria (see Figure 6.2). The complex patterns of population circulation have created difficulties in devising adequate monitoring and control strategies, especially since this circulation involves crossing the border into neighboring Ethiopia.

Migration and the incidence of disease and ill-health

Direct effects In some cases, migration may have other direct effects on disease incidence. This lies behind one of the possible explanations for "clusters" of childhood leukaemia arising near nuclear installations, such as that at Sellafield in West Cumbria, northern England (see Box 3.1, page 57). While some have argued that exposure to radiation is the cause of the raised incidence of leukaemia, the epidemiologist Leo Kinlen hypothesizes that

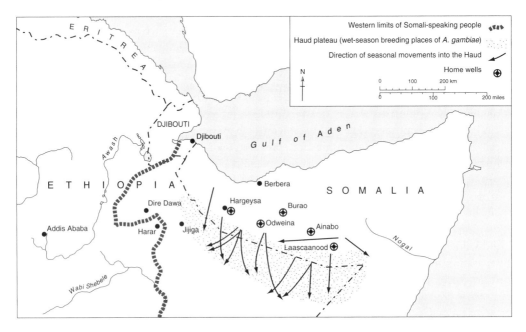

Figure 6.2   Movements of Somali pastoralists in the Horn of Africa

it results from exposure to an (as yet unidentified) virus, and that an outbreak of infection is most probable when there is an unusually high degree of population mixing. In essentially rural communities, such as that in West Cumbria in the later 1950s, or on the coast of northern Scotland (near the Dounreay nuclear power station), the local population is "challenged" by the rapid influx of a new population, brought in to manage and operate the nuclear installations. The immunity of the local population is low, and rapid immigration leads to a sudden increase in the exposure of the newly mixed populations to viruses. Leukaemia may be a rare response to a viral infection.

To test this hypothesis, Kinlen and his colleagues (1990) studied new towns in Britain, all of which had seen significant population growth after 1950. These towns were grouped into two classes: first, "overspill" towns (such as Basildon, Stevenage, and Welwyn Garden City), designed to provide accommodation for people dispersed from London after the Second World War; second, "rural" new towns (including Glenrothes, Corby and Cwmbran) that took population from a much more diverse set of origins. The hypothesis was that leukaemia incidence in the second set would be higher than among the first. The results bore this out. For example, between 1947 and 1965 the ratio of observed to expected deaths from leukaemia among children aged under five years in the rural new towns was 2.75 (a statistically significant excess), compared with 0.95 in the overspill towns (fewer than expected cases).

Other authors have conducted research that lends support to the Kinlen hypothesis. For example, Langford (1991) looked at deaths from childhood leukaemia (under 15 years of age) in England and Wales between 1969 and 1973, classifying these by geographical area

Table 6.3 Mortality from childhood leukaemia and population change in Britain

	Males	*Females*	*Total*
Population change >50%	37.7	39.1	38.4
Population change <50%	30.2	24.1	27.2
Relative risk (95% confidence interval)	1.25	1.62	1.41
	(0.9–1.7)	(1.2–2.2)	(1.1–1.8)

Numbers are mortality rates per million children aged less than 15 years
(*Source*: Langford, 1991)

(local authority). Areas are classified according to whether they have undergone substantial or more modest population change. His results (see Table 6.3) indicate that there are significantly higher death rates (especially among females) in the areas undergoing more population change.

A more sophisticated test of the Kinlen hypothesis requires data on the diversity of migrant origins, since it is the mixing of dissimilar populations that lies at the heart of this hypothesis. Accordingly, Stiller and Boyle (1996) construct a "migration diversity" measure, showing whether migrants tend to come from one source, or many sources. Those districts in England and Wales that had higher rates of migration, as well as more diverse sets of origins, were those in which leukaemia incidence was elevated.

Further work by Kinlen suggests that rural areas with a large influx of oil workers during the 1970s in Scotland are also those with excess rates of child leukaemia (Kinlen et al., 1993). The home addresses of over 17,000 such workers in Scotland were classified as either rural or urban, and the incidence of childhood leukaemia in both groups of areas was compared. Rural areas with the highest proportions of oil workers were those with the highest leukaemia incidence.

There is, then, quite a body of evidence (including that from other countries: see, for example, the French study by Rudant et al, 2006) to support the theory that it is migration and population-mixing, rather than direct or indirect exposure to radiation from the nuclear industry, that lies behind leukaemia incidence and leukaemia "clusters." However, it would be very premature to suggest that the case is now closed. Looking at other rare cancers, such as that of the eye (retinoblastoma), it seems that here possible exposure to nuclear radiation may be a causal factor. Morris and his colleagues (1993) recorded cases of people whose grandfathers had worked at Sellafield, but whose mothers had spent time living near the reprocessing plant. The suggestion was that the mothers may have been exposed as children to radionuclides in the general environment, or contaminated at home, and that they had had an increased rate of cell mutation, passed on to the next generation; thus, while the mothers themselves were healthy, their children were not. The children with retinoblastoma had never lived near Sellafield. This illustrates well the point that recording address at diagnosis may have no bearing on the origin of a disease, an issue to which we shall return.

Longer-term health impacts of migration There is a substantial literature on the differences in health status between migrants and non-migrants, specifically on the extent to which those moving from one area to another (and mostly from one country to another) ultimately "adopt" the health profile and "risk factors" for disease of those who have always lived there. As Elford and Ben-Shlomo (1997: 228) suggest, "migration provides a naturally occurring experiment which may establish the aetiological importance of factors acting at different points in the life course." In particular, it serves to establish the relative role of genetic, as opposed to environmental and socio-economic factors; if migrants from an area of low chronic disease incidence ultimately develop a higher disease incidence in their destination area this would suggest a more modest role for genetic factors.

We may look at cancer and risk factors for heart disease, as well as self-reported health, to illustrate this "acculturation" hypothesis. For example, English South Asians (those whose ethnic origin is in India, Pakistan, or Bangladesh) show higher rates of lung cancer and breast cancer than on the Indian sub-continent, results that are consistent with the notion of a transition of low cancer risk in the country of origin, to a higher risk in the country of residence. Winter and colleagues (1999) suggest this is due to changes in lifestyle among the migrants. Much of this work derives from a classical epidemiological tradition in which "ethnicity" is under-theorized, usually in terms of categories (such as "Indian") that may mask more than they reveal. As Elliott and Gillie (1998: 329) argue: "most studies serve to aggregate ethnic groups to such a level that individual differences in health and/or life beliefs become obfuscated in the name of adequate sample sizes; explanation and prediction take privilege over understanding." It is worth asking whether alternative explanatory frameworks might add to the existing literature.

The question arises as to the length of time which it takes for health status to be modified. A study of rural migrants to Nairobi, Kenya, found significant increases in blood pressure (see Box 6.2), as well as body weight, only ten months after migration (Elford and Ben-Shlomo, 1997: 234). Blood-lead levels among pregnant immigrants to South Central Los Angeles were significantly higher than among pregnant non-migrants (2.3 µg/dl, compared with 1.9), but this depended strongly on length of time since immigration. The greater the period women had been in the USA, the lower the blood-lead level. In Canada, obesity among immigrants increases with length of residence (Cairney and Ostbye, 1999). Obesity is usually measured using the BMI index (weight divided by the square of height; a body mass index of 25 kg/m^2 is the upper limit of the "normal" range). Women born in Canada have a mean BMI of 24.9 but while those living in Canada for less than five years have a mean BMI of 22.7; this rises to 24.8 for those living there for more than ten years. This acculturation effect remains even after adjusting for socio-economic and lifestyle factors. Porsch-Oezcueruemez et al. (1999) studied Turkish immigrants to Germany and found that total cholesterol levels were broadly similar to those in other western countries but very much higher than among those living in Turkey. In addition, blood pressure among men, and obesity in women, were also high in the immigrant population.

This kind of adaptation does not apply solely to those moving from one country to another. For example, the British Regional Heart Study has compared the blood pressure of non-migrants, born in the town where their blood pressure was examined, with the blood pressure of migrants. Men moving from southern England to Scotland had higher mean blood pressure than those staying behind (an average of 157/89, compared with 142/80), while those born in Scotland and moving south had lower mean blood pressure than those remaining in Scotland (143/80, compared with 148/85) (Elford and Ben-Shlomo, 1997: 234–5).

Box 6.2 Blood pressure

Since many studies of migration focus on changing blood pressure it is worth explaining briefly what blood pressure is and what is meant by *hypertension*.

Blood pressure is the pressure exerted on the artery walls by contraction of the heart's lower chambers (ventricles). Peak pressure is known as *systolic* pressure, while that between heart beats is *diastolic* pressure. Blood pressure is measured in millimetres of mercury. A normal reading would be approximately 120 mmHg (systolic) and 80 (diastolic), referred to as 120/80.

Hypertension is abnormally high blood pressure and is potentially dangerous in that it can lead to heart disease and stroke (Box 4.2, page 93). Body mass, consumption of alcohol, and intake of potassium are thought to influence high blood pressure. In middle age an increase in systolic pressure of 18 mmHg and 10 mmHg in diastolic pressure doubles the risk of stroke and increases the risk of heart disease by 50%. Regular monitoring of blood pressure, and reduction of high blood pressure, is therefore essential.

See Whincup and Cook (1997) for further details.

Research in central Africa has compared the growth of Hutu children whose families had migrated from Rwanda to Zaire to work in the copper mines, with that of non-migrants (Little and Baker, 1988). In Rwanda itself, the dominant Tutsis tended to be taller and heavier than the subordinate Hutus, but the migrant Hutus were taller and heavier than those remaining in Rwanda (see Figure 6.3). This was perhaps due to improved hygiene, diet, and health care in the Katanga province of Zaire.

In a series of studies, epidemiologists have looked at the health of those living in the Tokelau Islands, a series of atolls in the South Pacific, north-east of Fiji. Many islanders migrated to New Zealand after a hurricane in 1966 caused much damage, and studies of the health of migrants and non-migrants have focused on blood pressure, as well as chronic disease such as diabetes and asthma (Little and Baker, 1988; Elford and Ben-Shlomo, 1997: 229–30). In the mid-1970s, 812 adults were examined in Tokelau (532 non-migrants) and New Zealand (280 migrants); all had been examined five years earlier, before the migrants had moved. At that stage, there were no significant differences in blood pressure, but after adjustment for age and other factors the blood pressure of the migrants had risen significantly. Such longitudinal work has clear advantages over studies that simply look at migrants and non-migrants at one fixed point in time. Further work has indicated that Tokelauan children living in New Zealand had significantly higher blood pressure than children still living in the islands, even after taking into account weight and height, while other evidence suggests raised incidence of diabetes (especially among women) and asthma in the migrant group (see Figure 6.4).

It is fair to suggest that much of this literature is of a descriptive nature rather than offering convincing explanations of differences between migrants and non-migrants. The implication in many studies is that migrants simply adopt "lifestyles" (particularly diets) of the populations among whom they come to live. But the extent to which changing

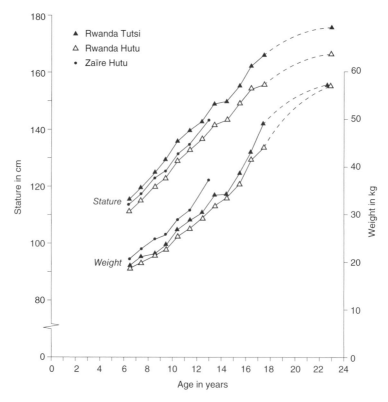

Figure 6.3   Growth in stature and weight of sedentary Rwanda Tutsi and Hutu, and Hutu children whose families had migrated to Zaire

(often reduced) material circumstances, or poorer stocks of social capital, play a role has yet to be fully explored.

Migration as a confounding variable   A conventional assumption in geographical epide-miology is that current place of residence is an adequate measure of exposure to possible risk factors. This may well be a reasonable assumption for diseases that have a short latency period, but for chronic diseases of adulthood (such as cancer and heart disease), which may have taken many years to develop, people may have changed residence several times. This problem is particularly serious when looking at disease incidence in small areas, since people are more likely to move short distances. But, regardless of geographic scale and resolution, current health may well reflect previous places of residence rather than the present one. Here, we examine some studies that illustrate this point well.

Consider, first, geographical variation in the incidence of primary acute pancreatitis in the Greater Nottingham area, part of the east Midlands region of the UK. In this disease enzymes attack the pancreas (a body organ which regulates sugar levels in the blood) and

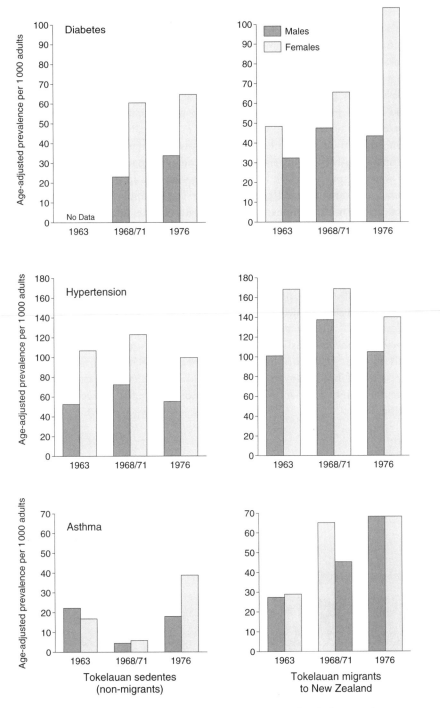

Figure 6.4 Rates of morbidity among Tokelau islanders and Tokelauan migrants to New Zealand

while the first attack (hence "primary acute") results in only about 26 new cases per year in the average district hospital in the UK, as many as one quarter of such cases may die during that attack.

In a classic study (Giggs et al., 1980) disease incidence was mapped by electoral ward, small areas of local government containing (very approximately) some four to five thousand people. The incidence map showed some evidence of a concentration of cases on the eastern side of the city of Nottingham, a feature that was confirmed when the probabilities of significantly more or fewer cases than on average were mapped (see Figure 6.5). A map of water supply areas (see Figure 6.6) suggests that areas of high disease incidence correspond to the area covered by a particular water supply area (Burton Joyce), and analysis of the chemical composition of the water in this and other areas reveals that it has far higher concentrations of calcium carbonate and magnesium than elsewhere. However, the second study (Giggs et al., 1988) looked at two cohorts of patients (214 presenting between 1969 and 1976; 279 presenting between 1977 and 1983) and reallocated patients according to their addresses five years prior to their first attack. The results (see Table 6.4) reinforce the finding of significantly high incidence in the Burton Joyce area. Thus, the earlier evidence linking incidence to water quality is strengthened when migration is accounted for. The study suggests that the association between disease incidence and contrasting drinking water supplies needs to be explored in more detail; this requires laboratory-based and clinical research. More generally, the study raises the issue of how many marginally significant links between health status and environmental quality have suffered through not being able to account for the effect of migration.

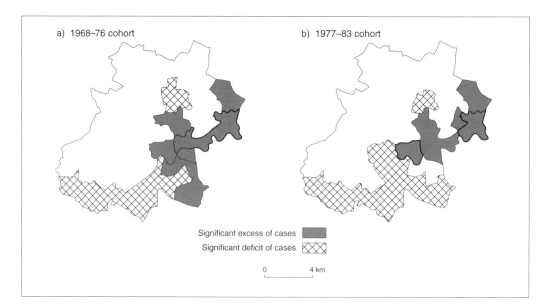

Figure 6.5   Areas of significantly elevated, and significantly lower, primary acute pancreatitis in Nottingham, 1968–83

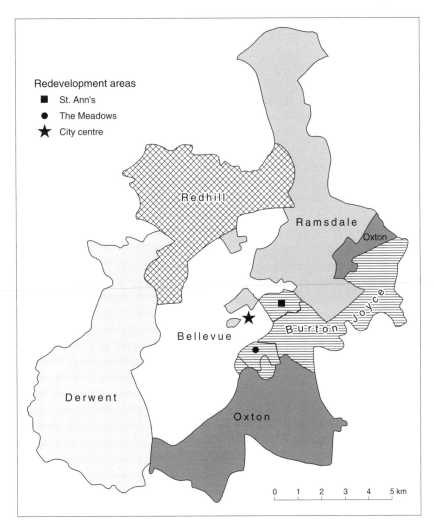

Figure 6.6 Water supply areas in Nottingham

Two other studies, both of neurological disease, make the same point. Riise and his colleagues (1991) studied 381 people who had developed multiple sclerosis (MS) in the Norwegian county of Hordaland between 1953 and 1987. They obtained data on place (community) of residence at birth, and at various ages up to the age of 25 years. They looked at the observed, and expected, number of pairs of patients who were resident in the same area and whose years of birth were within one year of each other. Results indicated that, until the age of about 15 years, there was little evidence of significant *space-time clustering* (see Box 6.3 below); however, between the ages of 16 and 20 (and for 18-year-olds in particular) there was very marked evidence of clustering. In other words, MS patients of a similar age lived close together in late adolescence to a much greater degree than chance

Table 6.4   Primary acute pancreatitis in Nottingham, England

| | | Average annual incidence rates (per million) | | | |
| | | All cases | | By former address | |
Water supply area	CaCO₃ (ml/l)	1969–76	1977–83	1969–76	1977–83
Derwent	197	62.8	92.2	62.8	82.4
Redhill	196	66.7	129.2	64.5	85.2
Oxton	151	45.7−	57.3−	29.7−	52.6−
Ramsdale	173	75.8	82.8	62.1	67.1
Bellevue	164	85.2	79.2	79.8	70.6
Burton Joyce	300	129.4++	187.5++	170.5++	273.9++

+/− means p < .05; ++/— means p < .01
(*Source*: Giggs et al., 1988)

> ## Box 6.3   Space-time clustering
>
> We considered in Chapter 3 how to detect "clustering" of disease and illness using the addresses of those diagnosed. Often epidemiologists wish to know whether cases that are close in space are also close in time; we may need to know if there is *space-time interaction*. This is particularly important if we wish to establish whether or not a disease is infectious. If cases that live nearby were also diagnosed at, or about, the same time this would strengthen the argument for an infectious aetiology. Where there is debate about disease causation (for example, in leukaemia and lymphoma, or multiple sclerosis) some researchers have used statistical methods to detect such clustering or interaction.
>
> A simple test is to form a table with two rows and two columns, the rows of which refer to whether or not a pair of cases are "close," and the columns denote whether or not a pair of cases are temporally "close." We then count how many pairs are close in time and space, close in space but not in time, close in time but not space, and not close in either. Statistically, we then determine whether the number of pairs close in time and space is significantly greater than expected on a chance basis. This is known as a *Knox test*. One difficulty is that a distance and time threshold must be fixed in advance. Are those living within 5 km (3 miles) "close" pairs, or should this be 10 km (6 miles)? Are those diagnosed within two days "close" in time, or should this be five days? There is some arbitrariness here, as well as dependence on the scale of investigation. In practice, researchers use a range of possible thresholds, or turn to other methods for detecting space-time interaction.
>
> The research described in the text, on multiple sclerosis in Norway, takes these ideas one step further, since the authors look not at address at time of diagnosis but rather at where people lived at different ages. See Thomas (1992) and Bailey and Gatrell (1995: 122–5) for further details.

would dictate. The tentative explanation is that MS is a delayed response (since patients will tend to present first in their thirties or forties) to a viral infection acquired, possibly via the exchange of saliva, in late teenage years. Merely mapping current place of residence would have failed completely to reveal this finding; a detailed knowledge of migration histories is required.

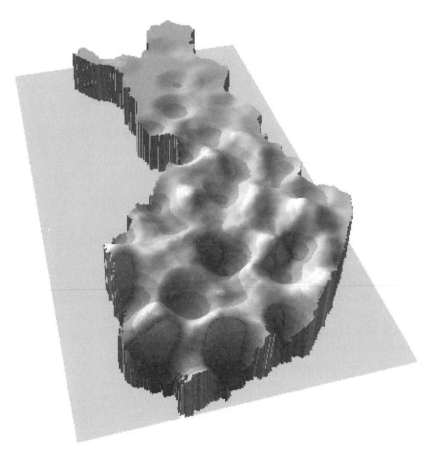

Figure 6.7   Relative risk of motor neurone disease in Finland

A third example comes from work by Sabel et al. (2000) on spatial variation in the incidence of motor neurone disease, or MND (also known as amyotrophic lateral sclerosis, or ALS) in Finland. MND is a progressive and ultimately fatal neurodegenerative disease, whose aetiology is unknown. It is a disease of low incidence (approximately 1–2 per 100,000) and affects primarily older age groups. Data were collected on 1000 deaths from MND between 1985 and 1995, matched by age and sex to population controls. Since the Finnish population register identifies all changes of address, it was possible to look at where both cases and controls had lived since the mid-1960s. A relative risk surface (see Chapter 3) was constructed according to current place of residence, and also former place of residence. It appears that those subsequently diagnosed with MND have, relative to people unaffected by the disease, spent many years living in the south-east of Finland, as well as inland from the city of Oulu and on the west coast, south of Vaasa. These are predominantly rural areas. Figure 6.7 illustrates this in a three-dimensional surface representation of the country, with light-shaded "peaks" denoting areas of increased risk, and darker "troughs" denoting areas of lower risk. Further, detailed epidemiological investigation is needed to shed light on the areas of elevated relative risk.

Table 6.5 Percentage population in Britain that is permanently sick or disabled, by origin of migration

Age group	All residents	Within-district	Between-district	Between-region	Outside GB
21–24	0.7	0.3	0.3	0.2	0.1
25–29	0.8	0.4	0.2	0.3	0.1
30–34	0.9	0.7	0.5	0.5	0.2
55–59	6.3	11.1	6.6	8.8	2.8
60–64	11.2	18.2	11.9	12.0	4.8

(*Source*: Bentham (1988), based on analysis of 1981 census data)

Impact of Health Status on Migration

Here we consider other links between migration and health. Rather than looking at how migration affects health we examine how health status affects the propensity to move. We consider the "health selection" hypothesis in more detail, and ask what evidence there is that only the healthy may migrate. Evidence is also reviewed that suggests people may migrate in order to gain access to health care or social support.

The selectivity of migration

Bentham (1988) considers the health selection hypothesis in some detail. He notes that migration rates tend to be higher for young, wealthy people employed in non-manual occupations; these people are likely to be healthier than the population at large, so that areas which are losing population tend to be those with higher mortality and morbidity. In the absence of information on migration, therefore, we do not know whether associations between ill-health and deprivation may be due in part to the selective loss of healthier people.

Bentham examines British data from the 1981 census on permanent sickness and disability and relates levels of morbidity to migration (where address has changed during the previous year). His results (see Table 6.5) suggest that, regardless of age, those moving the furthest (notably from overseas) have lower morbidity than the population as a whole; it is the "fitter" people who are moving. This is the "healthy migrant" effect. Among older adults, internal migrants tend to have higher levels of morbidity than the older population as a whole. Those moving shorter distances (within districts) have particularly high morbidity, suggesting that ill-health may be a possible factor in migration. This is evidence of "reverse" selectivity, in that it is the unhealthy rather than the healthy that appear to be moving. In more recent work, using individual data from the Longitudinal Study of England and Wales, Norman and his colleagues (2005) have shown that those moving from more to less deprived areas (using Carstairs deprivation scores) are healthier than those migrating from less to more deprived areas.

Table 6.6 Stroke mortality in the USA: migrants and non-migrants (1979–81)

Race	Sex	US-born	US-born interregional migrants	Immigrants
White	Both	73.1	71.6	65.0
	Male	79.8	75.4	68.5
	Female	68.3	68.3	62.1
Black	Both	108.4	95.9	51.4
	Male	121.4	103.3	55.2
	Female	98.8	90.1	48.6

Rates are deaths per 100,000 persons per year
(*Source*: Lanska, 1997)

What other evidence can be marshaled in support of health selection? Lanska (1997) showed that stroke mortality was significantly lower among immigrants (both white and non-white) to the USA than among the US-born resident population. Table 6.6 reveals that age-adjusted rates of stroke mortality are lowest among immigrants, but that those moving from one region to another also have lower rates than those who had not moved to another region. He argues that the costs, both financial and personal, involved in migration, operate to ensure that only the relatively healthy are "selected" for migration. No data are available, however, on the age at which migration occurred, or how many moves have been involved. Research undertaken in Western Australia (Moorin et al., 2006) suggests that the onset of serious disease in rural areas is associated with a reduced likelihood of migration to urban areas, even though access to health services in the cities is much greater. Either the economic means to migrate are lessened, or there is existing social support in the rural areas that those who are seriously ill need to draw upon. This is evidence of "unhealthy selection;" the unhealthy are *not* moving.

Other research in the USA, on maternal health, indicates that infant mortality is lower among the children of Puerto Rican-born and Hispanic Caribbean women than among those of Hispanic women born on the US mainland. For example, Landale and her colleagues (2000; 2006) have demonstrated a clear gradient for infant mortality among children born to Puerto Rican immigrants, (with the odds of mortality rising as the number of years lived in the USA increases; those who have lived for five years in the USA have a 9% higher risk than those living there for only a year, while those living in the USA for 20 years are 48% more likely to have an infant death. "Although infants of migrant women from Puerto Rico generally have lower mortality risks than do infants of native-born mothers, there is strong evidence that infant mortality risks rise with years of exposure to US society" (Landale et al., 2000: 901). Others have suggested that immigrant women are less likely to smoke or drink alcohol during pregnancy, though evidence indicates that these healthy behaviors disappear after arrival in the USA (Thiel de Bocanegra and Gany, 1997: 34). Wingate and Alexander (2006) also look at maternal health and birth outcome among migrants; but they compare Mexican-born women who had migrated to the USA before subsequently giving birth in the USA, with infants of US-born women of Mexican descent who had moved between regions, or between states, or within state. There was evidence, though not very strong, that infants of "non-mover" mothers had poorer outcomes (e.g. low birth weight, preterm delivery) than those who had moved from one region to another.

In general, the authors claim, "mothers who are healthy and able to move, regardless of whether the movement is from one country to another, or from one region to another, have healthier babies" (Wingate and Alexander, 2006: 497).

Kington et al. (1998) looked at the "functional status" (impairment or health problem that limits daily activity) of people aged 60 years and over, who had moved from the southern states of the USA. The health of blacks who were born in the South but lived elsewhere was better than that of those born in the South who had not migrated. There was no difference in health status between Southern-born whites who migrated and those who stayed. The contrasting fortunes of the black Americans could not be explained by differences in socio-economic status. Kington and colleagues put this down to selective migration: the "healthy migrant" effect. The findings contrast with studies of mortality, which show that African-Americans born in the South and migrating elsewhere, as well as those remaining in the South, have higher mortality rates than those born outside the South.

The selection hypothesis has been of particular relevance to studies of mental ill-health, especially in urban areas (Jones and Moon, 1987, Chapter 5). It has been argued that living in some inner-city areas "breeds" or causes mental health problems, while others contend instead that the mentally ill migrate or "drift" to such areas. This latter hypothesis is one of selective migration. Loffler and Hafner (1999) study people suffering from schizophrenia in the German cities of Mannheim and Heidelberg. In both, there have been concentrations of schizophrenics in low-status inner-city areas for well over 25 years. But a close examination of individual biographies reveals that schizophrenics have suffered downward mobility into poor areas as the illness has disrupted their lives. Selective migration is therefore confirmed here.

Migration for health care and social support

This is a somewhat neglected area of research. The emergence of HIV and AIDS has produced some findings, while the extent to which older people retire in order to seek out social support and health care has been given some attention. We consider each briefly.

Ellis and Muschkin (1996) have examined the migration patterns of people with AIDS, looking specifically at moves from different states in the USA, to Florida. Their particular interest centers on whether such people are moving to be close to informal sources of support, in particular elderly parents. As of late 1989 there were 9555 cases of AIDS diagnosed in Florida, of whom 535 had moved from outside the state. While 28% of these migrants had moved to Miami, and others to urban areas, 20% had moved to small town and rural areas. Related work (Graham et al., 1995) points to high levels of in-migration to rural areas by HIV/AIDS sufferers. Statistical analysis of the Florida data shows that, even after adjusting for the effects of population size and existing AIDS cases (since people with AIDS are more likely to move to areas where there are existing gay communities), the proportion of the population that is elderly has a significant influence on migration to particular counties. Although there is no direct evidence, "the most likely alternative motivation is to seek the care and support of family, primarily parents of post-retirement age" (Ellis and Muschkin, 1996: 1113). However, there are no data on individuals and so their conclusions must inevitably be rather cautious. Their own data suggest there is only a weak association between the concentration of an elderly population and the proportion of people with AIDS moving to particular counties (r = 0.13).

A more recent study (Wood et al., 2000) of people with HIV in British Columbia, Canada, shows that their migration to Vancouver is largely to be explained by their search for health care, although the culture of the city, including the comparative absence of prejudice there compared with other parts of the province, is also an explanatory factor. As in the Florida study, urban-rural migration was also accounted for by the need for social and family support. But in both studies the quantitative nature of the research means that explanation remains somewhat speculative. This is a research area that demands a qualitative perspective.

A good deal is known about patterns of migration among older people (Rogers et al., 1992), though detailed empirical work on the reasons for such moves is still required. Migration of the elderly is generally divided into that which is voluntary and that which is more constrained by personal circumstances, including health. In the first category are those who are recently retired, generally well-off and healthy, who seek a move for reasons of "amenity;" the classic example is the stream of migrants to Florida. The second type of older migrants comprises those less able to care for themselves, or who anticipate a need for care. This group includes those moving to be nearer to children and other relatives, perhaps when losing a partner, as well as those who develop a disability or chronic illness that necessitates institutional care. Warnes and colleagues (1999) have looked at Britons retiring to southern Europe (Tuscany in Italy, Malta, the Algarve in Portugal, and the Spanish Costa del Sol) and reveal that the onset of chronic illness and disability was likely to encourage a return to Britain. However, this varied from place to place; for those on Malta, with relatively well-developed primary care services and a predominantly English-speaking population, a return due to illness was less likely than from other areas. In general, though, the expatriates painted a very positive picture of retiring to southern Europe.

The Relationship between Migration and the Delivery of Health Services

Last, we consider the association between migration and the provision of health care to those moving from one place to another, usually from one country to another. To what extent do migrants get a "raw deal" from health care services?

Reijneveld (1998) looked at first-generation immigrants in Amsterdam, showing that those from Turkey, Morocco, and the former Dutch colonies made more demands on health care than non-immigrants. This reflected their poorer health status and was not to be explained simply in terms of lower income or occupational status. Other immigrant groups seem to make less use of preventive services, whether for screening or for ante-natal care, and where they do, the communication difficulties between primary health care providers and immigrants act as a barrier to effective health care. Part of the reason for non-use of services lies in the illegal status of some immigrants and their fear of losing employment (for example, Moroccan agricultural and construction workers in Spain: Ugalde, 1997). Culturally sensitive programs designed to address the health needs of migrants are likely to pay dividends. For example, Verrept and Louckx (1997) demonstrate that health "advocates," Moroccan and Turkish women recruited and trained as health workers, have improved the health status of immigrants to Belgium from Morocco and Turkey.

Access to health care in the USA among the foreign-born is limited by their lack of health insurance (Thamer et al., 1997). In 1990, the foreign-born population was twice as likely as the US-born population to be uninsured (26% compared with 13%), with those of Hispanic origin having an uninsured rate of 41%. However, in some parts of the USA, these figures are markedly worse. For example, in New York City, survey research (Sun et al., 1998) suggests the proportion of uninsured foreign-born people may be as high as 77%. The lack of a regular income and health insurance, and therefore inability to pay, means that immunization uptake is low (only 46% among the foreign-born). But efforts are being made by public health authorities to "target" such vulnerable immigrant populations, with immunization being offered free to children of immigrant families in New York City (Sun et al., 1998). In addition, the New York Task Force on Immigrant Health (NYTFIH) brings together academics, community activists, and health care providers to increase access to culturally sensitive health care (Thiel de Bocanegra and Gany, 1997). The NYTFIH has developed outreach strategies to make contact with immigrants likely to be suffering from tuberculosis, for instance by approaching them through community groups or on the street (where many will be street vendors).

In other settings, explanations for a low uptake of health services have little to do with material resources. For example, women moving to Israel from the former Soviet Union seem to avoid making use of preventive services, such as screening for breast or cervical cancer, even though they were using such services before they moved (Remennick, 1999). This may reflect difficulties in adapting to a new culture and environment, as well as linguistic and cultural barriers to accessing services. The pattern is also observed among those from Vietnam now living in London (Free et al., 1999). Qualitative (focus group) research was used to study community groups and this revealed that a lack of knowledge of out-of-hours arrangements in general practice, arising from communication difficulties, meant poor access to primary health care. Research on Filipino immigrants to the USA (Yamada et al., 1999), also using focus groups, suggests that the cost constraints in getting help for the treatment of tuberculosis are less serious than the wish to deny or hide the disease or to put faith in traditional treatments. Thus, linguistic barriers, cultural differences, and relative social and geographical isolation all serve to reduce the ability of migrants to gain access to health care services.

Other qualitative work has explored issues of gender and "race" in a more sophisticated way, drawing on feminism in particular to shed light on the health-seeking behaviors of first-generation women immigrants to Vancouver (Dyck, 1995b). The women studied by Dyck are from Hong Kong and the Chinese mainland, as well as from India and Fiji. But Dyck cautions against the simple coding of "race" or ethnicity as an explanatory variable, arguing that it is a social construct. The attachment of simple labels, she argues, "objectifies" these groups and serves both to stereotype individuals as passive victims of cultural change (rather than people with identity and human agency) and to "deflect attention away from the contributions of political and ideological processes, including the power relations of health service provision" (Dyck, 1995b: 249). Dyck considers in detail the case of one woman who seeks help from a variety of sources of health care, both traditional Chinese healers and a Chinese-Canadian family doctor (physician), but who, like others, is able to access informal social networks for health care. Such health care may be accessed locally, but some women draw on help and advice from those still living in China and the Indian sub-continent. Clearly, health care may be sought from a variety of "places."

Warfa and colleagues (2006) examine the impact on health service usage (and mental health status) that residential mobility has for Somali refugees living in London. Their qualitative research involved discussion groups with 21 Somali lay people (and, separately, with Somali professionals). The high frequency of moves, including those across health service boundaries, reflect poor housing quality, racism and discrimination; even in the new host country the respondents had "the experience of a 'forced move'" (Warfa et al., 2006: 508). Access to health services, including registration with a general practitioner, was problematic, compounded by language barriers and by the different cultural expectations of health care. As one lay female explained when attending an accident and emergency (A&E, or ER) service: "in Africa doctors examine the patient even if they see the person has no problem but here doctors look at their computer and decide to give you paracetamol without even touching your body" (Warfa et al., 2006: 511).

It is worth noting that the migration of health professionals themselves has a potential impact on the availability of health care. In Britain, there are long-standing concerns about the retention of a qualified medical (and nursing) workforce, and particularly about how movements of health professionals might cause inequalities in the supply of health care. Drawing on qualified staff from overseas, particularly the developing world, simply denudes such countries of their own much-needed health professionals.

Concluding Remarks

We have considered a number of different themes in this chapter, concerning the relationship between migration and health status, and between migration and health care delivery. It is quite clear that while there is a considerable literature on the former, research on the associations between migration and service provision has been much less substantial. What conclusions can we draw from the material presented here?

It is fair to say that much of the work reviewed here draws on essentially positivist approaches, creating statistical models of social epidemiology in which the "risk" of migration is examined after attempting to adjust for other factors. Clearly, it is important to do this, otherwise we will obtain bogus, or at least counter-intuitive conclusions. For example, Wei et al. (1996) present evidence that Mexican-Americans born in the USA have higher mortality rates than those born in Mexico. But this surprising finding is due largely to socio-economic confounding, since members of the latter group were of much lower socio-economic status. Once adjustment is made for this, the differences lessen substantially.

There is clearly scope for alternative approaches to migration and health. We have touched upon some of these here. For example, an understanding of the "experience" of migration and the health impacts of the event itself requires qualitative methods that may be located within a social interactionist or structurationist framework (Dyck and Dossa, 2007). And studying the health impacts of labor migration becomes impossible without engaging with a structuralist approach that looks at the political economy of health.

Regardless of the explanatory framework we adopt, many of the studies considered here have been cross-sectional, perhaps studying samples of migrants and non-migrants at one point in time. We saw some benefits to be gained by studying the health status of people before and after their moves (illustrated by the Tokelau Islanders), but this kind of longitudinal study is relatively rare (though see Newbold, 2005). It is certainly required in order to resolve debates about health "selection."

Health and migration are intimately linked. Given that migration is an inherently social and geographical process, and that health and health care are socially and geographically patterned, this is hardly surprising. Yet much more work needs to be done to clarify the relationships, and in particular to flesh out some of the ways in which migration impacts upon, and is affected by, health care delivery.

Further Reading

A good introduction to geographical research on migration is Boyle et al. (1998). This book also considers in more detail some of the types of migration touched on above, such as forced migration. For a recent overview of migration and its links to public health see Allotey and Zwi (2007) who consider briefly the health consequences of human trafficking, which we have not considered.

Gellert (1993) offers a good overview of the impact of international migration on the spread and control of communicable diseases. The chapters by Little and Baker, and by Kaplan, in the book edited by Mascie-Taylor and Lasker (1988), are well worth reading. There is a considerable literature on the differences in mortality among migrant groups. For a comprehensive picture for England and Wales, see Harding and Maxwell (1997). For a review of research on migration and health in Europe see Carballo et al. (1998).

PART III

Health and Human Modification of the Environment

Chapter 7

Air Quality and Health

The main aim in this chapter is to review some of the associations between air quality and ill-health. In speaking of air "quality" we are, of course, speaking mostly about air "pollution;" but reference to air quality permits us to discuss some of the interesting work on radon, a natural radioactive gas. Much of the literature in this area tends to adopt a classical positivist epidemiological stance. However, we shall see that some researchers call for a better understanding of "lay beliefs" about such links (a social interactionist perspective), while a strong case can be made for a structuralist interpretation of the most serious pollution occasioned by industrial "accidents."

We organize the discussion in terms of the *source* of pollution. Some gases, such as radon and low-level ozone, are rather diffuse in origin; they may be widespread and can be considered therefore as "areal" sources. Others, such as the pollutants accompanying vehicle exhaust emissions, arise from "linear" sources; principally the roads on which traffic flows. We thus consider the growing evidence linking proximity to major thoroughfares and ill-health. Last, we review a selection of research studies that examine "point" sources of industrial pollution; to what extent are those living near incinerators, coking plants, power stations, and other industrial sites more at risk of disease and ill-health than those living elsewhere? As with any classification, this three-fold division is somewhat simplistic; for example, vehicular emissions diffuse over a wide area and are hardly confined to the roads themselves! Nonetheless, it serves as a convenient organizing framework.

A variety of health effects will figure here. Of particular interest is respiratory disease such as asthma, since plenty of evidence exists that the incidence of asthma has increased over the past 50 years. Data for England and Wales suggest that the percentage of 12-year-old children diagnosed with asthma rose from about 4% in the early 1970s to almost 25%, while hospital admissions for those aged 5–14 years increased from about 4 per 10,000 in 1958 to over 20 per 10,000 in 1985. More recent data suggest that such rates of hospital admission have dropped since 1990 (all data from Anderson et al., 2007). In the USA, evidence suggests that the prevalence among children aged 5–14 years in 1980 was 4.4%, rising to 8.2% in 1995 (Moorman et al., 2007). Some caution needs to be exercised in interpreting temporal change, since it may be the case that "asthma" has been used as a label for respiratory conditions that were previously recorded otherwise. But asthma is far from being a disease of affluent countries; indeed, the World Health Organization suggests that the problems are considerably worse in low and middle-income countries, which bear the burden of over 80% of global asthma mortality.

Other forms of air pollution have been implicated in lung cancer and in cardiovascular disease, the hypothesis being that the pollutants enter the blood stream from the lungs and are carried to the heart; some heart attacks may, therefore, be triggered by air pollution. From a geographical point of view we are particularly interested here in studies that ask whether where you live affects exposure to, and health damage from, poor air quality. But other important studies adopt a time series approach, looking at temporal data on air pollution and corresponding patterns of morbidity and mortality.

Types of Pollutants

Early work on air pollution and health in the developed world focused on sulphur dioxide (SO_2) and smoke, since these were the main pollutants in urban areas before the 1970s. The principal source of SO_2 is the burning of fossil fuels, but the reduction in burning coal for domestic use has reduced emissions over the past 40 years. However, coal-fired power stations continue to be a major source of the pollutant. More recently, and with the growth of motor vehicle traffic, concern has shifted to pollutants such as nitrogen oxides (NOx) and volatile organic compounds (VOCs). The principal source of NOx is from motor vehicles, though power stations and other industrial processes contribute to the load. VOCs include benzene, emitted from petrol fumes and vehicle exhausts. The presence of sunlight causes NOx and hydrocarbons (the residues of partly burnt fuel) to combine with other gases in the atmosphere to form ozone (O_3), which we consider below.

Particulate air pollution refers to a mix of solid particles and liquid droplets. In contemporary society, motor vehicle exhaust emissions (especially from diesel engines) are a significant source. Interest centers on particles that can be inhaled (with a diameter less than 10 μm: so-called PM_{10} particles), and especially on the finer particles (diameter less than 2.5 μm: $PM_{2.5}$) that can be breathed deep into the lungs.

In the present chapter, as well as the following two chapters, we look at aspects of the field known as *environmental epidemiology*; this deals with the relationships between environmental quality and health outcomes. In this chapter we examine some of the evidence concerning the impact of air quality on human health (particularly respiratory disease and cancers). A key issue here, as in all environmental epidemiology, is to characterize the nature of the exposure. Levels of air quality need to be measured, or estimated. We need, ideally, to look at how a pollutant disperses from its source, and the extent to which people come into contact with it. How is the pollution distributed in time and space? These issues are considered briefly in Box 7.1.

We focus our attention here mainly on outdoor sources of air pollution, mostly because the "spatial" effects (proximity to source) have been so thoroughly studied. However, indoor pollution is a major health risk in some poorer countries where the use of biomass fuels is common and poor ventilation causes severe respiratory problems in small children (see Box 7.2).

> **Box 7.1** Exposure assessment
>
> A simple way of estimating exposure is simply to record whether someone lives in the same area as the source of air pollution. For example, if a county contained a suspected point source of pollution we could suggest that all those living in the county were potentially exposed to pollution. But this is clearly rather crude, since the movement of polluted air will not respect administrative boundaries! We could look at proximity to pollution sources, and distance is often used as a surrogate measure of exposure; as we shall see, some work looks at proximity to main roads, or to point sources of pollution. Better still would be to identify areas of risk using an air dispersion model, since the pollutant will not disperse uniformly in all directions. Or, rather than modeling air pollution, we could monitor levels of air quality at a set of locations. Issues then concern where to locate monitoring sites, and how many to use. Finally, in an ideal world, we would measure directly the exposure of a set of individuals, using personal monitors; sophisticated (and well-resourced) studies do so. There are different ways of characterizing "exposure," but the more effectively we try to do this the more costly it becomes. Many studies have tried to consider whether proxy or surrogate measures of exposure are adequate; for example, if there is a high correlation between pollution estimates derived from sophisticated environmental models and those based on simple distance, why not use the latter, cheaper alternative as a marker of exposure?
>
> Local meteorological conditions can affect the dispersion and subsequent concentration of outdoor pollutants, while human spatial behavior over the course of the day and week will serve to affect personal exposure and subsequent "dose." Our primary concern here is what the health effects of such doses might be. While laboratory work can assess the possible effects of single pollutants, and provide controls for a host of potential confounding variables, "air pollution" rarely consists of exposure to a single chemical; rather, we are invariably exposed to a mix of pollutants.

Area Sources

We look first at one classic American study, and then focus on two specific sources of air contamination. The former is research conducted by Dockery and his colleagues (1993), examining the association between mortality and air pollution in six American cities. It was a prospective study, following adults from the mid-1970s until death (where this had occurred). Access to individual-level data allowed the researchers to examine the effects of air pollution on mortality, while adjusting for the effects of age, sex, occupation, and, most crucially, smoking status. The six communities were: Watertown (Massachusetts); Steubenville (Ohio); Portage (Wisconsin); Topeka (Kansas); Harriman (Tennessee); and specific parts of St Louis (Missouri). Over 8000 individuals were recruited to the study in 1974, of whom 1430 had died by late 1989. Outdoor concentrations of air pollutants (total and fine particulates, SO_2, O_3, and sulphates) were measured at a central air-monitoring site in each community, at various dates.

Box 7.2 Health risks of indoor air pollution

Estimates suggest that nearly three billion people use biomass (charcoal, wood, crop waste, and animal dung) as their main source of domestic energy (Ezzati and Kammen, 2002). A wide range of pollutants is associated with the burning of biomass; these include: nitrogen dioxide, carbon monoxide, and particulates. A range of studies (reviewed by Ezzati and Kammen) implicates these pollutants in the domestic environment as causes of acute respiratory infections and ear infections in children, as well as asthma, chronic obstructive pulmonary disease, various respiratory cancers, and some eye disease. The scale of the problem dwarfs any of the health risks from air pollution – reviewed in this chapter – in high-income countries. For one thing, the level of exposure is an order of magnitude higher; while air quality standards suggest a daily average concentration of PM_{10} is 150 $\mu g/m^3$, exposure in homes using biofuels can be up to 5000 $\mu g/m^3$ throughout the year. But the real "scaling up" of the problem comes with the health burden; Ezzati and Kammen suggest that up to two million deaths a year in 2000 were attributable to indoor air pollution. The burden is not distributed evenly, either by income group or age and gender. Better-off families may use liquid petroleum gas (LPG), while poorer households use solid fuels that generate more pollution. Further, the major concentrations of pollutants are in kitchen areas, where women and children are the most exposed to fires and stoves. For these groups the exposure varies during the course of the day as well as seasonally, but it is they who are the most heavily exposed. Jin et al. (2006) report detailed empirical work on exposure to indoor air pollution in four poor, rural provinces in China, as well as illustrations of the different kinds of technologies used by households and differential exposures. For example, in Gansu and Shaanxi provinces there is usually a separate room for cooking, while in Guizhou and Inner Mongolia provinces cooking is usually performed in the living area. Improvements in ventilation can clearly assist. China has an ambitious program to introduce improved stoves in order to reduce indoor air pollution. There is clear evidence that this works; Chinese farmers using vented stoves had a significantly lower risk of lung cancer than those using open fires (Ezzati and Kammen, 2002: 1065). However, introducing new technologies requires cultural adaptation and behavioral change, as well as the political will and resources to cover the costs. The prizes are huge gains in public health. As Jin et al. (2006: 3174) put it, interventions to reduce the health hazards can be considered "as possibly paralleling those of water and sanitation in the last quarter of the twentieth century."

Results (see Figure 7.1) show mortality rates (adjusted for age, sex, smoking, education, and body mass) plotted against average pollution levels in each city. The correlation with fine particles is particularly striking. Results are essentially unchanged when looking at specific causes, such as cardiovascular and respiratory disease. But these data and results relate to the 1980s; what of data for the 1990s? In a follow-up study of the same cities Laden and her colleagues (2006) show that particulate levels have declined. But while the associations with all-cause, cardiovascular, and lung cancer mortality remain statistically significant it is interesting to observe that the reduction in exposure (albeit rather crudely measured) has led to a significant reduction in total mortality. In policy terms, the message is quite clear: we need to reduce urban air pollution in order to improve population health.

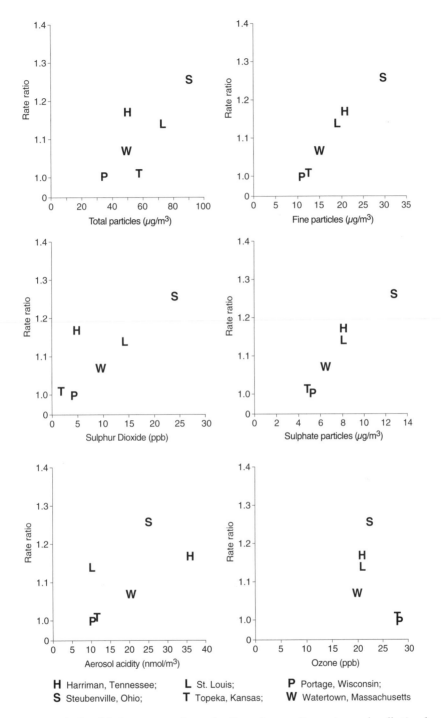

Figure 7.1 Relationship between estimated adjusted mortality ratios and pollution levels in six US cities

Radon

Radon is a radioactive gas formed during the decay process of uranium-238 and thorium-232, naturally occurring nuclides found in minerals in the earth's crust. Radon decays to produce further isotopes (known as radon "progeny"), which may attach themselves to water or to aerosols such as dusts. Upon reaching the open air the gas mixes with the atmosphere, but if trapped in underground mines or in houses, concentrations can build up. Since uranium is found in some igneous rocks (especially granites) and also some sedimentary rocks, radon levels are, on a broad scale, very variable from place to place. Within Britain, high concentrations (measured in Becquerels per cubic meter, Bq/m^3) are found in much of Cornwall, parts of west Devon and west Derbyshire, central Northamptonshire, and central Grampian in Scotland. In some of these areas 30% or more of the housing stock has radon levels in excess of 200 Bq/m^3 (the "action" level set by the National Radiological Protection Board in the UK, averaged over the year). In the USA (where the action level is 148 Bq/m^3), high levels of radon are found in some counties within Iowa, North Dakota, New Mexico, Minnesota, Nebraska, and Pennsylvania. Levels are elevated especially where the soil or rock is permeable and the gas can move through to the surface. At much more local scales radon levels are highly variable, depending on building construction and ventilation. Where a building has cracks or fissures the gas finds a route of entry and may be trapped if there is a low air exchange rate and the building is well sealed. Thus, radon concentrations may vary between adjacent dwellings. Levels also vary temporally, both diurnally (peaking in the early morning) and also seasonally (being elevated in winter months).

The primary danger of the gas lies in the radioactive decay process and the subsequent production of the radon progeny (such as polonium 214 and 218, bismuth 214, lead 214); the aerosols may lodge in lungs, exposing the lung tissue to alpha and gamma radiation. It is for this reason that the main health risk is considered to be lung cancer. Some studies have sought to demonstrate a link using aggregate data for sets of areal units, while other, more satisfactory, studies consider individual risk via case-control studies. The problem with the former is that they fail to provide adequate controls for confounding. As Goldsmith (1999) has shown, lung cancer rates for US counties are inevitably *negatively* correlated with radon levels measured for such counties. Residential radon levels are higher in low-density, suburban residences (where income is higher and people are willing to pay for measurements) and are therefore negatively associated with population density. Since population density is well known to correlate positively with lung cancer, it follows that aggregate levels of radon will necessarily be inversely correlated with lung cancer. Thus, population density confounds the relationship between the independent variable of interest (radon) and the dependent variable (lung cancer). There is much more to be gained from detailed, individual-level, case-control studies, since these can, in principle, adjust for smoking, the obvious and well-established risk factor for lung cancer.

Perhaps the most impressive of all studies is that by Darby and a group of collaborators across nine European countries (Darby et al., 2005). The quality of the study lies in its pooling of data from a set of 13 case-control studies, comprising, in total, over 7000 cases of lung cancer and over 14,000 controls. Further, a measured radon concentration is obtained for every individual, based on the length of time spent in each home, and data are available on smoking histories in order to assess the separate risks of radon and smoking. Results (see Figure 7.2) indicate a clear linear relationship between mean radon levels and the relative risk of lung cancer. Although rather few people are exposed to the highest radon concentrations (over 800 Bq/m^3) – hence the wide confidence interval – there is a doubling of relative risk. Put differently, and referring to the graph, for every 100 Bq/m^3 increase in exposure the risk of lung cancer increases by about 9%. However, for smokers the relationship is amplified; in other words, for life-long smokers exposed to cumulative high levels of radon the risk of lung cancer is substantially greater than for non-smokers. Darby and her colleagues conclude that radon levels in homes across Europe account for about 9% of deaths from lung cancer, and about 2% of all cancer deaths. Given the relatively low costs of reducing radon concentrations in homes, the potential payoffs, in lives saved, are highlighted in this important study.

Ozone

Ozone (O$_3$) is a summer pollutant, at its worst in more rural areas and when temperatures are high and wind speeds are low. In Britain, concentrations tend to be higher in the south of England and in Wales. There is very substantial temporal (both daily and seasonal) variation, with ozone "episodes" in summer months; the maximum concentrations are in the afternoon, with lower levels in early morning. Air quality goals for O$_3$ vary from country to country; in the UK an hourly average of 90 ppb is regarded as "poor." In Australia the hourly air quality goal is 120 ppb, while in Japan it is 60 ppb. In California the figure is 90 ppb.

Some estimates of population exposure to high levels of O$_3$ have been made. In the UK, research (UK Photochemical Oxidants Review Group, 1993) indicates that large numbers

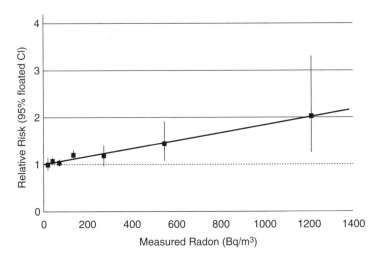

Figure 7.2 Relative risk of lung cancer according to measured residential radon concentration

Table 7.1   Estimate of population exposure (in millions) to ozone episodes in Britain (1987–90)

Average hours per year >90 ppb	*Rural areas*	*Urban areas*	*Total*
4	20.8	24.4	45.1
8	20.2	24.4	44.6
12	19.6	23.8	43.4
16	18.0	21.3	39.3
20	8.1	7.8	16.0
24	1.8	0.4	2.2

(*Source*: UK Photochemical Oxidants Review Group, 1993: 89)

of people may be exposed to more than 16 hours of high ozone concentrations during a year (see Table 7.1). There are marked urban-rural differences, with 1.8 million people in rural areas experiencing more than 24 hours above 90 ppb, compared with only 400,000 in urban areas.

Animal studies suggest that ozone damages the respiratory tract, although this damage is likely to be repaired after short-term exposure has passed. Controlled experiments on humans show that lung function (measured by forced expiratory volume, or FEV, the maximum volume of air that can be breathed out in one second) is impaired as duration and concentration of exposure increase. Vigorous exercise worsens the effect, since this increases the volume of air inhaled and allows the ozone to penetrate further. Studies of children attending summer camps in the USA have suggested that exposure to 100 ppb of O_3 leads to a reduction of about 3% FEV (UK Photochemical Oxidants Review Group, 1993) while adult studies suggest equivalent, or worse, effects on lung function.

Particular interest has centered on the possible link between asthma and ozone. When asthmatics are exposed to O_3 they show a greater response to allergens (such as pollen and house dust mite) than those exposed to filtered air; it is thus thought that O_3 increases sensitivity to pollen. Woodward and his colleagues (1995) report on a study in Houston, USA, which showed that the probability of an asthma episode increased as ozone exposure increased. Ozone does not appear to "cause" asthma; what it does is to allow allergens (such as house dust mite, pollens, and animal proteins from cats) to penetrate the airway wall by damaging its lining. The airways are then sensitive to stimuli such as cold air, irritant fumes and smoke, and even exercise.

Anderson and his colleagues (1996) have examined associations between ozone levels and mortality (all causes, respiratory, and cardiovascular) in London between 1987 and 1992. They look at the increase in mortality associated with increases in maximum hourly ozone concentrations, over a range of about 2 to 45 ppb. Regardless of season there are striking increases in respiratory mortality of about 5–6%. They argue that ozone exposure hastens death in people who are already seriously ill. Other work by the same research group has shown that high ozone levels on a given day lead to increased rates of hospital admissions for respiratory disease on the following day. Research on children in London seeking urgent care for acute episodes of wheeze (Buchdahl et al., 1996) demonstrates a U-shaped relation between attendances and ozone levels. Compared with controls

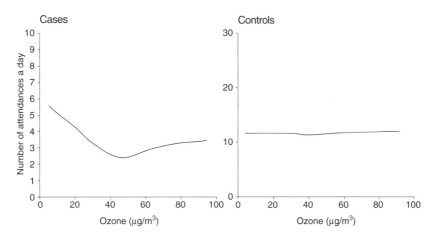

Figure 7.3 Relationship between number of children attending hospital accident and emergency services for acute wheeze (cases) and other reasons (controls), with ozone levels

(see Figure 7.3) the number of cases increases after exposure of about 50 ppb (mean 24-hour concentration). More recent work across 36 cities in the USA confirms an association between respiratory hospital admissions and exposure to ozone (Medina-Ramon et al., 2006), while an examination of daily variations in mortality in Shanghai, China between 2001 and 2004 (Zhang et al., 2006) suggests that ozone exposure is significantly associated with all-cause and cardiovascular mortality, particularly during the cold season.

Linear Sources

Although pollution from industrial and domestic sources in the developed world has decreased over the last 30 years, that from mobile sources has undoubtedly increased. Data for the USA reveal substantial growth in the volume of highway traffic over the past 40 years (see Table 7.2). There has been a virtual doubling of automobile (car) traffic since 1970, a 700% increase in journeys made by vans and SUVs, and a 250% increase in truck traffic; clearly, the potential for heavy exposure of people to the emissions from such vehicles has risen dramatically.

We examine here the evidence that links vehicle exhaust emissions to health problems. Such exhaust emissions include carbon monoxide, nitrogen oxides, and unburned hydrocarbons (all "primary" pollutants). We have already noted that the presence of sunlight causes the hydrocarbons and oxides of nitrogen to produce ozone, a "secondary" pollutant. In addition, while diesel engines emit no lead, and much less carbon monoxide than petrol engines, they emit small particulates; it is these that have emerged as strongly associated with health effects.

Table 7.2   Growth in vehicle traffic in USA (1970–2004)

	1970	1980	1990	2000	2004
Cars	920	1222	1418	1600	1705
Vans, pickups, SUVs	123	291	575	923	1014
Trucks	62	108	146	206	227

Note: Figures are in billions of vehicle miles
(*Source*: www.census.gov/compendia/statab/tables/07s1080.xls)

Table 7.3   Exposure to traffic-related air pollution and otitis media in Dutch and German children

	Netherlands		Germany	
	Adjusted O.R. (C.I.)	Sample size	Adjusted O.R. (C.I.)	Sample size
Exposure				
Fine particulates	1.13 (1.00–1.27)	2970	1.24 (0.84–1.83)	605
Nitrogen dioxide	1.14 (1.03–1.27)	2970	1.14 (0.87–1.49)	605

Note: O.R. is odds ratio; C.I. is the 95% confidence interval
(*Source*: Brauer et al., 2006: Table 3)

It is difficult in this type of work to capture adequately "exposure" to traffic-related air pollution. Personal monitoring is expensive and resource-intensive; surrogate measures include proximity to main roads, or traffic density, or modeled air pollution based on these and other variables. In a recent study conducted on German and Dutch children, quite sophisticated measures of exposure to traffic-related air pollutants were used. Taking monitored data on nitrogen dioxide and fine particulates, for 40 sites in each country, and data on road and dwelling density as variables to explain variation in pollution, Brauer and his colleagues (2006) were able to estimate the pollution burden for the home addresses of their child samples. They explore the association between pollution load and otitis media (ear infection), after controlling for possible confounders (such as indoor damp or mold, exposure to tobacco smoke, presence of pets). Their results (see Table 7.3) are suggestive of an association in both countries, though only significantly so among the Dutch cohort.

Gauderman and his colleagues (2007) have conducted a longitudinal study to assess the impact on child health of proximity to freeways in California. Over 3600 children were followed up from the age of 10 to 18 years, with their lung function measured each year. Two measures of exposure were used; first, simple proximity of the child's residence to the nearest major road and, second, estimates of traffic-related pollution at such residences, based on air dispersion models. Three measures of lung function were taken annually for each child, producing an extremely rich set of data. After adjusting for socio-economic factors, it was clear that those who lived within 0.5 km of a freeway had reduced lung function compared with those living further away. Of course, for various reasons it was not possible to assess each child each year, but for 1445 children complete data were available and the authors reveal quite strong evidence that for those having lived within 500 meters of a major road lung function was significantly lower than a reference group of children living more than 1.5 km from major roads (the horizontal line shown in Figure 7.4).

Other recent work on adults has suggested different consequences for poor health from exposure to particulates. Miller and her colleagues (2007) look at a large sample of over 66,000 women, enrolled into their study between 1994 and 1998, and living in one of 36 cities in the USA. Their interest focuses on the 1816 women who had had a cardiovascular "event" up to 2003; such an "event" might be a fatal heart attack, a stroke, or major surgery for heart disease. Miller's group seeks to determine if there is any association between such events and exposure to fine particulates ($PM_{2.5}$). Their measure of exposure is a little crude (they use the average annual concentration at the closest monitoring station, which could be several miles from place of residence). Nonetheless, their results suggest that, having adjusted for potential confounders, there is a more-than-doubling risk of death from cardiovascular disease for every 10 $\mu g/m^3$ increase in $PM_{2.5}$. The graphical relationship between risk and exposure (see Figure 7.5), in which 11 $\mu g/m^3$ is taken as a reference value, is quite linear, though the 95% confidence intervals are wide for higher levels of exposure, suggesting some uncertainty in the estimates. Further, the distribution of pollution estimates indicates that relatively few women would be exposed to high concentrations. Looking at both men and women exposed to traffic-related pollution, Tonne and her

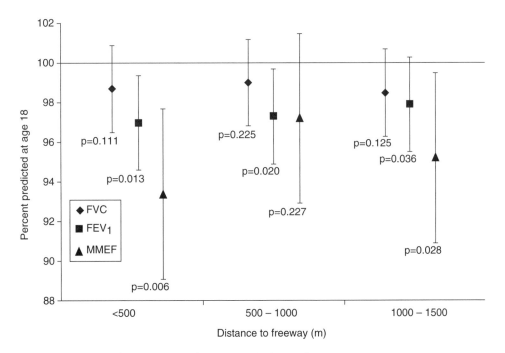

Figure 7.4 Lung function according to proximity to freeway

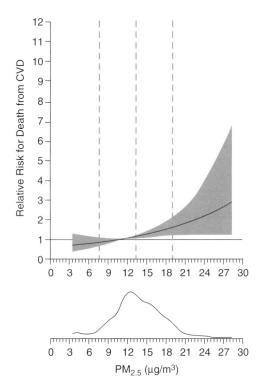

Figure 7.5 Exposure to fine particulates and death from cardiovascular disease

colleagues (2007) also suggest a statistically significant association with heart disease. Specifically, their study of over 5000 confirmed heart attacks (acute myocardial infarction) in Worcester, Massachusetts, between 1995 and 2003, finds a 5% increase in risk for those living near a major road.

While environmental epidemiologists have begun to study the way exhaust emissions affect those *living* near roads, other researchers have attempted to assess the exposure of road *users* to air pollutants. Concentrations of nitrogen oxides, carbon monoxide and VOCs inhaled by car users may be two or three times higher than those inhaled by pedestrians or cyclists. Benzene concentrations in vehicles are 18 times higher than background levels, and exposures among those people who use petrol (gasoline) pumps very frequently may be particularly high.

It should also be noted that since some air pollution is more common along very busy roads and road junctions, and these are usually places which are disadvantaged in other ways (poorer quality housing, for example) then the burden of pollution may fall disproportionately on poorer groups, including those who do not necessarily contribute to the problem. Issues of environmental health "equity" arise here (pp. 70–3 above), as they have in other areas considered in this book.

Point Sources

Here we assess the health risk of living near possible point sources of air pollution. Such point sources may come in a variety of forms. They can include: incinerators, such as those used for domestic (municipal), industrial, hazardous, and nuclear wastes; landfill sites (though these may be more a source of water pollution than of air pollution); industrial and manufacturing plants such as smelters, cement works, and fertilizer and pesticide factories; and power stations, such as those fueled by nuclear materials. We consider a small set of empirical studies drawn from a larger set, with particular attention to those where the pollution results from an acute, sudden episode. We begin with an account of one such disaster.

Point sources in the developing world

During the night of Sunday December 2, 1984, there was a leak of methyl isocyanate and hydrogen cyanide gases from the Union Carbide pesticides plant on the outskirts of the city of Bhopal, in the central Indian state of Madya Pradesh. Two thousand people were dead the following morning, while about 500,000 people were caught by the gas cloud, many of whom have continued to suffer since from emphysema, asthma, tuberculosis, and eye disease. Some 550,000 people had been awarded compensation by the end of 2003, with a maximum payment of under $10,000 and an average of $2200, sourced from a fund of $470 million which Union Carbide agreed to pay, five years after the event (Mukerjee, 1995; Broughton, 2005).

Health impacts include imprints etched on the eyes of those running during the night to escape the gases. As the author of a report commissioned by the Indian Council of Medical Research, M.P. Dwivedi, stated, "as they ran, the gas hurt their eyes, so they shut them as tight as possible, just leaving a slit to see through. Today, that slit, an opaque line on their corneas, is permanent" (in the *Guardian* newspaper, August 13, 1998). No one knew that, since methyl isocyanate reacts readily with water, a wet cloth placed over the face would have prevented the gas from penetrating the lung. Fifteen weeks after the disaster, researchers found that 38% of 260 people still living within two kilometers of the plant had "burning eyes," 19% had poor vision, and 6.5% had corneal opacity of the sort described by Dwivedi (Mukerjee, 1995). Three months afterwards, 39% of those living nearby had some form of respiratory impairment. Other work has compared pregnancy outcome among Bhopal women with those in an unexposed area (Bhandari et al., 1990). The rate of spontaneous abortions (miscarriages) in the exposed group was 24.2%, compared with only 5.6% in the control area. In a random sample of 454 adults, stratified by distance from the plant, Cullinan et al. (1997) found severe respiratory problems ten years after the event, with the frequency of symptoms declining with distance.

There have therefore been several classical epidemiological studies into the effects of the Bhopal disaster. However, one can argue that while this is of "scientific" interest it has had little impact on compensation claims and sheds no light on why the disaster happened. To do this, it is much more illuminating to reflect on the nature of pesticide production in India and wider issues of the political economy of development in the third world, issues discussed fully in Bogard (1989) and Shrivastava (1992); we might, in other words, gain a better understanding from the kind of structuralist perspective outlined in Chapter 2.

The license to import and use methyl isocyanate at the Union Carbide plant was based on its role in the manufacture of pesticides needed to increase food production as part of India's "Green Revolution." But we can also see the plant's location as being determined more by the need to transfer, wherever possible, hazardous production technologies from the developed world to developing countries. Increasing state regulation of such technology in the core leads to falling profits and pressure to relocate where cheap labor is available (the population of Bhopal rose dramatically after the Union Carbide plant was built). Yet those living near the plant knew little or nothing of its function or the potential hazard, while those employed there did not know how to handle non-routine events. "Not in my backyard" was not an option that the poor of Bhopal could afford, while the fact that the plant was relatively unimportant to the parent company meant that investment, particularly in safety and emergency planning, was very limited. Uncertainty was embedded in the fabric of the plant and the locality, and was used to tolerate risk. In addition, the hazards associated with pesticide production were mitigated (compensated) by the need to lessen the impact of food shortages and unemployment. For one author at least, the world system of capitalism is to blame. "Political tradeoffs to enhance legitimacy and global economic demands generated by the capitalist imperative of accumulation narrowed the chances for detecting the dangers at Bhopal" (Bogard, 1989: 104). To speak of Bhopal as an "accident" neglects the role played by human agents operating thousands of miles away, or the culpability of the global chemical industry. More generally, Broughton (2005) cautions that "the Indian economy is growing at a tremendous rate but at significant cost in environmental health and public safety as large and small companies throughout the subcontinent continue to pollute."

Our focus on this one major incident should not be taken to signal that point sources of air pollution in countries of the "south" are confined to major explosions. Indeed, in very rapidly developing economies there is plenty of evidence to suggest that pressures for growth are running ahead of those for maintaining environmental quality. For example, in China the demand for coal as a fuel for power stations has led to severe problems of air pollution in regions such as Shanxi Province (the main coal-producing region); the World Bank has recently indicated that 16 of the world's most polluted cities are in China, with Linfen in Shanxi perhaps the worst polluted city in the world.

Point sources in the developed world

No industrial disaster in the developed world matches the impact on death and morbidity of the Bhopal explosion. The potential health risks of a major nuclear accident are, of course, a serious concern, and we begin this section by considering what, if any, were the health impacts of an accident at a nuclear facility. Attention focuses on the contested nature of the evidence concerning the impact on cancer.

On March 28, 1979, there was a nuclear accident at the Three Mile Island plant near Harrisburg, Pennsylvania, resulting in a release of ionizing radiation. A presidential commission which reported shortly after the accident suggested that the maximum dose received by a person would have been less than the average annual background level, and that no detectable health effects would arise. However, public pressure led to research (Hatch et al., 1990; 1991) which sought to find out whether cancer rates near the plant had increased after the accident. In this context, proximity was defined by studying 160,000 people living within 10 miles (16 km) of the plant. Cancer and socio-economic data were collected for 69 census tracts. Atmospheric dispersion models provided estimates of possible doses delivered to people in each of those tracts, though these estimates were very uncertain.

The initial study (Hatch et al., 1990) looked at childhood cancer, and leukaemia in particular, during both the pre-accident period (1975 to March 1979) and following the accident (up to 1985) but since there were only four cases of leukemia, and relatively few of other cancers, little confidence could be placed in the estimates of association between cancer incidence and exposure to radiation. As far as these authors were concerned, there is no convincing evidence that radiation releases led to an increased risk of cancer. Later work by the same group detected an increase in all cancer in 1982 and 1983, but this was ascribed to factors other than radiation, such as the increased propensity, following the accident, to seek care. This research has been criticized by another group of researchers (Wing et al., 1997), who examine the association between cancer incidence and radiation dose, with controls for age, sex, socio-economic status, and pre-accident variation in incidence. Their work suggests that, following the accident, cancer incidence increased more in areas estimated to have been in the pathway of radioactive plumes than in less exposed areas. The fact that many years may need to elapse before any further increase in risk is detected, because of the long latency periods for many cancers, means that the story remains incomplete.

The Three Mile Island accident was an incident of short duration, in contrast to the longer-lasting impacts of more "chronic" air pollution or, indeed, other nuclear disasters such as that at Chernobyl in the Ukraine, in April 1986. More than twenty years after the event there is still considerable uncertainty over the longer-term health impacts (Baverstock and Williams, 2006). The WHO estimates that about 4000 cases of thyroid cancer – typically, very rare – in the former Soviet Union are due to childhood exposure to the radioisotopes of iodine. Of course, this number may increase as the population exposed to radiation increases. In addition, about 50 workers at the plant, or involved directly in the clear-up, were exposed to very high whole-body doses of radiation, and died as a result.

One major study that illustrates the health problems of living near long-term point sources of pollution involves the Monkton Coking Works in South Tyneside, north-east England, a plant emitting smoke and SO_2. The plant has been the focus of both traditional geographical epidemiological investigation (Bhopal et al., 1994) and also analysis of popular or lay beliefs about respiratory disease, using a mix of quantitative and qualitative techniques (Moffatt et al., 1995). In both studies, three areas were chosen for detailed survey work. Two were near the coking plant, one in the immediate vicinity (the "inner" zone), the other ("outer" zone) immediately to the north (see Figure 7.6). The third was a control area located between six and ten kilometers from the plant. The authors show how all three zones are broadly comparable in terms of socio-economic status. Bhopal and his colleagues (1994) looked at a large variety of health outcomes, including: mortality data (both all-cause and specific causes, such as lung cancer and circulatory disease); cancer registrations (respiratory and other cancers); tests of respiratory function among children; general practitioner (GP) consultations for respiratory and non-respiratory conditions; and a

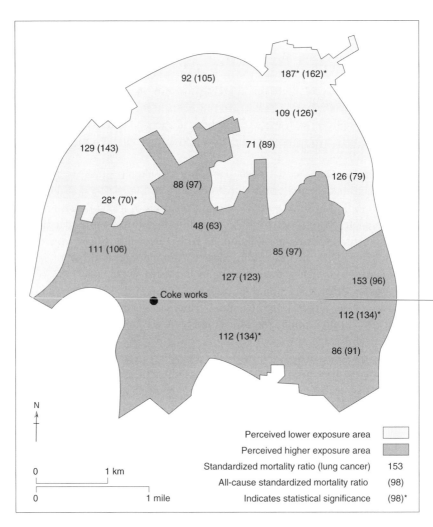

Figure 7.6 Standardized mortality ratios for lung cancer (all causes in parentheses) in relation to location of Monkton coking works

Table 7.4 Air pollution and general practice consultations (Monkton coking works study)

Air pollution (SO₂) level	Consultation rate in inner zone	
	Respiratory conditions	Non-respiratory conditions
Level 1 (highest)	752	2182
Level 2	713	2448
Level 3	625	2361
Level 4	560	2880
Level 5 (lowest)	424	2399

Note: Annual rates per 1,000 population
(*Source*: Bhopal et al., 1994: 244)

community survey of ill-health. Results indicated no significant difference in adult mortality or cancers among the three zones, and rates of self-reported asthma and bronchitis were broadly similar. For self-reported wheeze, cough, and sinus trouble there was a gradient of ill-health, with higher prevalence in the inner zone, closest to the coking works. This is an interesting finding, as it suggests some impact of the pollution on the upper respiratory tract, but not the lower tract or lung. The authors also had measurements of SO_2 available to them from a small number of monitoring sites. Analyzing GP consultations by pollution level in the inner zone (see Table 7.4) indicated a clear and significant increase in the consultation rate for respiratory problems as SO_2 levels increase, while consultations for non-respiratory problems did not vary with pollution. As the authors note, an analysis of readily available and routine sources of data reveals much less than one based on the harder-to-collect data on primary care consultations and self-reported morbidity. Bhopal and his colleagues conclude that "[T]he excess of respiratory problems observed in those living close to the works can best be explained as a result of their exposure to its emissions" (Bhopal et al., 1994: 246).

This study was originally sought by the local government, who, along with the coking works, had been pressed for a scientific study of the health problems perceived and suffered by local residents. These residents had their own strongly held beliefs about the link between pollution from the plant and their own experience of ill-health. Moffatt and her colleagues (1995) looked at these health perceptions via a postal survey that included both closed questions and an open section for comments. Interestingly, the three areas analyzed earlier (inner zone, outer zone, and control area) generated similar concerns about housing and family problems and difficulties with neighbors; problems that one would not expect to be associated with proximity to the coking works. On the other hand, there were clear differences among the zones in terms of complaints about dust, dirt, noise, smell, and air pollution. The quantitative evidence on health perceptions was reinforced by qualitative data. One parent of a 12-year-old commented: "If the plants and trees were dying, the air pollution couldn't have done me much good," while a 42-year-old woman reported that "I have always believed that the cokeworks were to blame for all my sinus problems as I never had any symptoms before I exchanged houses to my present address. I lived at my previous home for seven years without any sinus problems." A 53-year-old woman had "lived for many years with polluted air, the smell of sulphur carried with the wind, smoke billowing out day and night and window sills and cars covered with soot and grit" (all quotes from Moffatt et al., 1995: 888–9). The authors conclude that local anxieties are justified.

Popular ("lay") beliefs are often ignored or dismissed by public health professionals and epidemiologists as irrelevant distractions from the real task of uncovering the "facts." It is quite easy to belittle such beliefs as "sensitization bias:" to suggest that those living near sources of pollution are so aware of the existence of the plant that any attempt to interview them about it will inevitably encourage them to come forward with a long list of non-specific health problems. For Moffatt et al. (1995: 884), however, "epidemiological research that attempts to put to one side local concerns and popular beliefs thereby *distorts the very reality* it so assiduously seeks to describe" (our italics). For them, public concerns are as much a source of "data" as that emerging from large-scale surveys, perhaps more so, since the latter are divorced from the social context. And the dismissal of popular beliefs as sensitization bias is readily countered by noting that the perceptions of ill-health were quite specific, and restricted to respiratory complaints. Ironically, however, when the results of

the Monkton study were reported to a residents' action group they, like some local government officials, felt the study to be of limited value since it "only" produced evidence of
quite minor respiratory problems rather than the "serious" ill-health (such as cancer) which
was of more concern.

We have spoken quite often in this chapter of "associations" between exposure to air
pollution and ill-health. To what extent can we be less coy and speak of causal effects? One
way of assessing this is to ask if there are circumstances that allow us to mimic a natural
scientific experiment. A good specific illustration of how this might work is the workers'
strike (labor dispute) that closed, for 14 months in 1986–87, a large steel mill in the Utah
Valley. Pope (1989) shows that the levels of PM_{10} were roughly twice as high in winter
months when the steel mill was operational, compared with when it was temporarily
closed. Clear differences in hospital admissions of children, for bronchitis and asthma, were
observed. For example, in the fall (autumn) and winter months of 1985–86, and 1987–88,
such admissions numbered 127 and 133 respectively, while in the same period 1986–87
such admissions were only 46, a three-fold reduction.

Although of a totally different order of magnitude we can see broad parallels between
studies of point sources in both the developed and developing worlds. At one level we can
point to the wider structural economic forces at work, which give more weight to profit
and less to community health. But we can also point to the locations of these point sources,
which are, partly at least, shaped by the inability of local people to object to their being
sited amongst them. Noxious facilities tend to be located in areas where political opposition
is likely to be minimal. Geographers and public health professionals need to give much
greater attention to issues of environmental equity and to ensure that those "point sources"
most likely to impact upon health are located as far as possible from any population,
including one which lacks the political "muscle" to fight location decisions.

Concluding Remarks

There are good grounds for suggesting that there are many clear links between air pollution
and human health. The evidence indicating that radon gas emissions add to the burden of
lung cancer, particularly among smokers, is quite convincing. Research suggests that a pollutant such as ozone triggers asthmatic attacks among those suffering chronically from the
disease. Studies of the impact of vehicle exhaust emissions are more equivocal, though
other literature implicates particulates and nitrogen oxides as damaging to respiratory
health. Some point sources of pollution have had devastating effects, while others have
been harder to assess. Partly this is because, as with all epidemiological work, studies are
bedeviled by problems of confounding and exposure assessment. Aggregate studies of
health outcomes in geographical areas suffer from both problems, and we need to use
quite sophisticated environmental models, or better still, reasonably dense networks of
monitoring sites, in order to get better characterization of exposures. This applies to all
sources of pollution, whether point, linear, or areal.

Individual studies, if well-designed, can collect data on potential confounders (such as
smoking) and offer a better chance of detecting impacts on health caused by the exposure
(air pollution) of interest. But if studying disease with long latent periods, such as cancers,

these, too, will be found wanting unless they can assess exposure at previous places of residence or work. Further, studies investigating mortality, or serious chronic disease, need to be set against other studies which recognize that death, or even hospitalization, is the tip of a well-submerged iceberg.

As is clear from most of the studies reviewed in this chapter, research has tended to adopt a classic positivist stance. Apart from work by Moffatt and her colleagues (1995), few studies look at lay beliefs about sources of air pollution and their effects. Classically trained epidemiologists will dismiss the value of this, pointing to sensitization bias and arguing that those living near possible point sources of pollution will, literally, see the source and be keen to attribute ill-health to such sources. Yet, as Moffatt and her colleagues suggest, it is surely misguided to ignore those with the most immediate experience of pollution, those with direct exposure.

While there must continue to be well-designed epidemiological studies, ideally of a longitudinal nature (examining health outcomes before and after pollution incidents or episodes) we must also pursue other approaches to scientific investigation. Some of these need to be informed by a better understanding of the views of those affected, or potentially affected, by pollution sources. Equally, there is much to be gained from pursuing more structuralist accounts, ones that recognize the deeper structures underlying associations between air pollution and health, including the interests of business that may put profit before occupational and community health. We need to add a social and political model of health to classical epidemiological accounts.

Further Reading

For good overall reviews of the evidence on air pollution and human health see Brunekreef and Holgate (2002) and Kampa and Castanas (2008). Dunn and Kingham (1996) discuss some of the methodological difficulties in assessing the links between air quality and health.

The journal *Health Physics* is a good source of material on the links between radon gas and human health. For an overview of work on traffic and public health see Kjellstrom and Hinde (2007).

Full accounts of the Bhopal tragedy are to be found in Shrivastava (1992) and Bogard (1989). The former gives more information on the health consequences, while the latter is a stunning interpretation of the politics and political economy of risk and hazard as exemplified by the Bhopal disaster. Broughton (2005) reviews more recent findings concerning the disaster. An account of the Chernobyl disaster and some of the consequences is in Gould (1990).

For a very clear argument in favor of a social interactionist perspective on air pollution see Phillimore and Moffatt (1994), but also Bickerstaff and Walker (2003).

Proximity to hazardous waste sites can expose local populations to both air and water pollution. For a comprehensive picture of the public health impacts of such sites, primarily from a US perspective, see National Research Council (1991). This also has valuable general material on the principles of environmental epidemiology.

Chapter 8

Water Quality and Health

We looked briefly in Chapter 5 at the important contribution of safe water, and sanitation, for health. This chapter will look in more detail at these links, which are particularly acute in large parts of the developing world, and so we give some attention first to two of the most serious water-borne diseases occurring in these areas, cholera and schistosomiasis. We turn next to other gastrointestinal diseases, some of which are of increasing importance in the developed world, before looking at the hardness of water, which has been the focus of some research attention. Last, we examine the possible health consequences of contamination of water by chemicals and the contamination of water supplies by hazardous wastes. All of these have understandably generated considerable public concern and research effort in both the developed and developing worlds.

Most of the work examined here is drawn from the environmental epidemiology research literature. Much of this is necessarily "positivist" in orientation, seeking to employ large sample sizes to test hypotheses concerning exposure and outcome. We hope we make clear that there is a role for other perspectives, and that, as in the previous chapter, we should not neglect the individual's perception of risk. Most important, we are increasingly persuaded that we cannot secure a full understanding of the associations between disease/illness and water quality, without addressing wider structural and societal determinants.

Water-borne Diseases

Cholera

Cholera is a disease caused by the ingestion of a bacterium, *Vibrio cholerae*, present in water contaminated by faecal matter; the course of the disease is described graphically in Watts (1997: 173). There are over 100 serotypes of *V. cholerae*, two (O1 and O139) being responsible for the major disease epidemics. Symptoms are a massive loss of water and salts, leading to watery diarrhoea. If the depleted body fluids are not replenished (which is possible via simple oral rehydration therapy) then kidney and heart failure are a consequence.

We begin by examining some of the historical, as well as contemporary, research on cholera, which was the first disease shown by epidemiological methods to be water borne. There are two reasons for considering cholera here. First, the person who established the link between cholera and contaminated water, John Snow, produced a study in 1854 that has assumed almost mythological status as a piece of geographical epidemiology. This work by Snow, followed later by the microbiological research of Robert Koch that implicated the bacterium, played a major role in producing the sanitary reforms that helped control the disease in the developed world. The second reason for an extended discussion of cholera is quite simply because it continues to be a major cause of mortality across the globe.

The bacterium *V. cholerae* survives best in moderately saline waters, and thus prefers estuarine conditions, ideally where the temperature remains at 10°C (50°F) for several weeks at a time. Since it resides in the gut of apparently healthy individuals who may be traveling widely, it can spread via other transport routes. Thus, French soldiers working in early colonial Algeria transported the disease there in the early 1830s (Watts, 1997: 172). Britain suffered five cholera epidemics during the course of the nineteenth century, killing an estimated 130,000 people (see Learmonth, 1988: 143–52 for maps of its diffusion in the early to mid-nineteenth century). Over the same period India lost about 25 million people from the disease.

Before 1817, large parts of India were thinly populated by semi-settled and nomadic pastoralists in an "ecologically sound, cholera-free rural order" (Watts, 1997: 181), but British policies of compulsory resettlement onto fixed village sites disrupted this balance and did little to mitigate the effects of poor harvests. Malnutrition contributes to a pre-disposition to cholera since the body's defence mechanisms are weakened. The major epidemic of 1817, which originated in Jessore in Bengal (now Bangladesh) and spread to central India (helped on its way by the movements of the East India Company army), was preceded by famine. In the later-nineteenth century, other human intervention aided the spread of the disease, not least the role of the Public Works Department in over-seeing famine relief by concentrating huge numbers of people into camps supplied with contaminated drinking water, and by constructing a network of railways and irrigation canals unaccompanied by the drainage ditches needed to remove surplus water. "Impacting most heavily on the ability of cholera to slaughter millions were gentlemanly capitalists' very substantial investments in irrigation, in railways and port facilities and, equally decisively, the near absence of investment in public health" (Watts, 1997: 168). Watts also points a finger at the role of medical professionals who did little to halt the spread of cholera among the native Indian population, while for others the disease "was associated with much that European medical officers and administrators found outlandish and repug-nant in Hindu pilgrimage and ritual – so much so that the attack on cholera concealed a barely disguised assault on Hinduism itself" (Arnold, 1988: 8). And while some degree of medical intervention was essential in order to extract maximum output from the workforce – providing this did not eat into profits – this served mainly to bolster capital-ism's enterprise.

Following this first recorded pandemic in the early-nineteenth century (1817–23), others followed frequently into the early part of the twentieth century. All originated in the Ganga (Ganges) Brahmaputra delta in what is now Bangladesh and spread throughout much of Asia, the Middle East, Africa, and the Americas. In the early 1960s a new strain (El Tor) emerged from Indonesia before spreading worldwide. More recently, the O139 strain originated in India in 1991 and has spread into south-east Asia. In 1994 nearly 11,000 deaths and nearly 400,000 cases were reported to WHO, though these figures must be treated with caution as there is likely to be massive under-reporting; publicizing an out-break of cholera is unlikely to do much for fragile tourist industries. By 2005, both the number of cholera cases as well as the number of countries reporting cases had decreased. However, as indicated in Table 8.1, it is the poorest countries that remain most greatly impacted (World Health Organization, 2006). There have been a number of studies in high-incidence areas, such as Bangladesh. These show that the disease is socially patterned, with better-off families having lower disease incidence. In addition, higher incidence has been reported close to a hospital dealing with the disease, since local residents have used canal water polluted by effluent discharged by the hospital into the canal (Levine et al., 1976). Studies in South Africa have confirmed that incidence is elevated in areas of lower social status, while work in Mali found that those drawing water from one well located within ten meters (three yards) of three pit latrines were more likely to contract the disease (Hunter, 1997: 110). Cholera appeared in Latin America during the 1990s, killing over 11,000 people in the first five years of the decade. It first appeared in a coastal village in January 1991 but only five weeks later had diffused to 24 out of 29 Peruvian departments (Mintz et al., 1998: 86). In Mexico, the spatial patterning shows a north-south gradient, with higher incidence in the poorer southern states, while in Brazil it is the poorer northern states where incidence is highest (Mintz et al., 1998: 76). While it may well be the case that "health education and simple personal hygiene such as washing with soap can be signifi-cantly protective" (Hunter, 1997: 114), it is surely the case that to halt disease incidence and spread requires an attack on underlying structural determinants of poverty and poor sanitary infrastructure. As Mintz and colleagues argue, "the crowded populations of the developing world's peri-urban slums . . . still bear the burden of cholera disease" (Mintz et al., 1998: 65). WHO denoted the 1980s as the Drinking Water and Safe Sanitation Decade, but the provision of basic needs had been countered by rapid population change, to the extent that approximately half the world's population still lack access to adequate sanitation and safe water (United Nations, 2006). Indeed, the UN has declared 2008 the year of sanita-tion, with the catch phrase: sanitation is dignity. From a historical geographical perspective there have been a number of interesting studies of cholera. For example, Howe (1972) mapped the spread of the second pandemic (1826–51) from India to Britain, and within

Table 8.1   Reported cholera cases, 2005

Region	Total cases (%)	Total deaths (%)
Africa	125082 (95%)	2230 (98%)
Americas	24 (0.02%)	0 (0%)
Asia	6824 (5.2%)	42 (1.9%)
Europe	10 (0.01%)	0 (0%)
Oceania	3 (0.002%)	0 (0%)
TOTAL	131943	2272

(*Source*: World Health Organization, 2006)

Britain (see also Learmonth, 1988: 143–8). The likely route was from India to Russia and then, via the Baltic port of Danzig to Sunderland in October 1831. From there it spread both north into Scotland and south to Yorkshire, the Midlands, and then south-west England. A spatial diffusion study by Pyle (1969) explored three epidemics (1832, 1849, and 1866) in the nineteenth-century USA (see also Cliff et al., 1998: 272–3). Pyle was particularly interested in the relative importance of contagious and hierarchical diffusion processes. He showed that, in 1832, spread was dictated by distance (contagious diffusion), with towns near the origin of the outbreak registering the disease much earlier than those far away. In 1849 and 1866, improvements in transport meant that towns far from the origin in terms of geographical distance had become better connected in terms of travel time; spread was more hierarchical, with larger towns and cities reporting the disease much earlier than smaller settlements near the origin.

The classic epidemiological work – which shows a geographical imagination – is, as noted earlier, due to Dr John Snow working in London in the middle of the nineteenth century as the second cholera pandemic was under way (Cliff and Haggett, 1988: 7–11). In the second edition of his book, *On the Mode of Communication of Cholera*, published in 1854, Snow produced a map which showed that most of the deaths from cholera were to be found in a small area of Soho, and were of people who had taken water from a pump located in Broad Street; the water had become contaminated by a leaking cesspool. Snow saw the spatial patterning and its relationship to proximity to the pump: "the deaths are most numerous near to the pump where the water could be more readily obtained. The wide open street in which the pump is situated suffered most, and next the streets branching from it, and especially those parts of them which are nearest to Broad Street" (quoted in Cliff and Haggett, 1988: 53–5).

Accounts of Snow's work appear in numerous books on epidemiology and medical geography and focus on the detective work that led to the postulating of water-borne transmission (see Paneth et al., 1998 for further background). But as Snow himself made clear, this is only part of the story and links to material circumstances and the structural shortcomings of some of the contemporary water companies are also relevant. Snow (cited in Cliff and Haggett, 1988: 7–8) indicated that cholera in London "has borne a strict relation to the nature of the water supply of its different districts, being modified only by poverty, and the crowding and want of cleanliness which always attend it." Cholera was associated in particular with two companies drawing water directly from the River Thames, a "kind of prolonged lake . . . receiving the excrement of two millions and more inhabitants" (Snow, in Cliff and Haggett, 1988: 8). A contemporary cartoon (reproduced in Cliff and Haggett, 1988: 9) indicated where blame lay; against a drawing of industrial effluent was the following verse:

> *These are the vested int'rests, that fill to the brink,*
> *The network of sewers from cesspoool and sink,*
> *That feed the fish that float in the ink-*
> *-y stream of the Thames, with its cento of stink,*
> *That supplies the water that John drinks.*
> > (reproduced with kind permission of
> > Professor Andrew Cliff from
> > Cliff and Haggett, 1988: 97)

As we shall see in the next chapter, the vulnerability of certain marginalized populations is worsened in the face of increasing impacts on natural ecosystems resulting from global climate change.

Schistosomiasis

Schistosomiasis (also known as bilharzia) is a family of diseases resulting from infection with flatworms or "flukes" of the genus *Schistosoma*. The life cycle involves three factors: the human host, the worm, and an intermediate host, water snails (genus *Bulinus*). The worm's eggs are released via the urine or faeces of an infected person into freshwater, where the eggs hatch and penetrate the snail. After four to six weeks the cercariae (the worm's larvae) are released into the water and penetrate the skin of a human host. The worms migrate to the liver or lungs and, once mature, to the veins of other body organs. Different species of *Schistosoma* affect different parts of the body, so that some cause bladder and kidney disease, others cause disease in the liver and intestines. As is clear from this brief description, the water snail is a crucial element. This requires warm, preferably still or slow-moving and muddy, freshwater, with optimum temperature in the range 20–30°C (68–85°F).

Schistosomiasis is endemic in 74 developing countries and presently affects over 200 million people worldwide and although mortality is relatively modest (20,000 deaths per annum, due to subsequent development of cirrhosis and bladder cancer) it is a very debilitating disease in the tropics, second only to malaria in terms of prevalence (World Health Organization, 2007). Interestingly, more and more tourists are now contracting schistosomiasis given the rise of ecotourism and risk-based adventure tours. Agriculture, fishing and recreational water use all provide opportunities for disease transmission, and this explains the higher incidence among children and young adults. More than 80% of those infected are in sub-Saharan Africa. Praziquantel is the only available and effective (usually with a single dose) treatment against all forms of human schistosomiasis with few and only transient side effects (WHO, 2007). The cost of the average treatment is only about US$ 0.20 to 0.30. However, as we saw in earlier chapters, diagnostic procedures and safe drugs do not necessarily reach those most in need.

Explanations of disease incidence reveal the classic tension between behavioral and structural determinants. Given that the disease cycle has a clear link to poor disposal of faeces and urine, one argument for disease control would be for better education about hygiene. On the other hand, it is clear that alterations to water regimes have added to health hazards in some areas. While major irrigation schemes and flood control measures have had positive outcomes (in terms of increased food availability and therefore improved nutrition) they offer new habitats for the snail hosts and therefore new opportunities for the disease to spread. Such effects were recognized as long ago as 1971, when WHO suggested that schemes such as Aswan High Dam on the river Nile (and the large Lake Nasser behind it) were compounding the problem of infection. For example, one species of the worm, *Schistosoma mansoni*, affected only 3% of the population of one nearby Egyptian village in 1935, but this had risen to 73% by 1979, an increase that Abdel-Wahab et al. (1979) attributed to the dam construction. More recently, the construction of the Diama Dam on the Senegal River has introduced intestinal schistosomiasis to Senegal and Mauritania.

Geographical research on schistosomiasis illustrates nicely the range of approaches to the geography of health reviewed in Chapter 2, including broadly positivist, spatial analytic work and more humanistic qualitative research. For example, there is a wealth of research on the spatial distribution of infection at detailed spatial levels, research that examines the relationship between intensity of infection (egg counts) and a range of explanatory variables, including distance from water sources; predictably, distance to water sites is inversely associated with infection, though the strength of this association varies among study areas (Kloos et al., 1997). Such research uses detailed field work, as well as GIS and remote sensing (see Malone et al., 1997, for example), to map infection and to suggest areas in which control strategies should be focused. Others, however, prefer much more intensive and qualitatively grounded approaches. For example, earlier research by Kloos et al. (1983), some of whose detailed maps are reproduced in Learmonth (1988: 245–9), used an ethnographic approach to observe and record the behavior of boys aged between five and sixteen living in a small village, El Ayaisha, in upper Egypt. Home addresses were mapped, as were trips made to bathing places (usually the River Nile, but occasionally to canals), but time and duration of bathing, together with proportion of body immersed, were also noted, as were the kinds of activity undertaken (washing, swimming, fishing, and so on). Detailed maps of egg counts detected in urine samples were also prepared. Swimming seemed to be the predominant source of exposure to infection. Watts et al. (1998) adopt the framework of time geography, and methods of participant observation, to explore the detailed daily use of water sites by one household in Morocco. They argue, too, for a more sensitive awareness of the gendered use of space and water resources, suggesting that "this strategy will develop insights which would not be yielded by a conventional water study which looks at individual water contact activities dislocated from their social and cultural context" (Watts et al., 1998: 756).

The deeper underlying causes of the disease are of course the poor living conditions associated with poverty (Yi-Xin and Manderson 2005; Tchuenté, 2006). Hunter (1997) reports on a Brazilian study which found the highest egg counts in faeces from children of servants rather than children of household heads, and that low family income was also associated with high egg counts. Those households with flush latrines had lower egg counts than those with only pit latrines or none at all. Population displacement due to the movement of refugees has also played a part; for example, political instability in the Horn of Africa has led to the disease spreading to Somalia. This factor, coupled with the need for resources to be directed at the underlying causes of the disease (poverty, and lack of infrastructure) suggests that explanations of disease prevalence, and solutions to the problem, demand a structural or political economy approach.

Gastroenteritis

Three common sources of gastroenteritis around the world (though much studied, particularly in the developed world) are due to the protozoan parasites *Giardia* and *Cryptosporidium* and the bacterium *Campylobacter*. In all cases, the main symptom is diarrhoea, while animal and human sewage are key sources of environmental contamination. Giardiasis is quite widespread throughout the world and is a common cause of diarrhoea among international travelers, particularly among those who have consumed local tap water. The parasite survives in unfiltered and unchlorinated water supplies and therefore some outbreaks are due to the consumption of untreated water, such as that from streams (which may be contaminated by animal waste). Hunter (1997, Chapter 7) reports on a number of cases in the literature. In other instances, local water supplies are accidentally contaminated by overflows of sewage. Cryptosporidiosis outbreaks have, since the 1980s, been responsible for a considerable disease burden in the developed world. As with giardiasis there is a close correlation between attacks of diarrhoea and the consumption of contaminated tap water, such contamination again arising from flaws in filtration systems in water treatment works. One of the largest outbreaks was in Milwaukee in 1993, when over 400,000 people were affected. Here, water is drawn from Lake Michigan, and although this is treated by chlorination, filtration and other procedures it proved difficult to identify the precise area of system breakdown (Mackenzie et al., 1994). In other cases, consumption of groundwater, possibly contaminated by infected cattle, is a likely route for disease transmission.

Campylobacter transmission occurs though the consumption of undercooked meat (especially chicken) and untreated (unpasteurized) milk, as well as via contaminated water. Several studies implicate unboiled tap-water or groundwater supplies as sources of infection. Reservoirs that are open to animal wastes, or storage tanks that may be contaminated by roosting birds, are possible sources. The problem is particularly acute in rural areas. For example, those living in a rural community in Norway developed severe gastroenteritis following consumption of tap water fed by upland lakes; during spring, snow-melt flooding washed sheep faeces into the lakes from a path running alongside (Hunter, 1997: 140). The research evidence on campylobacter comes predominantly from the developed world. This might lead to the conclusion that the problem is unknown, or limited, in the developing world. This is grossly mistaken, as some studies (Hunter, 1997) indicate. For example, campylobacter is a major source of diarrhoeal disease in sub-Saharan Africa, again due mainly to the absence of a proper water supply.

There is little doubt that explanations for geographically localized outbreaks of water-borne disease must be sought not at the individual, but at the macro, level (Hrudey et al., 2002; Schabas 2002; Ali, 2004), as illustrated in the structural analyses of the Walkerton tragedy. Walkteron is a small town of approximately 4800 people located just north of Toronto, Ontario, Canada. In May of 2000, 2300 individuals experienced gastroenteritis, 65 were hospitalized, 27 developed haemolytic uremic syndrome (a serious and potentially fatal kidney disease), and 7 died. Suffering was not limited only to those who reported illness; others struggled with the psychosocial impacts of the lack of access to safe water in a developed world context, while others had to cope with the illness and deaths of their loved ones. The responsible pathogens were *Escherichia coli* O157:H7 and *Campylobacter jejuni*. The Government of Ontario established a public inquiry to determine both the causes and the responsibility for this tragedy. In essence, wells providing public water supply to the community were contaminated by faeces from cattle on nearby farms; heavy rainfalls played a role. Unfortunately, negligence on the part of public works employees meant that the water was not treated and was distributed in a contaminated state to local households. The negligence was deemed criminal by the courts and the public works employees were jailed. Many, however, blamed the situation on a neo-conservative provincial government led by a premier, Mike Harris, who had spent the previous five years cutting environmental and public health spending to the bone in Ontario (Schabas, 2002); the premier, on the other hand, blamed the incident solely on the incompetence of the public works employees (Ali, 2004).

Ali (2004) has undertaken a stunning structural analysis of the Walkerton tragedy by employing a framework informed by both sociological and ecological theories wherein he shows how large-scale structural determinants (e.g. globalization, climate change) interact with micro-scale determinants of the local milieu to produce significant changes in the ecology of disease. In the case of Walkerton, for example, he links international pressures for agribusiness with local farming practices by way of provincial policies of a neo-conservative government to explain the backdrop for the outbreak.

Hrudey and colleagues (2002) were involved in a range of ways in the inquiry process, and subsequently undertook a comparative analysis between this and other water-borne disease outbreaks in the USA, Canada, Denmark and England in order to discern the clear take home messages that would reduce the potential for future outbreaks. The result was a multiple barriers approach; that is, multiple barriers are/should be established to enhance the likelihood of access to safe water. These barriers are set up at source, treatment, distribution, monitoring and response stages of an integrated water resource management (IWRM) framework. While a sound strategy, it is also very costly – too costly for the world's poorest nations.

Water Hardness

Research on water hardness (of which calcium and magnesium are the main components) has shown it to be both adverse as well as protective in the context of health. For example, it is thought to be implicated in the development of eczema, and a study of children aged between four and sixteen years in Nottingham, England, used a GIS approach to offer some confirmation (McNally et al., 1998). Children's residential addresses were assigned to one of thirty-three water supply zones, for each of which data on water hardness (divided into four categories) were obtained. After adjustment for potential confounders there was evidence of an elevated risk of eczema among younger children (under 11 years), whether reported within the past year or at any time in the child's life (see Table 8.2). The odds of reporting eczema within the previous year was 1.54 in the areas with the hardest water (that is, 54% higher than in areas with soft water). A possible explanation is that calcium in the water acts as a skin irritant; in addition, hard water makes it more likely that there is a greater need for soap and shampoo when bathing, in order to develop a sufficient lather. A follow-up ecological study of over 450,000 Japanese school children (aged 6–12 years) in Osaka, Japan, found a relative odds of 1.12 (95% confidence interval 1.06 to 1.18; Miyake et al., 2004).

In other contexts, harder water seems to be protective of health. A well-known example is cardiovascular disease, although results are often mixed. For example, the British Regional Heart Study (Pocock et al., 1980; 1982) suggested that cardiovascular mortality in soft-water areas was about 15% higher than in harder-water areas. As Figure 8.1 demonstrates, the unadjusted SMR in areas of Britain served by supplies of soft water is about 120, but

Table 8.2 Water hardness and prevalence of eczema among children under 11 years in Nottingham, England

Water hardness[a]	Lifetime reported eczema prevalence (%)	Odds ratio[b] (95% CI)	1 year reported eczema prevalence (%)	Odds ratio[b] (95% CI)
118–35	21.2	1.00	12.0	1.00
151–7	23.4	1.13 (0.88–1.45)	14.1	1.19 (0.88–1.62)
172–214	22.7	1.16 (0.92–1.46)	14.6	1.32 (1.00–1.76)
231–314	25.4	1.28 (1.04–1.58)	17.3	1.54 (1.19–1.99)

[a] In mg/l of calcium and magnesium salts
[b] Adjusted for age, sex, socio-economic status, and distance from nearest health center
(*Source*: McNally et al., 1998: 529)

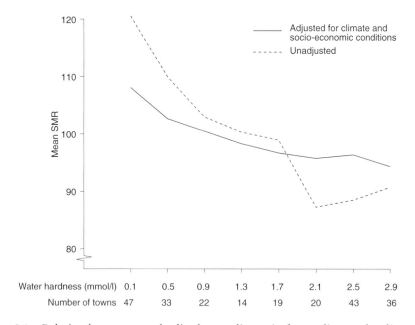

Figure 8.1 Relation between standardized mortality ratio for cardiovascular disease and water hardness, in a sample of British towns

only 90 in harder water areas, and the gradient, while attenuated, remains after adjustment for climate and socio-economic status. Such adjustment is critical, since areas of high mortality tend to be in the north of the country, where material deprivation (very broadly speaking) tends to be worse. There are, however, difficulties in interpreting this finding, because water hardness is correlated with other water parameters, including nitrates. As the researchers observe, "whether the lower CVD rates in hard water areas might be due to the beneficial effect from bulk minerals (e.g. calcium and magnesium) or the absence of harmful minerals or trace elements (e.g. sodium, lead and cadmium) remains unknown" (Pocock et al., 1982: 321). A recent meta-analysis (Monarca et al., 2006) reports that many – but not all – of the ecological studies reviewed found a protective association between cardiovascular disease mortality and water hardness, calcium, or magnesium. They hasten to add that results are not consistent This may in part be due to the difficulties – both conceptual and statistical – in undertaking these types of analysis. For example, the work reported above is an aggregate study; do the findings hold at the individual level? Work by Downing and Sloggett (1994) which studies premature mortality (all causes, for those under 65 years) demonstrates a significant relationship for men, but not for women. Adjusting for individual socio-economic circumstances, men living in hard water areas have a 10% reduction in mortality risk compared with those in soft water areas. However, the influences of unemployment, rented accommodation, and lack of a car increase the risk of premature mortality much more than does water hardness, a finding that will not surprise readers of Chapter 4.

Chemical Contamination of Drinking Water

Aluminium, fluoride, and arsenic

One of the most widely publicized public health incidents in Britain over the past 20 years was the accidental contamination of the local water supply in Camelford, Cornwall (south-west England). Here, in July 1988, a lorry driver tipped 20 tons of aluminium sulphate into a tank feeding the Lowermoor waterworks, exposing an estimated 20,000 residents to high levels of aluminium (and also lead and sulphate). Two reports were produced by a panel of scientists, chaired by Dame Barbara Clayton (Williams and Popay, 1994; Hunter, 1997: 246–8), and these concluded that while there were short-term effects on health there were no long-lasting consequences.

Residents continued to complain, however, 11 years after the incident. As one man in his seventies recalls (the *Guardian* newspaper, June 9, 1999): "We lose things or put them down in funny places, something we never did before. It's got to the state that I can't even repair the car. I pick up the manual and read it but by the time I have got the bonnet up I've forgotten what I have read." A critic might well counter that this is not parti-cularly unusual in a 70-year-old; however, the same man reported that shortly after the original incident: "I was sitting in a chair and found I couldn't move. My wife's hair turned red when she washed it and she had a rash up her arms. We had nausea, diarrhoea and mouth ulcers."

The Clayton Committee suggested that "it is not possible to attribute the very real current health complaints to the toxic effects of the incident, except inasmuch as they are the consequence of the sustained anxiety naturally felt by many people" (cited in Williams and Popay, 1994: 106). This is reminiscent of the dismissive "sensitization bias" in the Monkton coking works study of air pollution (see Chapter 7). However, a study by Altmann et al. (1999) suggests that the "scientific" conclusion of the Clayton Committee may be premature. The authors studied 55 people who were considering legal action and who had claimed to suffer organic brain damage. They also studied siblings of the cases, people who were broadly similar in age and who lived away from the Camelford area. A set of psychological tests to establish attention, coordination, visual skills, and memory were used to compare the cases and controls. A sensitive test for organic brain disease, the "symbol digit coding test," reveals that the performance of the cases is significantly worse (p = 0.03) than that of their siblings; box-plots revealing the lower and upper quartiles demonstrate this well (see Figure 8.2). The ability of cases (participants) to respond to visual stimuli is even worse (p = 0.0002). The authors conclude that "there are no other known causes for the effects that we have described in the people from Camelford" (Altmann et al., 1999: 810).

This example serves as a classical illustration of the tension between a "scientific" and a "lay" account of a pollution incident and is reminiscent of the case of the Sydney Tar Ponds discussed earlier (pp. 78–9 above). Here, unlike in many cases, there is no disputing that pollution occurred. The debate centers around its health impacts. A scientific study

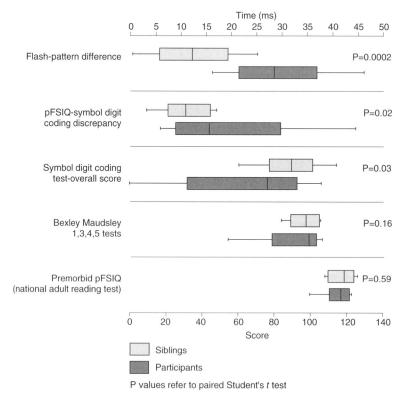

Figure 8.2 Comparison of cases (participants) and controls (siblings) in psychological tests following the Camelford water pollution incident

published more than ten years later points to significant psychological morbidity. Until this was published, the scientific evidence was weak and "experts" pointed to dangers of reporting and sensitization bias; if people know they are likely to have been exposed they are more likely to report symptoms. Should we have accepted the scientific account of the expert (Clayton Committee), or accounts of people who clearly were exposed to high levels of aluminium? Altmann and his colleagues have provided new evidence for accepting a link between exposure and morbidity, though this evidence remains contested.

There have in, fact been, concerns for many years of possible associations between raised levels of aluminium and the risk of neurological disease, including dementia, Alzheimer's disease, and motor neurone disease (amyotrophic lateral sclerosis). One geographical study in England (Martyn et al., 1989) considered rates of Alzheimer's disease among the population aged 40–69 years, in 88 local authority districts. Water quality data were obtained from the records of water companies. The risk of disease was 1.5 times higher in counties where aluminium levels were 0.02–0.04 mg/l, although at greater concentrations (>0.11 mg/l) the risk was no higher. The difficulty with this study, as with all aggregate studies, is that exposure is poorly characterized; can we really believe that all individuals currently living in a large area have had identical exposures to water of a particular chemical composition? A case-control study (Forster et al., 1995) in northern England failed to find a significantly elevated risk of Alzheimer's disease among those exposed to levels in excess of 0.15 mg/l. It remains difficult to know whether aluminium exposure "causes" these diseases or whether aluminium concentrations tend to increase more generally with aging. The Paquid study in France, for example, examined almost 4000 men and women aged 65 and over and found that aluminum on its own was not found to be a risk factor for cognitive impairment; however, when combined with calcium and pH, there was a statistically significant increase in cognitive impairment (Jacqmin et al., 1994).

The example of fluoride illustrates neatly some issues concerned with risk perception and risk communication. We saw in Chapter 4 how dental health among children improved in areas having fluoridated water, although material deprivation was also a factor. Fluoride reduces the solubility of tooth enamel and thus offers protection against decay (McDonagh et al., 2000). For example, a South African study (Du Plessis et al., 1995) indicated that there was a 50–80% reduction in tooth decay among black children when fluoride levels were between 0.2 and 0.9 mg/l. Evidence of this sort has led to calls from community dentists and public health professionals to have fluoride added to water supplies lacking the chemical. In Britain, the USA, and elsewhere there are vociferous groups who oppose such fluoridation; some cite evidence of links between fluoridation and cancer (bone cancer, or osteosarcoma, in particular), while others oppose a public health intervention over which they have no choice. One has some choice, they argue, over whether to exercise, and how much fatty food to consume, but fluoridated water is hard to avoid. Some cases of accidentally raised levels of fluoride (for example, due to faulty equipment) have produced acute episodes of gastrointestinal illness and nausea. Moreover, some studies in China, Kenya, Japan, and elsewhere indicate that where concentrations of natural fluoride are high (above 2 mg/l) there is a risk of fluorosis; here, calcium in bones is replaced by fluoride and bones begin to soften and crumble (Hunter, 1997: 257–8).

The association between fluoridated water and cancer was given publicity by a national newspaper (the *Guardian*, June 1999), which claimed that "in fluoridated areas of America, bone sarcoma rates among boys aged 9 to 19 are between three and seven times higher than in non-fluoridated areas." The newspaper article failed to point out that this study looked only at seven counties in one state, New Jersey, and simply compared aggregate rates of bone cancer in fluoridated and non-fluoridated communities. Indeed, osteosarcoma was identified in 12 males aged less than 20 years in fluoridated communities, and only 8 in non-fluoridated areas. This is hardly compelling evidence in favor of an association between exposure to fluoride and bone cancer. Other work has found that while osteosarcoma rates are higher in fluoridated areas, this bears no relation to time of fluoridation. Clearly, this level of detail cannot be discussed in a newspaper article, but for those reading it a doubtful association between "cancer" and fluoridated water is now embedded in their consciousness. For a full review of epidemiological studies see Cook-Mozaffari (1996).

A similar case can be made for the use of chlorinated disinfection by-products in drinking water. The use of chlorine to disinfect drinking water has been identified by many as the single greatest public health accomplishment. Without the use of chlorine in major drinking water supplies, we would see the return of many water-borne diseases discussed above (e.g. cholera). However, naturally occurring organic and inorganic materials present in raw water supplies combine with chlorine to form a variety of halogenated and non-halogenated compounds or disinfection by-products. There has been a substantial research focus on one family of these – trihalomethanes – given their suspected link to cancer, particularly breast cancer (Driedger et al., 2002). As is the case with fluoride, policy makers are faced with risk-benefit trade-off decisions; that is, do the risks of chlorination of public water supplies outweigh the benefits of keeping major outbreaks of water-borne diseases at bay? In making these decisions, policy makers have to rely on typically equivocal science, as we saw in the case with fluoridation. In the case of chlorine and breast cancer, the science is equally equivocal (Driedger and Eyles, 2001).

Arsenic is a highly toxic metal and its presence in groundwater means that those consuming such water are at risk of serious chronic disease. There are high natural concentrations of arsenic in water drawn from aquifers in the Ganges delta and this has led to serious skin conditions ("arsenical dermatitis"; see Smith et al., 2000). In south-west Taiwan the incidence of "blackfoot disease" (a peripheral vascular disease affecting the feet but sometimes hands, and leading to gangrene) is high and has been linked firmly to the consumption of groundwater contaminated by arsenic. Chen et al. (1988) reported results of a case-control study of over 300 patients which showed that those exposed to such water for over 30 years were 3.47 times more likely to develop the disease than those unexposed. Later investigations by Chen and Ahsan (2004) uncovered a doubling of lifetime mortality risk from liver, bladder, and lung cancers in Bangladesh owing to arsenic in drinking water. Other research in Taiwan indicates that there are high correlations between arsenic levels in water and the incidence of lung, skin, kidney, and bladder cancer (Chiou et al., 2001; Chen et al., 2003; Guo 2004). Concentrations of above 0.3 mg/l seem to be a risk factor for these cancers and other chronic disease. Thus Hunter (1997: 255) concludes that "arsenic in drinking water other than at very low concentrations has a significant adverse health impact." One solution to the problem is emerging from geochemists, who suggest that a relatively simple device can be used to bind iron hydroxide and arsenic; this can then be filtered out (Zhang et al., 2006). The system is now being tested in Bangladesh, along with several other intervention measures, including public education, posting of arsenic test results on private wells, and installing safe community wells (Opar et al., 2007).

Other Forms of Contamination

Hazardous waste sites

Consider, now, the health risks of exposure to contaminated drinking water, where the source of such contamination is sites disposing of hazardous wastes. This is not to deny that other sources of potential problems are air pollution from, or perhaps via the consumption of food supplied from areas polluted by, such sites. Of particular interest are organic contaminants, including pesticides, solvents, and petroleum products, but organics such as trichloroethylene (TCE) and trihalomethanes (THMs) have been studied in detail. TCE is used in dry-cleaning and as an industrial solvent, while THMs are a by-product of chlorination, formed by the action of chlorine on organic compounds.

While associations between hazardous waste disposal and human health are of global concern, great attention has been paid to this issue in the USA, where the Agency for Toxic Substances and Disease Registry (ATSDR) has a remit to consider the health effects of living near hazardous waste sites. In the USA as a whole, 50% of the population consumes groundwater, a figure that rises to 95% in rural areas. Given that the US Environmental Protection Agency (EPA) has estimated that 40 million people live within four miles (6.4 km) of "Superfund" sites (see Box 8.1) and four million within a mile of such sites (National Research Council, 1991) there is a clear geography to the possible association between exposure and health outcome. We need to bear in mind, as we consider some of the literature, that geographical proximity does not necessarily equate to direct exposure.

Many individual sites have been the focus of investigation. For example, Lagakos and colleagues (1986) carried out research on the town of Woburn in Massachusetts, where TCE and other organics had contaminated two wells. Twenty cases of childhood leukemia were recognized in the study area, compared with the 9.1 which would have been expected on the basis of national rates. Love Canal, in upper New York State, is one of the best known, and most intensively studied, waste sites in the USA. Here, water contaminated by benzene and other chemicals seeped into basements (and therefore exposure was by inhalation rather than direct consumption). One health outcome of interest here has been low birth weight (Vianna and Polan, 1984), results suggesting that between 1940 and 1953, when large volumes of chemicals were dumped in the site, there was a significantly increased number of infants of low birth weight in the most exposed site. Here, "exposure" was represented in terms of whether the house was on a low-lying natural drainage depression, or swale, in which water was more likely to percolate into the ground. Other work by Paigen and colleagues (1987) has monitored the growth of children born, and living at least 75% of their lives, in the Love Canal area, demonstrating that such children were significantly shorter than control children, differences that remained even after adjustment for possible confounding variables. Interestingly, by about 1995, the houses in the Love Canal neighborhood had been refurbished and sold to new families for virtually the same cost as similar homes in other areas of Niagara Falls, New York.

Others have looked at associations between exposure to contaminated water and more serious adverse pregnancy outcomes, such as congenital malformations. In particular, a number of studies (see National Research Council, 1991: 191–5 for a review) have

Box 8.1 Superfund

Growing concern about hazardous waste sites in the USA led to legislation to clean up the worst of those sites. The Superfund program was established in 1980 to locate, investigate, and clean up such sites. It is administered by the US Environmental Protection Agency (EPA). EPA conducts tests of water, soil, and air samples in order to assess the risk posed by a site. Sites are scored according to the risk they pose to public health. The worst such sites are put onto a National Priorities List and are eligible for a long-term program of remediation. This remedial action is funded using sums demanded from the companies responsible for the site contamination.

A key issue surrounding the Superfund sites is the cost-benefit analysis of site clean up. Essentially, site clean up is very expensive. Further, while the companies responsible for site contamination are theoretically liable for the costs of site clean-up, it is often difficult, if not impossible, to successfully extract payment from those companies. Therefore, researchers and policy makers have to decide: is it worth it? For example, Hamilton and Viscusi (1999) assert that the costs far outweigh the benefits. Undertaking a cost benefit analysis at a sample of 150 of the priority sites, they find that at the majority of sites the expected number of cancers averted by remediation is less than 0.1 cases per site and that the cost per cancer case averted is over $100 million. Is it worth it? We need to juxtapose the issue of cost benefit analysis with that of environmental justice. A growing body of evidence illustrates that people of color and low-income persons have borne greater environmental and health risks than the society at large; this is also the case with respect to residential proximity to Superfund sites in the USA (Bullard and Johnson, 2000). Like many environment and health issues, therefore, the policy response is a complex one. Lists of sites, together with maps, are available via the EPA's web-site.

examined links between exposure and heart malformations. For instance, Goldberg et al. (1990) studied heart malformations in the Tucson Valley, Arizona, part of which had been contaminated with trichloroethylene between 1950 and 1980. The study found that 35% of mothers with malformed infants had conceived their child, and lived during the first trimester (three months) of their pregnancy, in the contaminated area, three times as many as those without contact with polluted water. After the contaminated wells closed, in 1981, the odds ratio dropped from three to nearly one. Animal experiments confirm the plausibility of this relationship (Johnson et al., 2003). Assuming one can be confident that diagnoses, and record-keeping, have not changed over time, the kind of longitudinal work illustrated by Goldberg and colleagues has clear advantages over cross-sectional studies, since one can relate exposures during particular time periods to outcomes during the same, or later, periods.

Griffith et al. (1989) conducted a large-scale study of nearly 600 hazardous waste sites, spread across 339 counties throughout the USA. County-level mortality data were examined for 13 types of cancer and the rates in the 339 counties compared with 2726 "control" counties that did not have evidence of contaminated water. The authors reported significant excesses of deaths in the "case" counties, for both men and women, from cancers of the lung, bladder, stomach, colon, and rectum. However, this was a study of aggregate areal units that did not adjust for individual-level risk factors or confounders; for example, what evidence is there that smoking is more prevalent in the "contaminated" counties, which might have accounted for the excess risk of several cancers?

There are therefore several difficulties in trying to assess the association between hazardous waste and health outcomes. Most serious is the question of exposure: how do we measure this, and to what extent can we assume that the exposure has not varied over time? What is known about the residential histories of those affected, or other possible sources of exposure, as in the workplace? Definitions of study areas and study populations are also problematic; certainly, the first of these is inherently arbitrary. Studies of areal units may be suggestive, but detailed case-control studies are needed for a more convincing result, not least because such studies can adjust for the influence of confounding variables. Further, we are typically dealing with quite rare diseases of low incidence, and epidemiological studies require larger numbers of cases in order to achieve statistical power. All these issues, and more, are the very stuff of environmental epidemiology and have to be addressed by positivist approaches to the problem. Quite separate from these methodological issues, however, are more fundamental questions of epistemology. By this we mean once more the primacy of this kind of "scientific" knowledge over the accounts of illness given by those living near such sites. It is easy to make accusations of "scaremongering" and bias among local populations, yet it is, quite understandably, difficult to persuade those living near contaminated sites that three cases of a rare childhood cancer might be a statistical chance fluctuation rather than having a causal link to waste disposal. There is a deep irony in the fact that we need plenty of cases of disease to secure a respectable sample size, while also wishing to minimize the burden of any life-threatening condition on worried local communities (see Box 8.2).

Box 8.2   Lay epidemiology

As we have seen in both this chapter and the previous one, there is a tension between "scientific" investigations of disease "clusters" and the understandable concerns of ordinary people who either live near suspected sources of pollution or claim to have detected unusual local aggregations of disease or illness. In a sense, this is a conflict between "positivist" and "social interactionist" views of the world (see Chapter 2). There are numerous examples of communities claiming cancer "clusters" which public health specialists may or may not choose to investigate.

Some health professionals have shown how lay people can themselves undertake "scientifically rigorous" studies in order to persuade others that there are justifiable local health concerns. Marvin Legator and his colleagues collaborated on a guide to assist local communities in investigating suspected environmental hazards, such as hazardous waste sites. The book considers issues of experimental and questionnaire design, data analysis, and how to seek assistance from other agencies. It is an excellent resource for anyone, student or concerned citizen, who wishes to engage in their own epidemiological investigation.

See Brown (1992) and Legator et al. (1985).

Concluding Remarks

This chapter reveals sharp contrasts between the burden of disease in the developed and the developing worlds. In parts of Africa, South America, and the Indian sub-continent, issues of water quality revolve around the lack of safe drinking water (see Table 5.1, above, page 132) and the consequent outbreaks of severe infectious disease that are exacerbated by frequent natural disasters such as monsoon flooding. Vast numbers of people are at risk as a result. Although people in North America and Europe faced these risks in the nineteenth century, diseases such as cholera are now, in the developed world, at a distance. Periodic flooding in the developed world may result in heavy financial losses to business and the individual, but the public health risks are tiny in comparison with those in the developing world. While those living in quite comfortable material circumstances may be at risk from infections due to *Campylobacter* and *Cryptosporidium*, the health impact is in terms of morbidity and not, in general, mortality. Similarly, we have seen that, in the developed world, public health concerns about water quality relate to elevated levels of some metals, or to water hardness, or to possible exposure to organic and other contaminants from hazardous waste sites. Here too, we are not dealing with thousands of deaths.

In Chapter 4, we spoke of the Millennium Development Goals. Target 10 of MDG goal 7 (to enhance sustainability) is to cut by half, before 2015, the proportion of people in the world without sustainable access to safe drinking water and basic sanitation. Progress toward this target, of course, affects many other targets and goals: reduction of childhood mortality, reduction of major infectious diseases, increasing the quality of maternal health, empowering women, increasing school attendance (particularly of girls), reducing poverty and hunger, contributing to workforce health and contributing to economic growth of communities. Researchers have recently reported an analysis of the cost of achieving this target; Hutton and Bartram (2008) estimate the cost at $42 billion for water and $142 billion for sanitation. Key decisions will have to be made about the willingness of developed countries to contribute aid to achieve this goal.

As in the previous chapter, the tension between "scientific" (positivist) investigations of disease and illness, and public perceptions of risk, has emerged as an issue. What is missing is a fuller account of the structural factors that determine whether or not safe, potable water is supplied to the community, whether in the developed or developing world. Whether this relates to the privatizing of water companies and their wish to maximize profits and perhaps to under-invest in improving water quality, or to the political and economic constraints in the third world that prevent millions from enjoying access to safe supplies of water, an investigation of these wider-scale structural determinants is surely merited.

Further Reading

By far the most comprehensive review of literature relating to water quality and health is that by Hunter (1997); this contains a lengthy bibliography that provides an excellent starting point for further research.

Cliff and Haggett (1988) consider Snow's work on cholera and apply a battery of spatial analytical methods to his classic Soho data. See Paneth et al. (1998) for a more detailed description of John Snow's contribution to public health.

Mintz et al. (1998) paint a comprehensive global picture of contemporary and recent work on cholera. See Hrudey et al. (2002) for a good meta-analysis of investigations of water-borne disease outbreaks in the developed world.

There is a large literature on the health risks of hazardous waste disposal. A good overview is in the report written for the National Research Council (1991). See also the EPA website as well as Vrijheid (2000) for a review of the epidemiologic literature.

Chapter 9

Health Impacts of Global Environmental Change

The previous two chapters have focused on broad areas of environmental hazard: air and water quality. These hazards, along with others (such as sources of radiation, contaminated food, and noise, for example), have been a common area of research for environmental epidemiologists over several decades. But they are specific hazards acting in a particular setting. In this chapter we turn attention to broader processes of environmental and ecological change, those whose consequences for health are in part direct (as are air and water pollution) but also indirect. In particular, many of these processes are global in their scale and while, as we shall see, some of the research they spawn has tended to be more local in scope, the health impacts will be widespread.

Two broad areas of environmental change will be considered. First, the health consequences of ozone losses from the upper atmosphere, and, second, the impacts of climate change – the extent to which climate warming is likely to have negative consequences for human health. The various health impacts are summarized in Table 9.1, and the discussion in this chapter follows this framework. Since we have only relatively recently come to appreciate the nature and scale of these environmental changes there is still considerable uncertainty attached to the predicted health consequences. The "error bars" around some of the quantitative estimates of risk provided by researchers are therefore inevitably wide, and this needs to be borne in mind. However, the fact that climate change is happening and will have devastating consequences for the earth's population is no longer in dispute (Gore, 2006; Monbiot, 2007). Nor is the fact that the earth's poor will be among those most greatly impacted (Flannery, 2005; Monbiot, 2007).

Before we consider our first topic, we quote from Al Gore, perhaps the most visible politician seeking to get climate change into the public and political consciousness.

Undeniable tragedies have been unfolding in the part of Africa that includes southern Sudan to the east of Lake Chad, where genocidal murders have been commonplace in the region of Darfur. In Niger, just to the west of Lake Chad, the region wide drought has contributed to the famine conditions that put millions at risk. There are many complex causes of the famine and genocide, but a little-discussed contributing factor is the disappearance of Lake Chad, formerly the sixth largest lake in the world, in a period of only the last 40 years (Gore, 2006: 116).

The impacts of greenhouse gases on global climate have resulted in drastically reduced rainfall in the Sudan. As a result, Lake Chad has shrunk to one-twentieth its former size. With the lake's shrinkage, temperatures have soared, fish stocks have disappeared, agricultural production has collapsed and millions of human beings have been displaced. As Lake Chad dried up, periods of intense drought set the stage for the violence that has erupted in neighboring Darfur.

Yet Lake Chad is not the only example. The Aral Sea (Karakalpakstan, Central Asia) was once the fourth largest inland body of water in the world (Crighton et al., 2003). It was fed by two rivers, which flowed to the sea from the north and the south. Part of the Republic of Uzbekistan, this region was once part of the former Soviet Union, which decided – in the post-war era – to become a world leader in cotton production. Given that this area of the world is mostly desert, and that cotton is a water intensive crop, the two rivers feeding the Aral Sea had to be diverted to irrigate the crop. The environmental results were disastrous; by about 1960, the sea was one-third of its former size. The fisheries industry collapsed; hundreds of thousands of local residents were displaced and had no means of economic support; the salinity of the remaining sea waters increased to such a level as to pose serious health hazards to the local population; and, decades of excessive pesticide and herbicide use in the production of the cotton crop left the soil and groundwater highly toxic to humans and wildlife.

These are only two of numerous examples we could use to illustrate the impacts of humans on their environment, and the resulting impacts of that global change on human health. They also illustrate well the complexity of the range of factors – physical environment, economic, geopolitical, and cultural – that interact to influence health or ill-health.

Table 9.1 Health impacts of ozone depletion and climate change

Ozone depletion
   Skin cancer
   Cataract
   Immunosuppression

Climate change
   Direct effects
     • thermal stress
   Indirect effects
     • vector-borne diseases
     • food poisoning
     • sea level rise
     • agriculture

(*Source*: Based on Martens, 1998: 5)

Stratospheric Ozone Depletion

Ultraviolet radiation from the sun is usually divided into three classes, depending on wavelength. Long-wave radiation (UVA) occupies the range 315–400 nanometers (nm), while UVB and UVC radiation occupy 280–315 nm and 100–280 nm, respectively. The ozone layer in the upper atmosphere (stratosphere) ensures that none of the short-wave UVC radiation that can damage genetic material (DNA) reaches earth. It further absorbs much of the UVB radiation that also harms living things.

In 1985, research was published (Farman et al., 1985) that pointed to losses of ozone during spring in Antarctica, now popularly known as the "hole" in the ozone layer. But given the lack of human population in the region this is of less direct human concern than depletion at lower latitudes (Bentham, 1994). Specifically, as Madronich (1992) demonstrated, there have been losses of ozone amounting to about 6% at latitudes of 45°, in both hemispheres (see Figure 9.1), though this masks seasonal losses that are typically higher in winter months. Further, these are very broad regional averages, and losses may be much greater in certain areas.

Stratospheric ozone is depleted by chlorofluorocarbons (CFCs) and other so-called halocarbons, used in refrigeration, as aerosol propellants, and in solvents. On reaching the stratosphere they form very reactive chemicals, such as chlorine monoxide, which react with ozone to convert it back to oxygen. Despite agreements reached in Montreal in 1987 to reduce CFC usage (and subsequent toughening of this protocol in Copenhagen, in 1992, to phase out production and consumption of CFCs) it is anticipated that the atmospheric chlorine load will continue at a high level and that significant recovery of the ozone layer will not occur for at least another 50 years. Ironically, the production of

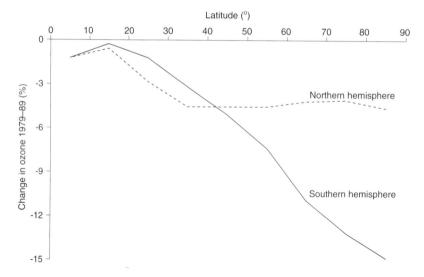

Figure 9.1 Changes in total ozone by latitude, northern and southern hemispheres, 1979–89

low-level (tropospheric) ozone due to pollution, particularly in Western Europe, helps to absorb some UVB radiation and diminish the effects of stratospheric ozone depletion.

What, then, of the consequences for health? Of the radiation that does get through to the earth's surface there is evidence that it can cause skin cancers and eye disease (cataracts). We need, then, to consider recent evidence on the incidence of such diseases. In so doing we shall, of course, need to give due weight to the usual epidemiological issues of quantifying exposure and assessing the relationship between radiation dose and health response, as well as issues of potential confounding variables.

Ozone depletion and skin cancer

There are three common forms of skin cancer, all of which are thought to be related to UV radiation (Department of the Environment, 1996). Two of these, basal cell carcinoma (BCC) and squamous cell carcinoma (SCC), are together referred to as *non-melanoma skin cancer* and account for about 90% of all skin cancers. The third, and most serious, is *malignant melanoma* (MM).

There is strong evidence that exposure to sunlight can cause non-melanoma skin cancer, but that this exposure occurred ten years or more prior to diagnosis. Cumulative lifetime exposure to sunlight seems to be a risk factor for SCC in particular, while adult risk of BCC is considered to be related to sun exposure in childhood and adolescence. Those working outdoors, in farming and fishing, are at risk, with cancers on exposed body areas, such as the head and neck. Incidence of non-melanoma is particularly high in Australia and New Zealand, where considerable proportions of the population have fair skin. The incidence is also higher among older people, but mortality rates from these cancers are generally low.

Malignant melanoma is much less common, but affects younger people as well as older age groups, and mortality rates are relatively high. What is known about international variations in MM incidence? As data indicate (see Table 9.2) the incidence of MM is quite high in Australia and New Zealand, Further, incidence is much higher in Northern Europe compared with countries from southern latitudes. For example, age-standardized rates in Denmark and Sweden are approximately six times those in Greece. Mortality from MM is much higher than for BCC and SCC, and is rising in the UK, as is incidence (see Figure 9.2). MM mortality has increased in Canada, where the estimated annual increase has been 4.8% for males and 3.1% for females (Lens and Dawes, 2004). In the USA, the lifetime risk of an American developing MM in 1935 was one in 1500 individuals; in 2002, the risk was 1 in 68 individuals (Lens and Dawes, 2004). Within New Zealand, Bulliard et al. (1994) have reported higher incidence and mortality with increasing proximity to the equator. Data on incidence reveal that, for both men and women, there have been long-term increases in incidence over 20 years, and that incidence is over 50% higher in the northern region of the country than further south (see Table 9.3). Recently, however, we have seen a leveling off and even a slight decline in MM in New Zealand (see Figure 9.3) due to aggressive public health education programs (Pearce et al., 2006); these are occurring internationally in areas of high incidence of MM (Garvin and Eyles, 2001).

Vacation and recreational behaviors, though not necessarily those involving skin protection, may go some way towards accounting for the social class gradient in MM, where the

Table 9.2 Incidence of cutaneous malignant melanoma (per 100,000) for 23 selected countries

Country	Male		Female	
	Crude	ASR	Crude	ASR
Australia	51.6	40.5	40.7	31.8
New Zealand	45.2	36.7	44.4	34.9
Sweden	19.8	12.6	19.9	13.3
U.S.A.	16.4	13.3	12.9	9.4
Denmark	14.8	10.6	17.6	13.0
Switzerland	12.5	9.3	15.0	11.1
The Netherlands	12.2	9.4	16.7	12.9
Austria	11.5	8.8	15.4	10.4
Canada	10.6	8.2	10.6	8.0
Hungary	10.3	7.6	10.3	6.8
Israel	9.7	9.4	11.0	9.8
Germany	9.3	6.5	11.4	7.1
France	8.6	6.8	11.1	7.9
U.K.	8.3	6.1	11.3	7.7
Poland	6.6	5.6	8.6	6.7
Italy	6.5	4.6	8.2	5.5
Russian Federation	6.3	5.4	6.4	4.7
Spain	4.0	2.8	6.8	4.5
South Africa	3.8	6.4	3.6	4.8
Brazil	2.9	3.5	2.0	2.2
Greece	2.5	1.9	3.2	2.0
Japan	0.63	0.40	0.49	0.29
China	0.21	0.22	0.17	0.17

ASR is Age-standardized incidence rate
(Source: Lens and Dawes, 2004)

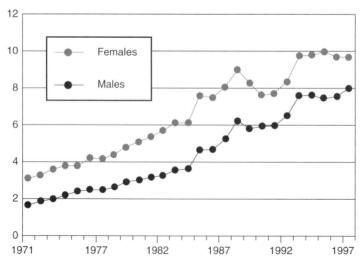

Figure 9.2 Incidence of cutaneous malignant melanoma in the UK, age-standardized to the European standard population (rates per 100,000)

Table 9.3 Incidence of melanoma in New Zealand (1968–89)

Region	Average latitude	Men 1968–73	Men 1984–9	Women 1968–73	Women 1984–9
Northern	36°S	12.2	23.2	19.1	26.5
Midland	38°S	9.1	24.9	14.4	28.1
Central	41°S	7.6	21.9	11.4	25.8
Southern	44°S	6.6	18.3	12.1	24.4

Rates per 100,000, standardized to world population, non-Maori population only
(*Source*: Bulliard et al., 1994: 236)

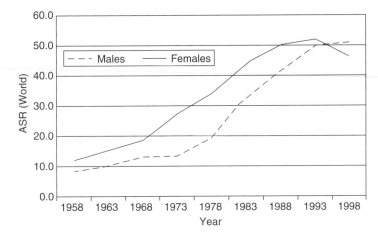

Figure 9.3 Age-standardized incidence rates of malignant melanoma in New Zealand, 1958–98

standardized mortality rate in professional groups is about 170, while for partly skilled and unskilled workers it is about 55. Gender differences in MM incidence are also of significance, with women being much more likely than men to have melanomas on their legs. Some public health experts therefore demand changes in health behaviors in order to reduce incidence. They point out that the contemporary preference for suntanned skin in many parts of the western world is socially constructed, in the sense that earlier centuries saw cultural value placed on white skin (Garvin and Wilson, 1999). They also indicate that people underestimate the risk of exposure to sunlight, in much the same way that other familiar risks controllable by the individual are under-estimated, compared with those that are "managed" by the state and corporate interests (Bentham, 1993). Whether lower levels of use of (expensive) high-protection sun creams among the less affluent will attenuate the social class gradient in MM will only be apparent during the next two or three decades. It may be that such creams protect against sunburn but are less effective in protecting against skin cancer (Office for National Statistics, 1997: 61).

Martens (1998) has conducted extensive research on the likely consequences of ozone depletion, with particular reference to the Netherlands and Australia. His approach is to set up a mathematical model of ozone depletion and to link estimates of UV skin dose to the incidence of skin cancer. The model includes the delay between exposure and tumour development, and various scenarios are explored on the basis of assumptions made about population size and age distribution, as well as rates of ozone loss. Modeling under conditions of uncertainty is fraught with danger, but nonetheless Martens envisages increases (by 2050) in non-melanoma of up to 60 cases per 100,000 in Australia, but quite modest increases (1 per 100,000) in MM incidence. Results indicate that although compliance with international agreements should see a reduction in chlorine concentrations and some stabilizing of ozone depletion in the early part of the twenty-first century, the lag between exposure and disease, coupled with an aging population, will lead to increases in all types of skin cancer until at least 2050. Slaper and his colleagues (1996) suggest that, even with the Copenhagen agreement to phase out ozone-depleting chemicals, by 2050 there will be an additional number of cases of skin cancer numbering (each year) about 33,000 in the USA, and 14,000 in Britain, Germany, Denmark, Belgium, and the Netherlands. This is a considerable health burden. But, as they point out, the impact of the international agreements to phase out CFCs should mitigate the effects on ozone depletion and therefore skin cancer. However, there is always a concern that certain countries that have or have not ratified the agreements will continue to produce or increase production of ozone depleting substances. Further, there is some concern about their production in developing countries, where alternatives are/may not be cost effective (de Gruijl and Leun, 2000). And yet, without these agreements, the numbers of deaths would likely have been many times higher.

Other health impacts of ozone depletion

Although the bulk of the research effort on the health risks of UV radiation has been devoted to skin cancer, some researchers have pointed to possible impacts on eye disorders, particularly damage to the lens (increasing opaqueness or cataracts). There are three main types of cataract, and UVB radiation seems most heavily implicated in cortical cataract, in that it damages proteins in the lens.

A study of over 800 fishermen (known locally as "watermen") in Chesapeake Bay, USA, demonstrated a positive association between UVB exposure and cortical cataract; a doubling in UVB exposure over the lifetime is associated with a 60% increase in the risk of this cataract (Taylor, 1995). This study sought a rigorous measure of exposure, using detailed data on ambient UVB levels and personal exposure histories (including protection from the wearing of hats and sunglasses). The effect of UVB exposure was unaffected by adjustment for other possible risk factors. This is important, since merely demonstrating an increase in cataract with decreasing latitude, without adjusting for poor diet and poor material circumstances, as some studies have sought to do, is unhelpful. Other work (Cruickshanks et al., 1992) has confirmed these findings. Men living in Beaver Dam, Wisconsin, who were exposed to higher levels of UVB radiation, were 36% more likely to have severe cataract, even after adjusting for other risk factors. Research in Japan (Hayashi et al., 1998) examined the association between the prevalence of cataract in 47 districts (prefectures) and estimated UVB levels, adjusting for the proportion of the population that is over 75 years of age. There was a weakly significant relationship for women but not for men, though the study is hampered by poor measurement of exposure.

Armstrong (1994) has considered evidence which suggests that UVB radiation affects the development of immunity to natural infections (see also Lucas et al., 2006). The longer-term consequences of this are hard to predict, but may lead to increases in infectious disease. Armstrong cautions that UVB could impair human response to immunization with live virus vaccines (such as that for measles), and that it can stimulate the replication of the AIDS virus (HIV) in the white blood cells ("T lymphocytes") that protect against viral infection. He also refers to convincing evidence that exposure to UVB radiation can reactivate previous infection with the herpes simplex virus (which causes cold sores).

On balance, there is plenty of evidence to suggest that UVB radiation causes skin cancer and some forms of cataract. It may also be implicated in other cancers, such as non-Hodgkin lymphoma; research by Langford et al. (1998) in Europe indicates that, after adjustment for socio-economic factors, estimated levels of UVB radiation predict variation in mortality from this disease. This work uses multi-level modeling, which we reviewed briefly in Chapter 3. It demonstrates a positive relationship between mortality and UVB in the UK and France, but in Italy the relationship is reversed. Interestingly, recent research in Australia and elsewhere indicates that exposure to ultraviolet radiation may be protective in the cases of some illness. For example, Staples et al. (2003) found an inverse relationship between latitude and ultraviolent radiation for type 1 diabetes in Australia.

From a health-promotion perspective the evidence suggests that exposure to the midday sun in summer months should be minimized and that wearing wide-brimmed hats is a simple but effective measure to counteract the effects of ozone layer depletion. A number of health campaigns along these lines are in place. Recent literature points to the potential detrimental effects of taking these messages too far; for example, lack of ultraviolet radiation results in vitamin D deficiency, which in turn has been linked to ill-health of adults and children (Lucas et al., 2006).

Global Climate Change

It is now well known that the additional burden of carbon dioxide, methane, and other gases, produced as the result of domestic, industrial, and agricultural activity, enhances the natural "greenhouse" effect, whereby such gases absorb and trap energy that is re-emitted by the earth's surface. The UN's Intergovernmental Panel on Climate Change suggests there has been a "discernible" human influence on the climate system (United Nations, 2007). The UN IPCC reports an increase in average global temperature of 0.2°C (UN IPCC, 2007). We have also seen an increase in precipitation; wet areas are getting wetter, dry areas are getting drier, severe weather events are increasing in number, polar ice is melting, species are in danger. Over the next century, we will see an exponential increase in a new class of human beings: environmental refugees (Haines et al., 2006).

While a decade or so ago there was some uncertainty around the potential impacts of climate change, that uncertainty is dissipating in light of the scientific evidence. The UN IPCC suggests that continued emissions of greenhouse gasses at or above current rates would cause further warming and induce many changes in the global climate system during the twenty-first century that will very likely be larger than those observed during the twentieth century. Nor will these impacts be felt uniformly around the globe; it will be those peoples in the coastal areas, and the poor and marginalized who will feel the wrath of climate change so much more than, say, North Americans. "A change in world climate would have wide-ranging, mostly adverse, consequences for human health" (Haines et al., 2000: 730) and will be felt as both *direct* effects – the impact of temperature increases on human physiology – and *indirect* effects, where the health impacts are mediated by the ways in which climate affects sea levels and ecosystem behavior (see Table 9.1 above, p. 219). Indeed, the World Health Organization estimates that, by 2000, the global burden of disease attributable to climate change had already exceeded 150,000 excess deaths annually (Frumkin et al., 2008).

Direct effects: thermal stress

Healthy people can cope well with changes in temperature; for example, leaving a warm house in the winter for a brisk walk on a cold evening will cause few problems for the fit person. But, outside a comfortable temperature range, as the body attempts to adjust to extreme cold or heat there can be risks to cardiovascular and respiratory health. These are likely to affect older people more so than younger, healthy people. We consider here, particularly, the evidence concerning changes in mortality due to temperature increases.

Broadly speaking, there is a U-shaped relationship between mortality and ambient temperature. While mortality is high during extreme cold, it improves as the temperature rises; where average temperatures are below the comfort level, mortality improves by about 1% for every 1°C increase in average temperature. But as temperature rises above the comfort level there is an estimated 1.4% increase in mortality (Martens, 1998). In New York City, for example, deaths from all causes rise sharply if the temperature is higher than 33°C (91°F) (Kalkenstein, 1993). Looking specifically at cardiovascular mortality among those aged over 65 years, a unit increase in temperature in cold conditions reduces mortality by about 4%, while in very warm conditions it increases mortality by 1.6%. For deaths from respiratory causes there is a 3.8% reduction, and 10.4% increase, respectively. It appears from this that respiratory disease, in particular, is relatively sensitive to changes in temperature. These studies (see also Eurowinter Group, 1997) are drawn from cities in temperate climates, located in the developed world.

Martens (1998: 118–25) models the impact on mortality of possible global warming: results suggest that for places such as Singapore, where the climate is warm all year round, mortality will increase, but that for cities in colder climates (London, for example) modest increases in mortality in warmer months will be substantially offset by reductions in winter mortality. Evidence for cardiovascular mortality in selected countries (for those aged over 65 years) is shown in Table 9.4. These scenarios are surrounded by a high degree of uncertainty and depend upon the ability of people to adapt physiologically to temperature change. Nonetheless, Martens' general conclusion is that global warming will probably reduce mortality, especially due to cardiovascular disease, because of warmer winters. Global climate change should therefore reduce excess winter mortality due to bronchitis, influenza, and heart disease.

However, global warming is projected to increase the frequency of heatwaves and decrease the frequency of winter cold spells. Modeling research has shown that heat-related deaths may increase substantially by 2050. For example, a study of ten Canadian cities suggests that, using Montreal as an example, heat related deaths would increase from

Table 9.4 Estimated change in cardiovascular mortality due to thermal stress (population aged over 65 years)

Country	Cold-related mortality change[a]	Warmth-related mortality change[a]
Singapore	0	43
Japan	−79	18
Netherlands	−181	19
UK	−250	10
USA	−184	32
Canada	−235	26
Spain	−129	33
Australia	−98	22

[a] Per 100,000 population
(*Source*: Martens, 1998: 123)

70 per year to 240–1140 in an average summer (Haines et al., 2000). The 2003 heatwave in France (close to 40,000 attributable deaths across Europe) is an example of the impacts of temperature increases even in developed countries (Haines et al., 2006).

Indirect effects

There may, of course, be other, more adverse, effects of global climate change. We should not neglect the interaction between global climate change and air pollution. Higher temperatures in summer months aid the photochemical reactions that produce low-level ozone (see Chapter 7). In addition, research suggests that climate change will be associated with an increase in extreme weather events, and a rise in the incidence of storms and flooding. The flooding of low-lying coastal areas will bring with it direct loss of life, as well as potentially devastating impacts on infrastructure. We consider in this section the more indirect influences of such climate change. There is likely to be a considerable health burden arising from the indirect effects of global climate change (McMichael and Haines, 1997; Haines, et al., 2000), because changes in temperature and rainfall will have major impacts on ecosystems. Ecosystem disturbance will manifest itself in terms of vector-borne disease, in infections leading to food poisoning, and in agricultural production, land use, and other change. We consider these in turn.

Impacts on infectious (especially insect-borne) disease

Models estimating the effects of climate change project substantial increases in the transmission of infectious diseases worldwide – malaria, dengue fever – while others – schistosomiasis for example – will see a decrease in transmissibility due to excessive warming of water as well as some regional drying (Haines et al., 2000). Increased prevalence of cholera is also likely, due to the impact of temperature increases on algal blooms, which potentiate the transmission of cholera. This is already happening in areas of Bangladesh (Haines et al., 2000). Research suggests that climate change will affect both the vectors and the infective agents that transmit infectious diseases such as malaria, dengue fever, and trypanosomiasis (sleeping sickness). Each is examined briefly.

Martens (1998) estimates that about 2.4 billion of the world's population is at risk from malaria, with between 300 and 500 million people suffering currently from the disease. Malaria is caused by the *Plasmodium* parasite (of which *P. vivax* and *P. falciparum* are most common in tropical areas and the latter is especially lethal), but the parasite is transmitted by species of the *Anopheles* mosquito, the saliva of which is injected into the human bloodstream. The ranges of the mosquito species are highly dependent on both temperature and rainfall, with an optimum temperature of about 20–25°C (68–77°F). Above this temperature mosquitoes will not survive. Similarly, while a minimum of 1.5 mm of rainfall a day is required, excess rainfall washes away mosquito larvae. But temperature also controls the survival of the parasite itself. Details of these climatic controls are given in Martens (1998).

Martens runs some global climate circulation models under different scenarios of climate change and finds that large parts of North America, Europe, Australia, and Asia may be at increased risk of malaria transmission; here, the *Anopheles* mosquitoes are already present but the *Plasmodium* parasite cannot currently survive because of low temperatures. Elsewhere, regional studies in Zimbabwe and the Andes of southern America suggest that the impact will be greatest in more upland areas within regions that are already malarial: malaria will spread into higher altitudes and therefore affect highland populations that are currently protected. Maps of the possible global impact, as well as scenarios in Zimbabwe, are shown in McMichael and Haines (1997). Martens' (1998) work suggests that the global population potentially at risk will probably rise from 2.4 billion to well over 3 billion, perhaps generating a further 220–480 million cases of the disease. The major burden will continue to be in tropical Africa, where Tanser and colleagues (2003) use global climate change models to predict a 5–7% increase in the incidence of malaria.

A detailed case study of Rwanda is instructive (Loevinsohn, 1994). Rwanda (see Box 6.1, p. 160) had, in 1991, a population of about seven million, most of whom were living in rural highland areas. Loevinsohn shows that there have been long-term increases in temperature, though not in rainfall, since 1960, and a parallel increase in the recorded incidence of malaria, from about 35 per 1000 people in 1982 to over 150 per 1000 in 1990. Loevinsohn uses data from one health center, serving a population of 38,000, to examine smaller-scale variation. He finds that, between 1984 and 1987, incidence rose most sharply in high altitude areas and among children aged less than two years. These are areas and a population group in which malaria had previously been rare. Monthly incidence between 1983 and 1990 is explained statistically by mean minimum temperature and rainfall, though the best-fitting model (explaining about 80% of the variation in incidence) is one relating incidence in any one month to climate data in the preceding two months. This is because the parasite takes 40–57 days to develop under the prevailing temperature conditions, and also because it takes time for the rainfall to collect in low-lying breeding sites. Loevinsohn attributes the increased incidence solely to these climatic influences; there were no other obvious explanations, such as a reduction in spraying programs to control mosquitoes or in-migration of non-immune people. The upsurge in malaria in Rwanda resulted from increases in temperature and rainfall.

In southern Asia, malaria outbreaks seem to be associated with periods of particularly heavy rainfall (Lindsay and Birley, 1996). Historically, there has been a close correlation

Box 9.1   El Niño

In order to understand the climate phenomenon known as El Niño, we need to know that, normally, off the west coast of South America cold air flows west to form the south-east trade winds, arriving in the west Pacific and then warming, rising, and at high altitudes flowing east to complete a circulation. But this pattern is disturbed every two to seven years. During the Christmas season (hence El Niño – "Christ child") the prevailing trade winds weaken and the warm surface waters that are normally driven west by the trade winds flow east towards South America. Strictly, El Niño refers to the warm ocean current flowing along the coast of Ecuador and Peru during December. The periodic, large-scale disturbance of ocean and air circulation is known as the El Niño Southern Oscillation (ENSO). This brings drought to countries bordering the Indian and Pacific Oceans and lengthy wet periods to Pacific regions. ENSO amplifies climate variability; in other words, periods of rainfall and drought are more intense than in unaffected areas. The opposite of El Niño is La Niña, characterized by unusually cool ocean temperatures in the eastern Pacific.

There is growing evidence that these events may lead to an increase in vector-borne disease. For example, the 1997–98 El Niño event resulted in widespread respiratory illness in Indonesia and Brazil due to haze from uncontrolled burning of tropical forests (Haines et al., 2000). Nicholls (1993) suggests that the incidence of Australian encephalitis is associated with climatic extremes. There are outbreaks of the disease in south-east Australia between January and May, following heavy rainfall and flooding (accompanying a La Niña episode) which promotes the mosquito vector responsible for disease transmission. An unusually wet summer in 1974, again a La Niña event, led to an epidemic of fever in southern Africa, while in the previous year extreme monsoon conditions in India produced an epidemic of Japanese encephalitis in the northern Indian state of Uttar Pradesh. The death toll was 5000, mostly children.

How does ENSO relate to global warming? Essentially, global temperature increases intensify or amplify the pattern of rainfall variability. An enhanced greenhouse effect may make ENSO events more frequent and more intense. As a result, the incidence of vector-borne diseases, where the vectors thrive on wet and warm conditions, is likely to increase.

between deaths from malaria and the rainfall peaks that follow El Niño Southern Oscillation (ENSO) events (see Box 9.1 and Figure 9.4).

Dengue fever is another vector-borne disease, the vector being the mosquito *Aedes aegypti*. It is a viral infection, the viruses (of which there are four serotypes) belonging to the genus *Flavivirus*. Infection varies from a relatively mild influenza-like illness to a severe form of haemorrhagic disease (with bleeding from soft tissues) that is often fatal. The disease is widespread throughout much of Asia and Central and South America, with over 100 million cases reported each year. WHO indicates that the haemorrhagic form affects children in particular and that mortality is about 5%, leading to 24,000 deaths per year. Rapid urbanization, population movement, the resistance of mosquitoes to insecticide, and the inadequate provision of piped water, are all implicated as factors in the increased incidence of dengue.

Temperature affects the distribution of the mosquitoes, the frequency with which they will bite, and also influences the incubation period of the virus, for example at 27°C (81°F)

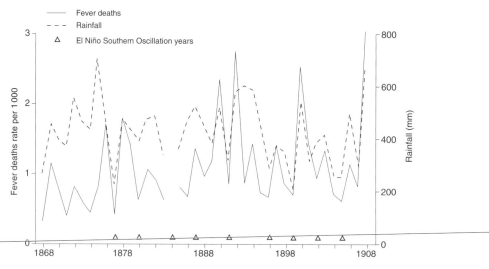

Figure 9.4 Relationship between deaths from malaria and rainfall in the Punjab, 1868–1908

the incubation period is ten days, but this drops to seven days at 34°C (93°F) (Martens, 1998: 43). This shortening of the incubation period leads to an increase in the rate of transmission of the disease. Work by Jetten and Focks (1997) indicates that a warming of the global climate by 2°C would see transmission of dengue in some parts of southern Europe (Spain and Greece) and in the southern states of the USA. At some times of the year, for example late summer, current climates in North America would permit transmission of dengue. But, as with malaria, there is evidence that dengue is spreading into more highland areas in places in which it is already endemic, especially in Central and South America (McMichael, 1998: 23). For example, Mexico City, which is currently free from the disease because of its altitude, is surrounded by lower-altitude, dengue-endemic areas and modest increases in temperature would permit the mosquitoes to survive at higher altitudes. Jetten and Focks (1997) suggest that Melbourne, Australia, is another city that might be affected. Hales et al. (2002) strongly suggest, however, that to truly understand the impacts of climate change on the incidence of dengue fever we must also take into account other factors, such as population change. On the basis of their empirical models, they indicate that the global population at greater than 50% risk for dengue fever will increase from 1.5 billion to 6 billion by 2085, compared to 3.5 billion if climate change did not occur.

Further, Githeko et al. (2000) review the mixed impacts of climate change. For example a short-term increase in temperature and rainfall (as seen in the 1997–98 El Niño) caused *Plasmodium falciparum* malaria epidemics and Rift Valley Fever in Kenya. This is partially due to accelerated parasite development and an explosion of vector populations. However, at the same time, these changes produced a *decrease* in malaria transmission in neighboring Tanzania.

Trypanosomiasis ("sleeping sickness") is carried by the tsetse fly (*Glossina morsitans*); the trypanosoma are parasitic protozoa whose hosts are wild and domestic animals. The disease, which is widespread in much of sub-Saharan Africa, is fatal if left untreated. It tends to affect young and middle-aged adults, since they spend more time in fields and are therefore more likely to come into contact with the fly. Factors that affect the resting sites for adult tsetse flies, such as long-term changes in rainfall patterns, can affect the epidemiology and transmission of the disease (Githeko et al., 2000). As with other vector-borne diseases, authors have sought to relate data on the spatial distribution of the vector to environmental data, and then to assess the possible implications of climate change on its spread. For example, Rogers and Williams (1993) use GIS to model the distribution of the tsetse fly in East Africa. One climate variable, the maximum of the mean monthly temperature, is a highly significant predictor. Assuming mean increases in temperature of 1–3°C, highland areas of Zimbabwe become suitable for *Glossina*, indicating that, as with other vector-borne diseases, sleeping sickness may well spread to previously unaffected regions in tropical areas. There is a very clear role to be played by GIS and remote sensing, if informed by a good understanding of disease ecology, in the prediction of disease risk and in disease surveillance. Interestingly, Rogers and Williams (1993: 89–90) make the point that epidemics of sleeping sickness in Uganda over 100 years ago were quite possibly due to the movements of a colonial army and its porters. As they put it, "catastrophic epidemics are often associated with the arrival of a conquering army closely followed by the parasites to which the army has already adapted."

Rodents are implicated in some newly emerging, and re-emerging infectious diseases, given that prolonged droughts deplete rodent predators (owls, snakes, coyotes) and rains provide new food supplies (Githeko et al., 2000; Haines et al., 2000). For example, there was an outbreak of flu-like illness in the south-western United States in 1993, with symptoms progressing to acute respiratory distress (Epstein, 1995). Investigations suggested that hantavirus was the culprit, a virus that is transmitted in the saliva and excreta of rodents (Duchin et al., 1994). But what caused the outbreak? Epstein suggests that heavy rainfall promoted the growth of vegetation and insects on which the rodents prey, causing a ten-fold increase in the rodent population over one year. He implies that climate change (El Niño events) may lie behind the outbreaks of rare infections.

Other health effects of climate change

The Inter-governmental Panel on Climate Change has forecast sea-level rises of up to 40 cm (16 inches), perhaps more, by the turn of this century. This will nearly double the number of people around the world who live in areas prone to flooding, to nearly 100 million (Haines et al., 2006). The consequence of this will be population displacement, some of the health impacts of which we reviewed in Chapter 6, as well as outbreaks of disease such as cholera (Chapter 8). In August 2000, we witnessed precisely these outcomes in coastal areas of India. As with most impacts reviewed here, the burden falls on those least able to cope materially; environmental "inequity" will therefore be an even greater evil in the years ahead.

Climate change will also be likely to impact on agriculture. In particular, there will be regional variations in temperature change, in precipitation, and therefore in the ability of crops to germinate and grow. In general, experts predict a net negative impact on food production, although production in some temperate zones, such as the Canadian prairies, may actually increase (Haines et al., 2006). Reduced agricultural yields will lead to increased malnutrition and hunger; we have already noted global inequalities in health outcomes and, more specifically, the impact on health in later life of low birth weight and poor nutritional status in early life. Estimates of the additional number of hungry people by the year 2060 range from 40–300 million (Monbiot, 2007).

Health Effects of Other Global Environmental Change

While the bulk of research evidence focuses on the health consequences of climate change, it is clear that the next 100 years will see other forms of global and environmental change that will also be significant. We can consider here processes of land use change, including deforestation and reforestation, as examples of other regional and global change processes (see Flannery, 2005; Haines et al., 2006; and Monbiot 2007 for full reviews).

To some extent the diseases that are the focus of considerable public and epidemiological interest are those regarded as "emerging infectious diseases" (see Box 9.2). The vector-borne Lyme disease may increase in incidence, either because of temperature increases or land use change. Lyme disease is caused by a spirochaete, *Borrelia burgdorferi*, which is transmitted by the tick vector *Ixodes ricinus* (Mawby and Lovett, 1998). This tick is a parasite on deer and mice. The disease was first recognized in the mid-1970s around the village of Old Lyme, Connecticut. Transmission of the spirochaete is influenced by temperature, but also by land use change, since where abandoned farmland reverts to woodland the latter provides an ideal breeding ground for the deer and mice hosts. As Mayer (2000) argues, deforestation of land on the outskirts of commuter suburbs in New England has created new habitats suitable for deer population. These ecological conditions are highly suitable for the transmission of the disease but, as Mayer suggests, the increasing incidence is very much a function of economic pressures to develop land. For example, Haines and colleagues (2006) report that there have been latitudinal shifts in ticks that carry encephalitis in northern Europe, although it is important to remember that there could be alternative explanations for such events (e.g. changes in confounding factors such as land use, as Mayer suggests).

During the 1990s, about 12,500 cases of Lyme disease were reported each year to the US Centers for Disease Control and Prevention (CDC), while between 2003 and 2005, over 64,000 cases were reported (CDC, 2007). Most cases occur, not surprisingly, in the summer months when people are more likely to spend time out of doors. Further, the majority of cases come from the north-east, but states such as Wisconsin and California also report large numbers of cases (see Kitron and Kazmierczak, 1997, for a study of the spatial distribution of the disease in Wisconsin using GIS and remote sensing; see also Cromley et al., 1998 for a similar study based in Connecticut). Those working outdoors, and more specifically in woodland and brush areas, as well as those participating in recreational activities such as camping and hunting, are at particular risk, though the burden of disease falls mostly on children and adults of middle age (peak incidences are in age groups 5–9 years, and 50–54 years). There is a strong geographical (west-east) gradient in risk, with fewer than one case per 100,000 in the British Isles, yet 69 per 100,000 in southern Sweden and 130 per 100,000 reported in Austria (1995 data).

Box 9.2 Emerging infectious diseases

Until quite recently it was assumed, in the developed world, that infectious disease was largely a thing of the past and that the attention of public health professionals needed to be focused almost exclusively on chronic disease such as cancer and heart disease. The emergence of HIV and AIDS in the 1980s served as a warning that this conclusion was premature. More recently still, the emergence of various viral diseases (Lyme disease, West Nile virus, SARS) has meant that the study of "new" infections has become a considerable focus of research activity. Several examples are mentioned in this chapter, and these and others are considered by Mayer (2000).

One recent example, which gained tremendous publicity around the world was the outbreak of the SARS virus in Canada in the winter of 2003. Severe Acute Respiratory Syndrome, or SARS, was introduced to Canada by a visitor returning from Hong Kong. Transmission to a family member, later admitted to hospital, resulted in a large outbreak and an acute public health scare across the entire country. As of July 10, 2003, a total of 438 cases were reported in Canada, 86% of which were in the province of Ontario. The syndrome is characterized by fever, headache, myalgia, cough and shortness of breath. It leads to pneumonia and occasionally to acute respiratory distress syndrome and death (Varia et al., 2003). A total of 128 cases were transmitted within hospitals, putting both the community as well as health service workers at risk. Seventeen deaths were reported and over one-third of cases were experienced by hospital personnel. With public health measures reminiscent of the influenza outbreak of the nineteenth century (e.g. attempting to quarantine entire schools of teenagers and confine them to their homes), SARS was eventually brought under control, but the entire public health community in Canada was traumatized by this new emerging infectious disease. The outcome was to establish a Royal Commission into the response by Canadian public health and the establishment of the newly minted Public Health Agency of Canada to deal with what might come next – perhaps drug-resistant TB or avian influenza, leading to a major pandemic in human populations. Researchers have recently mapped global hot spots for emerging infectious diseases (Jones et al., 2008). While these researchers insist that the major source will come from wildlife, it is the treatment of that wildlife by humans (deforestation, concentrating wildlife on smaller and smaller parcels of land) and the interaction of that wildlife with humans that pose the greatest risks. Our immune systems are not able to deal with exposures to these new emerging infectious diseases and the impacts are predicted by these researchers to be "extraordinarily lethal."

The perceived impact of many of these diseases is greater than their real contribution to the burden of disease, especially in the developed world. Chronic disease will indeed continue to be the main problem here. Nonetheless, vigilance is required over new viral infections, and especially resurgent infections such as tuberculosis. In the developing world, worries over rare infections such as Ebola pale into insignificance when we consider the unimaginable burden of severe diarrhoeal disease, itself largely a function of poor infrastructure.

The emergence of West Nile virus in North America in the early 2000s is yet another example of the potential impacts of global climate change on health and disease patterns. West Nile virus is by no means a *new* illness and has been endemic in many parts of the world (e.g. Egypt) for centuries. It emerged in North America in 1999, causing an outbreak of meningoencephalitis in the New York City area which resulted in seven deaths

(Elliott et al., 2008). Of the total 123 non-fatal cases detected in the USA in 1999–2001, the median age of patients was 65 years, with a range from 5–90 years; 60% were over 60 years of age and 63% were female. As of 2005, there were 2470 human cases of the virus reported to the Centers for Disease Control; in Canada, 1335 human cases were reported in 2003. The virus is transmitted to humans by infected mosquitoes. *Culex pipiens* is an important vector that breeds in underground standing water found in city drains, catch basins, and in any other pools of standing stagnant water. During a long, hot summer, these water sources become even richer in the rotting organic material that *Culex* needs for survival. The winter of 2002–2003 in southwestern Ontario/northeast USA was a particularly warm one, allowing the vector to survive through the winter and thrive in the long hot summer that followed. The result was a substantial number of dead birds infected by mosquitoes (recall the discussion of the study by Mostashari et al., 2003, in Chapter 3) and illness – sometimes quite serious – in humans.

An outbreak occurred in Oakville, Ontario in the summer of 2002. It was Canada's first experience with West Nile virus. Sixty cases occurred in a population of about 400,000, with onset during the months of August and September. Most cases occurred within two postal code forward citation areas (Loeb et al., 2005). A team of researchers undertook a seroprevalence study in the area to determine the proportion of the population who were bitten by an infected mosquito but did not develop symptoms. Results indicated that the seroprevalence was about 3%, which was similar to other studies done in North America around that time; further, it was found that respondents who practiced two or more protective behaviors had an approximately 50% reduction in the risk of infection (Elliott et al., 2008).

The clearance of forest and woodland in South America, and the consequent development of extensive monoculture has led to the emergence of new haemorrhagic fever viruses (arenaviruses). One example, from Venezuela, is Guaranito virus, confined to rural populations engaged in cattle ranching. Like hantavirus, this has rodents as its host. Mortality rates are high. More broadly, the cutting of roads through rainforests encourages the development of water-collecting ponds at the roadside; these provide good breeding grounds for mosquitoes and the subsequent emergence of malaria. In Thailand, loss of forest has been associated with a rise in *Anopheles* mosquitoes. Using GIS, Gomes et al. (1998) have shown that in some districts the proportion of land area devoted to forest has fallen from 37% to 25% in ten years; forest has been replaced by commercial crops in plantations that provide suitable niches for mosquitoes. The new mosquito colonies "act as a reservoir of intense, unchecked malaria transmission, from which the workers who come to work during harvest periods can carry the parasite to their homes across Thailand" (Gomes et al., 1998: 94).

Concluding Remarks

This chapter has ended with a discussion of emerging infections, a discussion that arises out of a wider analysis of the health impacts of global environmental change. But while the last three chapters all dealt with some aspect of health and the physical environment, and earlier chapters considered the social, it is appropriate that this concluding section points to the links between the environmental and the social.

These links are summed up admirably by Epstein (1995: 170): "social, political and economic factors ... are clearly integral to the condition and management of the environment, for therein lie the driving forces of global change ... and *the inequitable distribution of exposures, vulnerabilities, and access to treatment*" (our italics). Mayer (2000) takes a broadly similar view, arguing that a "political ecology" framework (synthesizing traditional disease ecology with a political economic, or structuralist perspective) is required to understand the unintended health consequences of human-induced environmental change.

Getting to grips with an understanding of many of these emerging or resurgent infections requires a new form of geographical epidemiology, one that draws on a historical perspective but also on many of the social sciences – anthropology, sociology, economics, and politics. For Farmer (1999: 42–3), as with Urry's (2007) manifesto on "mobilities," we need to recognize that the appropriate units of analysis are not necessarily geographical ones such as nation-states; rather, they are flows of people, money, and viruses. We need a critical and a political epidemiology that examines how global financial and political institutions, institutionalized racism, and (neo)colonial history are implicated in disease aetiology. For writers such as Farmer, many diseases are not newly "emerging" or "re-emerging;" diseases such as tuberculosis may have declined in incidence within many developed countries, but for the poor in such countries, and many in the developing world, the notion of resurgence is a myth. In advancing the research agenda on links between global environmental change and health, it is essential that the "social" be retained. Geographies of health, at this global scale, must continue to draw on both environmental and social science.

Further Reading

See the 2007 report of the UN Intergovernmental Panel on Climate Change for a thorough review of the causes and consequences of global climate change. Three key books consider the social, political and economic issues relating to climate change: Flannery (2005); Gore (2006); and Monbiot (2007). Gore's book, in particular, focuses on the public/human response to climate change.

Basu and Samet (2002) have reviewed the relation between elevated ambient temperature and mortality. Githeko et al. (2000) have undertaken a useful regional analysis of climate change and vector-borne diseases.

A useful introduction to the health effects of global climate change is provided in Haines et al. (2006). See also the edited volume by Martens and McMichael (2002). See Mayer (2000) for a useful overview of emerging infectious diseases. The text by Cromley and McLafferty (2002) on GIS and Public Health is a useful primer on the applications of GIS and spatial analysis to understanding re-emerging infectious diseases. Finally, as indicated earlier, Farmer (1999) has written an important book that, fueled by an anthropological as well as epidemiological imagination, merits a reading by anyone concerned with health, space, and place.

Chapter 10

Conclusions: Emerging Themes in Geographies of Health

In this concluding chapter we look first at some contemporary problems at a global level, before indicating what are (or, in our view, should be) some emerging themes at the national scale, then considering issues that are manifested at a local (neighborhood or domestic) scale. Although something of an over-simplification, it is fair to suggest that the volume of current and recent geographical research on health is inversely proportional to scale. In other words, relatively little work has yet been undertaken (at least, by geographers) that focuses on the international dimension of health and health care. Plenty is undertaken at a national level, albeit with a substantial focus on countries in the "north" rather than the "south." But in recent years, and with the "cultural turn" in human geography, most attention has been focused on local issues, whether health in small neighborhoods, in formal care settings, or in the home.

Methodologically, geographic research continues to draw on a wide range of research methods, both quantitative and qualitative, but with an increasing emphasis on the latter. This mirrors the widespread turn away from statistical methods to qualitative approaches in the discipline as a whole. But we would not wish to overstate this, and there is recent evidence of a re-engagement with quantitative and spatial analytic approaches. Indeed, in some cases high-quality research *demands* the mixing of both quantitative and qualitative techniques. Our own view is that methodological pluralism – and training – is important, if for no other reason than to understand the continuing engagement with multi-level models at one end of the spectrum and (for example) various forms of qualitative data analysis at another. But methodology cannot be reduced to technique; it encompasses the "approach" to a research problem or question, an approach that may, as we suggest, draw on the mixing of methods. To take just one recent example, Hanchette (2008) seeks to understand the geographic distribution of childhood lead poisoning in eastern North Carolina, first describing the patterns using spatial data analysis and then using historical and archival methods to dig deeper. Her overall approach is a political ecology one that, having identified a spatial association between areas of lead poisoning and areas with high proportions of African-Americans, goes on to explain this in terms of the history of tobacco production and the movement of African-Americans into dilapidated housing built many years earlier by middle-income white families who could afford lead-based paint. The children have suffered the consequences of exposure to the toxic lead.

The Macro-scale: Health and the "Global"

"Globalization" has become something of a watchword over the past 20 years, although its conceptual significance now seems to be waning. Briefly, it can be defined as a "process of greater integration within the world economy through movements of goods and services, capital, technology and (to a lesser extent) labor, which lead increasingly to economic decisions being influenced by global conditions" (Schreker and Labonte, 2007: 284). While there have been many gains resulting from these processes, including economic growth and new employment opportunities, some writers claim that these gains have been unequally distributed, both among and within countries. One area of sharp debate has been the interventions made by international agencies such as the World Bank, the International Monetary Fund, and donor agencies, to influence or direct policy in developing countries. For example, the World Bank introduced structural adjustment programs (SAPs) during the 1980s in order to shape such countries' macro-economic policies; implementation of these SAPs by the governments of such countries has been a condition for loans. Such policies include cuts in public sector health spending, increasing privatization (Turshen, 1999), imposition of fees on those requiring health care, and other forms of cost recovery. In addition, the global pharmaceutical companies prefer to invest in drugs that have significant commercial returns. As Schreker and Labonte (2007: 298) point out, of 1400 new drugs marketed between 1975 and 1999, only 16 were to treat tuberculosis and tropical disease, both of which account for a significant burden of disease.

But not everyone agrees that "globalisation" – at least when this is constructed as increasing openness to international trade – is bad for population health. Kunitz (2007) has reviewed considerable empirical evidence and concludes that "while openness has not resulted in the benefits promised by the optimists, neither has it had the deleterious consequences for the health of many populations that the pessimists predict" (Kunitz, 2007: 174). Further, in a wide-ranging review, Breman and Shelton (2007) conclude that structural adjustment programs can have both positive as well as negative impacts on health outcomes.

The literature on global social, economic and environmental processes and their links to human health is developing rapidly. We see little evidence that geographers are engaging with these debates, but consider that there is real scope for them to do so. Quite simply, if we are to understand the nature of the obesity "epidemic," the impact of smoking in developing countries, and the rise of drug-related health problems, we need to set these issues in a global context (see several chapters in Kawachi and Wamala, 2007). And while the "global" and health is about more than the problems of environmental change considered in Chapter 9, we simply have to do more to understand the health consequences of climate change and the impacts this may have.

We would make the same observation of engagement concerning human rights and the health impacts of major international conflicts. On the first, there is surely scope for those interested in global geopolitics, anthropology, and law to come together with health geographers to research such impacts and to argue for their amelioration. Such impacts are distributed unequally. In the developed world we take for granted human rights (of which the right to survive is the most taken for granted), but in much of the world these rights are violated and are symptoms of deep-seated "pathologies of power" (Farmer, 2005).

On the second, we looked briefly in Chapter 6 at the health of refugees, whose movements are often triggered by regional wars that spread across international borders. Such wars impact as much, if not more, on vulnerable civilians, especially those who do not have the resources to escape the conflict. Further, as we have seen all too often in Iraq and Afghanistan, supposedly "intelligent" weapons can hit civilian areas as much as the intended military targets (Allotey and Zwi, 2007: 161). Yet there is much important research to be done on the health and health care of international migrants who are not *forced* to relocate, but do so in order to improve their life chances. In the UK, much is made of the possible social and health impacts of "asylum seekers" and those migrating from new members of the European Union, particularly their demands on health services. Sadly, the media often portray such people as "problems." Further, and linked to discussions of environmental change in the previous chapter, we need to understand the possible human impacts of large-scale population movements that may arise as climates continue to change.

The Meso-scale

The country-level examples on which we have drawn in this book show clear geographic biases, reflecting primarily the relative preponderance of research. Consequently, the UK, the USA, Canada, and other countries in the developed world, have figured prominently in terms of the examples on which we have drawn. What is very striking is the dearth of reported research in large parts of Asia, including, in particular, China, Japan, and countries in South and South-east Asia. In part, this reflects the lack of translated work, but also the lack of access to data sources that we take for granted elsewhere. For example, while data (and, consequently, atlases) on mortality are common in the developed world, none has been published recently for China, and such data as there are concern large geographic areas (see, for example, Dummer and Cook, 2007) or small surveys.

As a rather simple, but nonetheless revealing, picture of the attention paid by geographers and others to certain countries of the world, we present data on the proportions of research papers published in *Health & Place* (the premier journal for health geography) since 2001 (see Table 10.1). Over the seven-and-a-half-year period (we only include data for half of 2008) we note that 84% of the published research has focused on the developed world, with almost two-thirds focusing on the UK, USA or Canada. It is interesting to note a simple temporal trend; research that has an empirical focus on North America and Australia and New Zealand has grown over the last 3–4 years, while that on Africa and Asia has declined. We would not wish to make too much of this – there are many other journals in which to publish! – but it does suggest where the balance of research effort lies.

We have noted elsewhere the "cultural turn" in health geography, with well-known health geographers such as Robin Kearns and Wil Gesler at the forefront of these developments. But we would wish to see more cultural research focusing on health care practices that most of us would find abhorrent. For example, what geographical research is being undertaken on female genital mutilation, whether in parts of sub-Saharan Africa or among certain Muslim communities in the western world? What attention is being paid by geographers to the sex-selection in certain parts of India and China in which illegal terminations (euphemistically called "miscarriages") lead to perhaps only 300–400 girls

Table 10.1   Papers published in *Health & Place* (2001–2008), by region/country (percentages)

	UK	US	Canada	Australia & New Zealand	Europe	Asia	Africa	S. America
2001–8	29	19	13	13	10	7	7	2
2001–4	39	14	9	9	7	11	11	1
2004–8	22	22	16	15	13	5	5	2

born for every 1000 boys, when about 950 is the worldwide figure. Quite properly, geographers have concerned themselves with health inequalities in the developed world, especially at a neighborhood scale, but we look for more research on major gender-based inequalities such as these.

The volume of literature on the geography of health inequalities in the developed world has expanded dramatically since the first edition of this book was published. Such research continues to address many of the themes considered in Chapter 4, whether on the extent to which such inequalities have widened, the relative role of context or composition, or the impact of social capital (see the various contributions in Boyle et al., 2004). Methodological pluralism characterizes all this work, whether using multi-level models (Sellstrom et al., 2008) or qualitative methods (Popay et al., 2003).

The Micro-scale

A rich body of work continues to emerge that looks at the relationship between "place" and health, and more specifically place and well-being. To a considerable extent this recent body of literature develops Gesler's original concept of therapeutic landscape (see page 9 above), which, as Williams' recent edited collection (2007) shows, has become a central feature of health geography. The concept has been extended, from the original concern with iconic healing places (such as Lourdes), to consider everyday environments (such as waiting rooms in health centres and domestic settings), zoological and community gardens, woodlands, as well as the interpretation of literary works.

Out of a concern with therapeutic landscapes has evolved a more generic geography of care – as with so much in current health geography, largely western in focus – that deals with care settings and the relative roles of public and private providers, as well as the voluntary sector. Research has focused on disabled people and older adults in particular, such that in the latter case we can begin to speak of a "geographical gerontology" (Andrews et al., 2007; Cutchin, 2007). Care settings include both the home and assisted or group residential care. Understanding how older adults cope (or do not cope) with the transition from home to a new space that has to be shared with others, has proved to be an important area of recent and contemporary research (Milligan, 2006). Closely related to this has been a concern with "emotional geographies" (Davidson et al., 2005), the intersection between place, people and emotion. People have both feelings *for* places as well as feelings engendered *by* them (Milligan, 2007). Davidson's work has focused particularly on those with phobias such as agoraphobia, while others look at specific settings such as hospices (settings for palliative care) and how – as fit and "proper" places to die – the particular place and the feelings it evokes are intertwined. In addressing these issues, whether from social interactionist, or feminist, or other perspectives, geographers should seek both to reflect deeply on the ethics of their research and the extent to which it "enables" marginalized groups (Valentine, 2003).

In conclusion, there seem to us to be an almost unlimited range of problems and issues for health geographers to tackle, at a range of geographic scales, and using a range of methods. As the disciplinary barriers in the human and natural sciences continue to dissolve, there is further scope for those trained in particular disciplines to share their skills and perspectives. A good recent example is Cutchin's (2007) attempt to bring together geography and social epidemiology in order to understand exposures to a petrochemical complex in Texas. On a larger scale, one of us (Susan Elliott) is involved in a Canadian National Centre of Excellence (NCE), AllerGen (www.allergen-nce.ca/), which involves over 60 scientists (clinical, natural, and social) from Canada and elsewhere, working to understand the broad range of determinants (genetic, environmental, social, cultural, and economic) of the emerging epidemic of allergies and asthma we are seeing in the developed world. In sum, we welcome any such attempts to pool expertise and this surely has to characterise future health geographies.

References

Aase, A. and Bentham, G. (1994) The geography of malignant melanoma in the Nordic countries: the implications of stratospheric ozone depletion, *Geografiska Annaler*, 76B, 129–39.

Aase, A. and Bentham, G. (1996) Gender, geography and socioeconomic status in the diffusion of malignant melanoma, *Social Science & Medicine*, 42, 1621–37.

Abdel-Wahab, M.F., Strickland, G.T., El-Sahly, A., El-Kady, N., Zakaria, S. and Ahmed, L. (1979) Changing pattern of schistosomiasis in Egypt 1935–1979, the *Lancet*, i, 242–4.

Adamson, J.A., Ebrahim, S. and Hunt, K. (2006) The psychosocial versus material hypothesis to explain observed inequality in disability among older adults: data from the west of Scotland Twenty-07 study, *Journal of Epidemiology and Community Health*, 60, 974–80.

Aggleton, P. (1990) *Health*, Routledge, London.

Akhtar, R. and Izhar, N. (1994) Spatial inequalities and historical evolution in health provision, in Phillips, D.R. and Verhasselt, Y. (eds.) *Health and Development*, Routledge, London.

Ali, S.H. (2004) A socio-ecological autopsy of the E. coli O157:H7 outbreak in Walkerton, Ontario, Canada, *Social Science & Medicine*, 58, 2601–12.

Allotey, P. and Zwi, A. (2007) Population movements, in Kawachi. I. and Wamala, S. (eds.) (2007) *Globalization and Health*, Oxford University Press, Oxford.

Alterman, T., Shekelle, R.B., Vernon, S.W. and Burau, K.D. (1994) Decision latitude, psychologic demand, job strain, and coronary heart disease in the western electric study, *American Journal of Epidemiology*, 139, 620–7.

Altman, D. (1991) *Practical Statistics for Medical Research*, Chapman and Hall, London.

Altmann, P., Cunningham, J., Dhanesha, U., Ballard, M., Thompson, J. and Marsh, F. (1999) Disturbance of cerebral function in people exposed to drinking water contaminated with aluminium sulphate: retrospective study of the Camelford water incident, *British Medical Journal*, 319, 807–11.

Anderson, H.R., de Leon, A.P., Bland, J.M., Bower, J.S. and Strachan, D.P. (1996) Air pollution and daily mortality in London, 1987–92, *British Medical Journal*, 312, 665–9.

Anderson, H.R., Gupta, R., Strachan, D.P. and Limb, E.S. (2007) Fifty years of asthma: UK trends from 1995 to 2004, *Thorax*, 62, 85–90.

Andrews, G.J., Cutchin, M., McCracken, K., Phillips, D.R. and Wiles, J. (2007) "Geographical gerontology": the constitution of a discipline, *Social Science & Medicine*, 65, 151–68.

Angus, J., Kontos, P., Dyck, I., McKeever, P. and Poland, B. (2005) The personal significance of home: habitus and the experience of receiving long-term home care, *Sociology of Health & Illness*, 27, 161–87.

Arcury, T.A., Gesler, W.M., Preisser, J.S., Sherman, J., Spencer, J. and Perin, J. (2005) Special populations, special services: the effects of geography and spatial behavior on health care utilization among the residents of a rural region, *Health Services Research*, 40, 135–56.

Armstrong, B.K. (1994) Stratospheric ozone and health, *International Journal of Epidemiology*, 23, 873–85.

Arnold, D., ed. (1988) *Imperial Medicine and Indigenous Societies*, Manchester University Press, Manchester.

Ashton, J., ed. (1992) *Healthy Cities*, Open University Press, Milton Keynes.

Bailey, T.C. and Gatrell, A.C. (1995) *Interactive Spatial Data Analysis*, Addison Wesley Longman, Harlow.

Banatavala, N., Roger, A.J., Denny, A. and Howarth, J.P. (1998) Mortality and morbidity among Rwandan refugees repatriated from Zaire, November, 1996, *Pre-hospital Disaster Medicine*, 13, 17–21.

Barker, D.J.P., ed. (1992) *Fetal and Infant Origins of Adult Disease*, BMJ Publishing Group, London.

Barker, D.J.P. (1994) *Mothers, Babies, and Disease in Later Life*, BMJ Publishing Group, London.

Barker, D.J.P. and Bagby, S.P. (2005) Developmental antecedents of cardiovascular disease: a historical perspective, *Journal of the American Society of Nephrology*, 16, 2537–44.

Barnett, R., Pearce, J. and Moon, G. (2005) Does social inequality matter? Changing ethnic socioeconomic disparities and Maori smoking in New Zealand, 1981–1996, *Social Science & Medicine*, 60, 1515–26.

Bartley, M., Blane, D. and Davey Smith, G., eds. (1998) *The Sociology of Health Inequalities*, Blackwell, Oxford.

Basu, R. and Samet, J.M. (2002) Relation between elevated ambient temperature and mortality: a review of the epidemiological evidence, *Epidemiologic Reviews*, 24, 190–202.

Baverstock, K. and Williams, D. (2006) The Chernobyl accident 20 years on: an assessment of the health consequences and the international response, *Environmental Health Perspectives*, 114, 1312–7.

Baxter, J. and Eyles, J. (1997) Evaluating qualitative research in social geography: establishing "rigour" in interview analysis, *Transactions of the Institute of British Geographers*, 22, 505–25.

Bell, J.F., Zimmerman, F.J., Almgren, G.R., Mayer, J.D. and Huebner, C.E. (2006) Birth outcomes among urban African-American women: a multilevel analysis of the role of racial residential segregation, *Social Science & Medicine*, 63, 3030–45.

Benach, J. and Yasui, Y. (1999) Geographical patterns of excess mortality in Spain explained by two indices of deprivation, *Journal of Epidemiology and Community Health*, 53, 423–31.

Bennett, S., Gilson, L. and Mills, A. (2007) *Health, Economic Development and Household Poverty: from Understanding to Action*, Routledge, London.

Ben-Shlomo, Y. and Chaturvedi, N. (1995) Assessing equity in access to health care provision in the UK: does where you live affect your chances of getting a coronary artery bypass graft? *Journal of Epidemiology and Community Health*, 49, 200–4.

Ben-Shlomo, Y. and Davey Smith, G. (1991) Deprivation in infancy or adult life: which is more important for mortality risk? the *Lancet*, 337, 530–4.

Ben-Shlomo, Y., White, I. and Marmot, M. (1996) Does the variation in the socioeconomic characteristics of an area affect mortality? *British Medical Journal*, 312, 1013–4.

Bentham, G. (1986) Proximity to hospital and mortality from motor vehicle traffic accidents, *Social Science & Medicine*, 23, 1021–6.

Bentham, G. (1988) Migration and morbidity: implications for geographic studies of disease, *Social Science & Medicine*, 26, 49–54.

Bentham, G. (1993) Depletion of the ozone layer: consequences for non-infectious human disease, *Parasitology*, 106, 39–46.

Bentham, G. (1994) Global environmental change and health, in Phillips, D.R. and Verhasselt, Y. (eds.) *Health and Development*, Routledge, London.

Bentham, G., Hinton, J., Haynes, R., Lovett, A. and Bestwick, C. (1995) Factors affecting non-response to cervical cytology screening in Norfolk, England, *Social Science & Medicine*, 40, 131–5.

Benzeval, M. and Judge, K. (1996) Access to healthcare in England: continuing inequalities in the distribution of general practitioners, *Journal of Public Health Medicine*, 18, 33–40.

Benzeval, M., Judge, K. and Whitehead, M. (1995) The role of the NHS, in Benzeval, M., Judge, K. and Whitehead, M. (eds.) *Tackling Inequalities in Health: An Agenda for Action*, The King's Fund, London.

Beral, V., Roman, E. and Bobrow, M., eds. (1993) *Childhood Cancer and Nuclear Installations*, BMJ Publishing, London.

Berland, G.K., Elliott, M.N., Morales, L.S., et al. (2001) Health information on the Internet: accessibility, quality, and readability in English and Spanish, *Journal of the American Medical Association*, 285, 2612–21.

Bhandari, N.R., Syal, K., Kambo, I., et al. (1990) Pregnancy outcome in women exposed to toxic gas at Bhopal, *Indian Journal of Medical Research*, 92, 28–33.

Bhopal, R.S., Phillimore, P., Moffatt, S. and Foy, C. (1994) Is living near a coking works harmful to health? A study of industrial air pollution, *Journal of Epidemiology and Community Health*, 48, 237–47.

Bickerstaff, K. and Walker, G. (2003) The place(s) of matter: matter out of place – public understandings of air pollution, *Progress in Human Geography*, 27, 45–67.

Blakely, T. (2002) Socio-economic position is more than just NZDep, *New Zealand Medical Journal*, 115, 109–11.

Boardman, J.D. (2004) Stress and physical health: the role of neighborhoods as mediating and moderating mechanisms, *Social Science & Medicine*, 58, 2473–83.

Bobadilla, J.-L., Cowley, P., Musgrove, P. and Saxenian, H. (1994) Design, content and financing of an essential national package of health services, *Bulletin of the World Health Organization*, 72, 653–62.

Bogard, W. (1989) *The Bhopal Tragedy: Language, Logic, and Politics in the Production of a Hazard*, Westview Press, Boulder, Colorado.

Bollini, P. and Siem, H. (1995) No real progress towards equity: health of migrants and ethnic minorities on the eve of the year 2000, *Social Science & Medicine*, 41, 819–28.

Boyle, P.J., Curtis, S., Graham, E. and Moore, E., eds. (2004) *The Geography of Health Inequalities in the Developed World*, Ashgate Press, Aldershot.

Boyle, P., Gatrell, A.C. and Duke-Williams, O. (1999) The effect on morbidity of variability in deprivation and population stability in England and Wales: an investigation at small-area level, *Social Science & Medicine*, 49, 791–9.

Boyle, P., Halfacree, K. and Robinson, V. (1998) *Exploring Contemporary Migration*, Addison Wesley Longman, Harlow.

Boyle, P.J., Kudlac, H. and Williams, A.J. (1996) Geographical variation in the referral of patients with chronic end stage renal failure for renal replacement therapy, *Quarterly Journal of Medicine*, 89, 151–7.

Brauer, M., Gehring, U., Brunekreef, B., et al. (2006) Traffic-related air pollution and otitis media, *Environmental Health Perspectives*, 114, 1414–8.

Breman, A. and Shelton, C. (2007) Structural adjustment programs and health, in Kawachi, I. and Wamala, S. (eds.) *Globalization and Health*, Oxford University Press, Oxford.

Brothwell, D. (1993) On biological exchanges between the Two Worlds, in Bray, W. (ed.) *The Meeting of the Two Worlds: Europe and the Americas 1492–1650*, Oxford University Press, Oxford.

Broughton, E. (2005) The Bhopal disaster and its aftermath: a review, *Environmental Health: a Global Access Science Source*, 4, (accessed as www.ehjournal.net/content/4/1/6).

Brown, P. (1992) Popular epidemiology and toxic waste contamination: lay and professional ways of knowing, *Journal of Health and Social Behaviour*, 33, 267–81.

Brunekreef, B. and Holgate, S.T. (2002) Air pollution and health, the *Lancet*, 360, 1233–42.

Buchdahl, R., Parker, A., Stebbings, T. and Babiker, A. (1996) Association between air pollution and acute childhood wheezy episodes: prospective epidemiological study, *British Medical Journal*, 312, 661–5.

Bullard, R. and Johnson, G. (2000) Environmentalism and public policy: environmental justice, grassroots activism and its impact on public policy decision making, *Journal of Social Issues* 56, 555–578.

Bulliard, J.-L., Cox, B. and Elwood, J.M. (1994) Latitude gradients in melanoma incidence and mortality in the non-Maori population of New Zealand, *Cancer Causes and Control*, 5, 234–40.

Burra, T.A., Elliott, S.J., Eyles, J.D., Kanaroglou, P.S., Wainman, B.C. and Muggah, H. (2006) Effects of residential exposure to steel mills and coking works on birth weight and preterm births among residents of Sydney, Nova Scotia, *The Canadian Geographer*, 50, 242–55.

Butler, R. and Parr, H., eds. (1999) *Mind and Body Spaces: Geographies of Illness, Impairment and Disability*, Routledge, London.

Cairney, J. and Ostbye, T. (1999) Time since immigration and excess body weight, *Canadian Journal of Public Health*, 90, 120–4.

Caballero, B. (2007) The global epidemic of obesity: an overview. *Epidemiologic Reviews*, 29, 1–5.

Carballo, M. and Siem, H. (1996) Migration, migration policy and AIDS, in Haour-Knipe, M. and Rector, R. (eds.) *Crossing Borders: Migration, Ethnicity and AIDS*, Taylor and Francis, London.

Carballo, M., Divino, J.J. and Zeric, D. (1998) Migration and health in the European Union, *Tropical Medicine and International Health*, 3, 936–44.

Carlson, P. (2004) The European health divide: a matter of financial or social capital? *Social Science & Medicine*, 59, 1985–92.

Carrasquillo, O., Himmelstein, D.U., Woolhandler, S. and Bor, D.H. (1999) Trends in health insurance coverage, 1989–1997, *International Journal of Health Services*, 29, 467–83.

Carr-Hill, R., Place, M. and Posnett, J. (1997) Access and utilisation of health care services, in Ferguson, B., Sheldon, T. and Posnett, J. (eds.) *Concentration and Choice in Healthcare*, Financial Times Healthcare, London.

Carr-Hill, R., Rice, N. and Roland, M. (1996) Socioeconomic determinants of rates of consultation in general practice based on fourth national morbidity survey of general practices, *British Medical Journal*, 312, 1008–13.

Centers for Disease Control (2007) Lyme Disease – United States, 2003 – 2005, *MMWR Weekly*, 56, 573–576.

Chapple, A. and Gatrell, A.C. (1998) Variations in use of cardiac services in England: perceptions of general practitioners, general physicians and cardiologists, *Journal of Health Services Research and Policy*, 3, 153–8.

Chen, C.-J., Wu, M.-M., Lee, S.-S., Wang, J.-D., Cheng, S.-H., and Wu, H.-Y. (1988) Artherogenicity and carcinogenicity of high arsenic artesian well water. Multiple risk factors and related malignant neoplasms of blackfoot disease, *Arteriosclerosis*, 8, 452–60.

Chen, Y. and Ahsan, H. (2004) Cancer burden from arsenic in drinking water in Bangladesh, *American Journal of Public Health*, 94, 741–4.

Chen, Y.-C., Guo, Y.-L.L., Su, H.-J.J., et al. (2003) Arsenic methylation and skin cancer risk in southwestern Taiwan, *Journal of Occupational and Environmental Medicine*, 45, 241–8.

Chiou, H.-Y., Chiou, S.-T., Hsu, Y.-H., et al. (2001) Incidence of transitional cell carcinoma and arsenic in drinking water: a follow-up study of 8102 residents in an arseniasis-endemic area in northeastern Taiwan, *American Journal of Epidemiology*, 153, 411–8.

Chouinard, V. (1999) Body politics: disabled women's activism in Canada and beyond, in Butler, R. and Parr, H. (eds.) (1999) *Mind and Body Spaces: Geographies of Illness, Impairment and Disability*, Routledge, London.

Chouinard, V. and Crooks, V.A. (2003) Challenging geographies of ableness: celebrating how far we've come and what's left to be done, *The Canadian Geographer*, 47, 383–5.

Clapp, R.W., Howe, G.K. and Jacobs, M. (2006) Environmental and occupational causes of cancer re-visited, *Journal of Public Health Policy*, 27, 61–76.

Clarke, K., Howard, G.C.W., Elia, M.H., et al.(1995) Referral patterns within Scotland to specialist oncology centres for patients with testicular germ cell tumours, *British Journal of Cancer*, 72, 1300–2.

Clauw, D.J., Engel, C.C., Aronowitz, R., et al. (2003) Unexplained symptoms after terrorism and war: an expert consensus statement, *Journal of Occupational and Environmental Medicine*, 45, 1040–8.

Cliff, A.D. and Haggett, P. (1988) *Atlas of Disease Distributions*, Blackwell, Oxford.

Cliff, A.D., Haggett, P. and Ord, J.K. (1986) *Spatial Aspects of Influenza Epidemics*, Pion, London.

Cliff, A.D., Haggett, P. and Smallman-Raynor, M. (1998) *Deciphering Global Epidemics: Analytical Approaches to the Disease Records of World Cities, 1888–1912*, Cambridge University Press, Cambridge.

Cliff, A.D., Haggett, P. and Smallman-Raynor, M. (2000) *Island Epidemics*, Oxford University Press, Oxford.

Cloke, P., Crang, P. and Goodwin, M. eds. (1999) *Introducing Human Geographies*, Arnold, London.

Cloke, P., Philo, C. and Sadler, D. (1991) *Approaching Human Geography*, Paul Chapman, London.

Cohen, D.A., Finch, B.K., Bower, A. and Sastry, N. (2006) Collective efficacy and obesity: the potential influence of social factors on health, *Social Science & Medicine*, 62, 769–78.

COMMIT Research Group (1995) Community intervention trial for smoking cessation (COMMIT): II. Changes in adult cigarette smoking prevalence, *American Journal of Public Health*, 85, 183–92.

Congdon, P. (2003) *Applied Bayesian Modelling*, John Wiley, New York.

Cook-Mozaffari, P. (1996) Cancer and fluoridation, *Community Dental Health*, 13, 56–62.

Cooper, H., Arber, S., Fee, L. and Ginn, J. (1999) *The Influence of Social Support and Social Capital on Health: A Review and Analysis of British Data*, Health Education Authority, London.

Corburn, J., Osleeb, J. and Porter, M. (2006) Urban asthma and the neighborhood environment in New York City, *Health & Place*, 12, 167–79.

Cornwell, J. (1984) *Hard-earned Lives: Accounts of Health and Illness from East London*, Tavistock Publications, London.

Cox, B.D., Huppert, F.A. and Whichelow, M.J., eds. (1993) *The Health and Lifestyle Survey: Seven Years On*, Dartmouth, Aldershot.

Craddock, S. (1995) Sewers and scapegoats: Spatial metaphors of smallpox in nineteenth century San Francisco, *Social Science & Medicine*, 41, 957–68.

Craddock, S. (2000) Disease, social identity, and risk: rethinking the geography of AIDS, *Transactions of the Institute of British Geographers*, 25, 153–68.

Craddock, S. (2001) Scales of justice: women, equity, and HIV in East Africa, in Dyck, I., Lewis, N.D. and McLafferty, S. (eds.) (2001) *Geographies of Women's Health*, Routledge, London.

Crang, M. (2003) Qualitative methods: touchy, feely, look-see? *Progress in Human Geography*, 27, 494–504.

Creese, A., Floyd, K., Alban, A. and Guinness, L. (2002). Cost-effectiveness of HIV/AIDS interventions in Africa: a systematic review of the evidence, the *Lancet*, 359, 1635–42.

Cresswell, T. (2004) *Place: An Introduction*, Wiley-Blackwell, Oxford.

Crighton, E., Elliott, S.J., van der Meer, J., Small, I. and Upshur, R. (2003) Impacts of an environmental disaster on psychosocial health and well-being in Karakalpakstan, *Social Science & Medicine*, 56, 551–67.

Cromley, E., Carter, M., Mrozinski, R. and Ertel, S.-H. (1998) Residential setting as a risk factor for Lyme disease in a hyperendemic region, *American Journal of Epidemiology*, 147, 472–477.

Cromley, E.K. and McLafferty, S.L. (2002) *GIS and Public Health*, The Guilford Press, New York.

Crooks, V.A. and Chouinard, V. (2006) An embodied geography of disablement: chronically ill women's struggles for enabling places in spaces of health care and daily life, *Health & Place*, 12, 345–52.

Crossley, N. (2001) *The Social Body: Habit, Identity and Desire*, Sage, London.

Cruickshanks, K.J., Klein, B.E. and Klein, R. (1992) Ultraviolet light exposure and lens opacities: the Beaver Dam Eye Study, *American Journal of Public Health*, 82, 1658–62.

Cullinan, P., Acquilla, S. and Dhara, V.R. (1997) Respiratory morbidity 10 years after the Union Carbide gas leak at Bhopal: a cross-sectional survey, *British Medical Journal*, 314, 338–42.

Cummins, S. and Macintyre, S. (2006) Food environments and obesity – neighbourhood or nation? *International Journal of Epidemiology*, 35, 100–4.

Curtis, S. (2004) *Health and Inequality: Geographical Perspectives*, Sage, London.

Curtis, S. and Jones, I.R. (1998) Is there a place for geography in the analysis of health inequality? in Bartley, M., Blane, D. and Davey Smith, G. (eds.) *The Sociology of Health Inequalities*, Blackwell, Oxford.

Cutchin, M. (2002) Virtual medical geographies: conceptualizing telemedicine and regionalization, *Progress in Human Geography*, 26, 19–39.

Cutchin, M. (2007) Therapeutic landscapes for older people: care with commodification, liminality, and ambiguity, in Williams, A. (ed.) *Therapeutic Landscapes*, Ashgate, Aldershot.

Daniels, P., Bradshaw, M., Shaw, D. and Sidaway, J., (eds) (2004) *An Introduction to Human Geography*, Pearson, London, second edition.

Darby, S., Hill, D., Auvinen, A., et al. (2005) Radon in homes and risk of lung cancer: collaborative analysis of individual data from 13 European case-control studies, *British Medical Journal*, 330, 223–9.

Davey, B. and Seale, C. (eds.) (1996) *Experiencing and Explaining Disease*, Open University Press, Buckingham.

Davey Smith, G., Whitley, E., Dorling, D. and Gunnell, D. (2001) Area based measures of social and economic circumstances: cause specific mortality patterns depend on the choice of index, *Journal of Epidemiology and Community Health*, 55, 149–53.

Davidson, J., Bondi, L. and Smith, M., eds. (2005) *Emotional Geographies*, Ashgate, Aldershot.

Dear, M. and Wolch, J. (1987) *Landscapes of Despair: From Deinstitutionalization to Homelessness*, Polity Press, London.

Department of the Environment (1996) *The Potential Effects of Ozone Depletion in the United Kingdom*, Stationery Office, London.

Department of Health (2005) http://www.dh.gov.uk/en/PublicationsAndStatistics/Statistics/StatisticalWorkAreas/StatisticalWorkforce/DH_4107332

Diggle, P.J. and Rowlinson, B.S. (1994) A conditional approach to point process modelling of elevated risk, *Journal of the Royal Statistical Society, Series A*, 157, 433–40.

Diggle, P.J., Gatrell, A.C., and Lovett, A.A. (1990) Modelling the prevalence of cancer of the larynx in part of Lancashire: a new methodology for spatial epidemiology, in Thomas, R.W. (ed.) *Spatial Epidemiology*, Pion, London.

Dissanayake, C.B. and Chandrajith, R.L.R. (1996) Iodine in the environment and endemic goitre in Sri Lanka, in Appleton, J.D., Fuge, R. and McCall, G.J.H. (eds.) *Environmental Geochemistry and Health*, the Geological Society, London.

Dobson, M.J. (1997) *Contours of Death and Disease in Early Modern England*, Cambridge University Press, Cambridge.

Dockery, D.W., Pope, C.A., Xu, X., et al. (1993) An association between air pollution and mortality in six US cities, *New England Journal of Medicine*, 329, 1753–9.

Dolk, H., Vrijheid, M., Armstrong, B., et al. (1998) Risk of congenital anomalies near hazardous-waste landfill sites in Europe: The EUROHAZCON study, the *Lancet*, 352, 423–7.

Dong, W., Ben-Shlomo, Y., Colhoun, H. and Chaturvedi, N. (1998) Gender differences in accessing cardiac surgery across England: a cross-sectional analysis of the Health Survey for England, *Social Science & Medicine*, 47, 1773–80.

Dorn, M.L. (1998) Beyond nomadism: the travel narratives of a "cripple," in Nast, H.J. and Pile, S. (eds.) (1998) *Places Through the Body*, Routledge, London.

Downing, A. and Sloggett, A. (1994) The use of the Longitudinal Study for environmental epidemiology, *Longitudinal Study Newsletter*, 11, 5–9.

Draper, G.F., Vincent, T., Kroll, M.E. and Swanson, J. (2005) Childhood cancer in relation to distance from high-voltage power lines in England and Wales: a case-control study, *British Medical Journal*, 330, 1290–5.

Driedger, S.M. and Eyles, J. (2001) Organochlorines and breast cancer: the uses of scientific evidence in claimsmaking, *Social Science & Medicine*, 52, 1589–1605.

Driedger, S.M., Crooks, V.A., and Bennett, D. (2004) Engaging in the disablement process over space and time: narratives of persons with multiple sclerosis in Ottawa, Canada, *The Canadian Geographer*, 48, 119–36.

Driedger, S.M., Eyles, J., Elliott, S.J., and Cole, D.C. (2002) Constructing scientific authorities: issue framing of chlorinated disinfection byproducts in public health, *Risk Analysis*, 22, 789–802.

Du Plessis, J.B., van Rooyen, J.J., Naude, D.A. and van der Merwe, C.A. (1995) Water fluoridation in South Africa: will it be effective? *Journal of the Dental Association of South Africa*, 50, 545–9.

Duchin, J.S., Koster, F.T., Peters, C.J., et al. (1994) Hantavirus pulmonary syndrome: a clinical description of 17 patients with a newly recognised disease, *New England Journal of Medicine*, 330, 949–55.

Dummer, T.J.B. and Cook, I.G. (2007) Exploring China's rural health crisis: processes and policy implications, *Health Policy*, 83, 1–16.

Duncan, C., Jones, K. and Moon, G. (1996) Health-related behaviour in context: a multi-level modelling approach, *Social Science & Medicine*, 42, 817–30.

Dunn, C.E. and Kingham, S. (1996) Establishing links between air quality and health: searching for the impossible? *Social Science & Medicine*, 42, 831–41.

Dunn, J.R., Frohlich, K.L., Ross, N., Curtis, L.J. and Sanmartin, C. (2006) Role of geography in inequalities in health and human development, in Heymann, J., Hertzman, C., Barer, M. L. and Evans, R.G. (eds.) (2006) *Healthier Societies: From Analysis to Action*, Oxford University Press, Oxford.

Dyck, I. (1995a) Hidden geographies: the changing lifeworlds of women with multiple sclerosis, *Social Science & Medicine*, 40, 307–20.

Dyck, I. (1995b) Putting chronic illness "in place": women immigrants' accounts of their health care, *Geoforum*, 26, 247–60.

Dyck, I. (1998) Women with disabilities and everyday geographies: home space and the contested body, in Kearns, R.A. and Gesler, W.M. (eds.) *Putting Health into Place: Landscape, Identity and Well-being*, University of Syracuse Press, Syracuse, New York.

Dyck, I. (1999) Body troubles: women, the workplace and negotiations of a disabled identity, in Butler, R. and Parr, H. (eds.) *Mind and Body Spaces: Geographies of Illness, Impairment and Disability*, Routledge, London.

Dyck, I. and Dossa, P. (2007) Place, health and home: gender and migration in the constitution of healthy space, *Health & Place*, 13, 691–701.

Dyck, I., Kontos, P., Angus, J. and McKeever, P. (2005) The home as a site for long-term care: meanings and management of bodies and spaces, *Health & Place*, 11, 173–85.

Dyck, I., Lewis, N.D. and McLafferty, S., eds. (2001) *Geographies of Women's Health*, Routledge, London.

Dyck, I. and Tigar McLaren, A. (2004) Telling it like it is? Constructing accounts of settlement with immigrant and refugee women in Canada, *Gender, Place and Culture*, 11, 513–34.

Eibner, C. and Sturm, R. (2006) US-based indices of area-level deprivation: results from HealthCare for Communities, *Social Science & Medicine*, 62, 348–59.

Elford, J. and Ben-Shlomo, Y. (1997) Geography and migration, in Kuh, D. and Ben-Shlomo, Y. (eds.) *A Life Course Approach to Chronic Disease Epidemiology*, Oxford University Press, Oxford.

Elliott, P.J., Hills, M., Beresford, J., et al. (1992) Incidence of cancer of the larynx and lung near incinerators of waste solvents and oils in Great Britain, the *Lancet*, 339, 854–8.

Elliott, S.J. (1999) And the question shall determine the method, *Professional Geographer*, 51, 240–3.

Elliott, S.J. and Gillie, J. (1998) Moving experiences: a qualitative analysis of health and migration, *Health & Place*, 4, 327–40.

Ellis, M. and Muschkin, C. (1996) Migration of persons with AIDS – a search for support from elderly parents? *Social Science & Medicine*, 43, 1109–18.

Elliott, S.J., Loeb, M., Harrington, D. and Eyles, J. (2008) Heeding the message? Determinants of risk behaviours for West Nile Virus, *Canadian Journal of Public Health*, 99, 137–41.

Elliott, S.J., Taylor, S.M., Walter, S., Stieb, D., Frank, J. and Eyles, J. (1993) Modelling psychosocial effects of exposure to solid waste facilities, *Social Science & Medicine*, 37, 791–805.

Elovainio, M., Kivimäki, M., Steen, N. and Vahtera, J. (2004) Job decision latitude, organizational justice and health: multilevel covariance structure analysis, *Social Science & Medicine*, 58, 1659–69.

Epp, J. (1986) Address: the honourable Jack Epp, *Health Promotion International*, 1, 413–7.

Epstein, P.R. (1995) Emerging diseases and ecosystem instability: new threats to public health, *American Journal of Public Health*, 85, 168–72.

Eurowinter Group (1997) Cold exposure and winter mortality from ischaemic heart disease, cerebrovascular disease, respiratory disease and all causes, in warm and cold regions of Europe, the *Lancet*, 349, 1341–6.

Evans, R.G., Barer, M.L., Lewis, S., Rachlis, M. and Stoddart, G.L. (2000) Private highway, one-way street: the deklein and fall of Canadian medicare? *Centre for Health Services and Policy Research*, University of British Columbia, Vancouver, British Columbia.

Eyles, J. and Donovan, J. (1986) Making sense of sickness and care: an ethnography of health in a West Midlands town, *Transactions of the Institute of British Geographers*, 11, 415–27.

Eyles, J., Taylor, S.M., Johnson, N. and Baxter, J. (1993) Worrying about waste: living close to solid waste disposal facilities in southern Ontario, *Social Science & Medicine*, 37, 805–12.

Ezzati, M. and Kammen, D.M. (2002) The health impacts of exposure to indoor air pollution from solid fuels in developing countries: knowledge, gaps, and data needs, *Environmental Health Perspectives*, 110, 1057–68.

Ezzati, M., Lopez, A.D., Rodgers, A., Vander Hoorn, S. and Murray, C.J.L. (2002) Selected major risk factors and global and regional burden of disease, the *Lancet*, 360, 1347–60.

Farman, J.C., Gardiner, B.G. and Shanklin, J.D. (1985) Large losses of total ozone in Antarctica reveal seasonal ClO$_x$/NO$_x$ interaction, *Nature*, 315, 207–10.

Farmer, P. (1992) *AIDS and Accusation: Haiti and the Geography of Blame*, University of California Press, Berkeley.

Farmer, P. (1999) *Infections and Inequalities: The Modern Plagues*, University of California Press, Berkeley.

Farmer, P. (2005) *Pathologies of Power: Health, Human Rights, and the New War on the Poor*, University of California Press, Berkeley.

Farmer, T., Robinson, K., Elliott, S.J. and Eyles, J. (2006) Developing and implementing a triangulation protocol for qualitative health research, *Qualitative Health Research*, 16, 377–94.

Ferguson, B., Sheldon, T.A. and Posnett, J., eds. (1997) *Concentration and Choice in Healthcare*, FT Healthcare, London.

Ferguson, D.E. (1979) The political economy of health and medicine in colonial Tanganyika, in Kaniki, M.H.Y (ed.) *Tanzania Under Colonial Rule*, Longman, London.

Fisher, J.B., Kelly, M. and Romm, J. (2006) Scales of environmental justice: combining GIS and spatial analysis for air toxics in West Oakland, California, *Health & Place*, 12, 701–14.

Flannery, T. (2005) *The Weather Makers*, Harper Collins, Toronto, Ontario.

Flowerdew, R. and Martin, D., eds. (2005) *Methods in Human Geography*, 2nd edition, Pearson Education Ltd, Harlow, Essex.

Fone, D.L. and Dunstan, F. (2006) Mental health, places and people: a multilevel analysis of economic inactivity and social deprivation, *Health & Place*, 12, 332–44.

Forster, D.P., Newens, A.J., Kay, D.W.K. and Edwardson, J.A. (1995) Risk factors in clinically diagnosed presenile dementia of the Alzheimer type: a case-control study in northern England, *Journal of Epidemiology and Community Health*, 49, 253–8.

Fox, N.J. (1999) *Beyond Health: Postmodernism and Embodiment*, Free Association Books, London.

Frank, J., Lomax, G., Baird, P. and Lock, M. (2006) Interactive role of genes and the environment, in Heymann, J., Hertzman, C., Barer, M.L. and Evans, R.G. (eds.) *Healthier Societies: From Analysis to Action*, Oxford University Press, Oxford.

Free, C., White, P., Shipman, C. and Dale, J. (1999) Access to and use of out-of-hours services by members of Vietnamese community groups in south London: a focus group study, *Family Practice*, 16, 369–74.

Frohlich, N. and Mustard, C. (1996) A regional comparison of socioeconomic and health indices in a Canadian province, *Social Science & Medicine*, 42, 1273–81.

Frumkin, H., Hess, J., Luber, G., Malilay, J. and McGeehin, M. (2008) Climate change: the public health response, *American Journal of Public Health*, 98, 1–11.

Gardner, M.J. (1992) Childhood leukaemia around the Sellafield nuclear plant, in Elliott, P., Cuzick, J., English, D. and Stern, R. (eds.) *Geographical and Environmental Epidemiology*, Oxford University Press.

Gardner, M.J., Snee, M.P., Hall, A.J., Powell, C.A., Downes, S. and Terrell, J.D. (1990) Results of case-control study of leukaemia and lymphoma among young people near Sellafield nuclear plant in West Cumbria, *British Medical Journal*, 300, 423–9.

Garvin, T. (1995) "We're strong women": building a community-university research partnership, *Geoforum*, 26, 273–86.

Garvin, T. and Eyles, J. (2001) Public health responses for skin cancer prevention: the policy framing of sun safety in Australia, Canada and England, *Social Science & Medicine*, 53, 1175–89.

Garvin, T. and Wilson, K. (1999) The use of storytelling for understanding women's desires to tan: lessons from the field, *The Professional Geographer*, 51, 297–306.

Gatrell, A.C. (1983) *Distance and Space: A Geographical Perspective*, Oxford University Press, Oxford.

Gatrell, A.C. (2002) *Geographies of Health: an Introduction,* 1st edition, Blackwell Publishing, Oxford.

Gatrell, A.C., Berridge, D., Bennett, S., et al. (2001) Local geographies of health inequalities, in Boyle, P., Curtis, S., Graham, E. and Moore, E. (eds.) (eds.) *The Geography of Health Inequalities in the Developed World*, Ashgate Press, Aldershot.

Gatrell, A.C., Garnett, S., Rigby, J., Maddocks, A. and Kirwan, M. (1998) Uptake of screening for breast cancer in South Lancashire, *Public Health*, 112, 297–301.

Gatrell, A.C., Popay, J. and Thomas, C. (2004) Mapping the determinants of health inequalities in social space: can Bourdieu help us? *Health & Place*, 10, 245–57.

Gatrell, A.C., Thomas, C., Bennett, S., et al. (2000) Understanding health inequalities: locating people in geographical and social spaces, in Graham, H. (ed.) *Understanding Health Inequalities*, Open University Press, Buckingham.

Gatrell, C.J. (2007) Secrets and lies: breastfeeding and professional paid work, *Social Science & Medicine*, 65, 393–404.

Gauderman, W.J., Vora, H., McConnell, R., et al. (2007) Effect of exposure to traffic on lung development from 10 to 18 years of age: a cohort study, the *Lancet*, 369, 571–77.

Gellert, G.A. (1993) International migration and control of communicable diseases, *Social Science & Medicine*, 37, 1489–99.

Gerhardt, U. (1989) *Ideas About Illness: An Intellectual and Political History of Medical Sociology*, Macmillan, London.

Geronimus, A.T., Bound, J., Waidmann, T.A., Hillemeier, M.M. and Burns, P.B. (1996) Excess mortality among blacks and whites in the United States, *New England Journal of Medicine*, 335 (21) 1551–8.

Geronimus, A.T., Bound, J. and Waidmann, T.A. (1999) Poverty, time, and place: variation in excess mortality across selected US populations, 1980–1990, *Journal of Epidemiology and Community Health*, 53, 325–34.

Gesler, W. (1993) Therapeutic landscapes: theory and a case study of Epidauros, Greece, *Environment and Planning D: Society and Space*, 11, 171–89.

Gesler, W.M. (1996) Lourdes: healing in a place of pilgrimage, *Health & Place*, 2, 95–106.

Gesler, W. (1998) Bath's reputation as a healing place, in Kearns, R.A. and Gesler, W.M. (eds.) *Putting Health into Place: Landscape, Identity and Well-being*, Syracuse University Press, Syracuse, New York.

Getis, A. and Ord, J.K. (1992) The analysis of spatial association by use of distance statistics, *Geographical Analysis*, 24, 189–206.

Getis, A. and Ord, J.K. (1999) Spatial modelling of disease dispersion using a local statistic: The case of AIDS, in Griffith, D.A., Amrhein, C.G., and Huriot, J.-M. (eds.) *Econometric Advances in Spatial Modelling and Methodology: Essays in Honour of Jean Paelinck*, Kluwer, Massachusetts.

Giacomini, M.K. and Cook, D.J. (2000) Users' guide to the medical literature XXIII. Qualitative research in health care: are the results of the study valid? *Journal of the American Medical Association*, 284, 357–62.

Giggs, J.A., Bourke, J.B. and Katschinski, B. (1988) The epidemiology of primary acute pancreatitis in the Nottingham defined population area, *Transactions of the Institute of British Geographers*, 5, 229–42.

Giggs, J.A., Ebdon, D.S. and Bourke, J.B. (1980) The epidemiology of primary acute pancreatitis in Greater Nottingham: 1969–1983, *Social Science & Medicine*, 26, 79–89.

Ginns, S.E. and Gatrell, A.C. (1996) Respiratory health effects of industrial air pollution: a study in east Lancashire, UK, *Journal of Epidemiology and Community Health*, 50, 631–5.

Githeko, A., Lindsay, S.W., Confalonieri, U. and Patz, J. (2000) Climate change and vector-borne diseases: a regional analysis, *Bulletin of the World Health Organization*, 78, 18–29.

Gittelsohn, A. and Powe, N.R. (1995) Small area variations in health care delivery in Maryland, *Health Services Research*, 30, 295–317.

Glassman, J. (2001) Women workers and the regulation of health and safety on the industrial periphery, in Dyck, I., Lewis, N.D. and McLafferty, S. (eds.) *Geographies of Women's Health*, Routledge, London.

Gleeson, B. (1999) *Geographies of Disability*, Routledge, London.

Glenn, L.L., Beck, R.W. and Burkett, G.L. (1998) Effect of a transient, geographically localised economic recovery on community health and income studies with longitudinal household cohort interview method, *Journal of Epidemiology and Community Health*, 52, 749–57.

Gober, P. (1994) On geographic variation in abortion rates in the United States, *Annals of the Association of American Geographers*, 84, 230–50.

Gober, P. and Rosenberg, M.W. (1999) Looking back, looking around, looking forward: a woman's right to choose, in Dyck, I., Lewis, N.D. and McLafferty, S. (eds.) *Geographies of Women's Health*, Routledge, London.

Goddard, M. and Smith, P. (2001) Equity of access to health care services: theory and evidence from the UK, *Social Science & Medicine*, 53, 1149–62.

Goffman, E. (1959) *The Presentation of Self in Everyday Life*, Anchor Books, Doubleday, New York.

Goldberg, S.J., Lebowitz, M.D., Graver, E.J. and Hicks, S. (1990) An association of human congenital cardiac malformations and drinking water contaminants, *Journal of the American College of Cardiologists*, 16, 155–64.

Goldsmith, J.R. (1999) The residential radon-lung cancer association in US counties: a commentary, *Health Physics*, 76, 553–7.

Goma Epidemiology Group (1995) Public health impact of Rwandan refugee crisis: what happened in Goma, Zaire, in July 1994?, the *Lancet*, 345, 339–44.

Gomes, M., Linthicum, K. and Haile, M. (1998) Malaria: the role of agriculture in changing the epidemiology of malaria, in Greenwood, B. and de Cock, K. (eds.) *New and Resurgent Infections: Prediction, Detection and Management of Tomorrow's Epidemics*, John Wiley, Chichester.

Gordon, D. (2003) Area-based deprivation measures – a UK perspective, in Kawachi I. and Berkman F. (eds.) *Neighbourhoods and Health*, Oxford University Press, Oxford.

Gore, A. (2006) *An Inconvenient Truth*, Rodale Publishers: Emmaus, Pennsylvania.

Gorman, B.K. (1999) Racial and ethnic variation in low birthweight in the United States: individual and contextual determinants, *Health & Place*, 5, 195–208.

Gould, P.R. (1990) *Fire in the Rain*, Polity Press, Cambridge.

Gould, P. and Wallace, R. (1994) Spatial structures and scientific paradoxes in the AIDS pandemic, *Geografiska Annaler*, 76B, 105–16.

Graham, R.P., Forrester, M.L., Wysong, J.A. Rosenthal, T.C., and James, P.A. (1995) HIV/AIDS in the rural United States: epidemiology and health services delivery, *Medical Care Research Review*, 52, 435–52.

Gray, A., ed. (1993) *World Health and Disease*, Open University Press, Buckingham.

Grbich, C. (1999) *Qualitative Research in Health: An Introduction*, Sage, London.

Greenwood, B. and de Cock, K., eds. (1998) *New and Resurgent Infections: Prediction, Detection and Management of Tomorrow's Epidemics*, John Wiley, Chichester.

Gregson, N. (2005) Agency: structure, in Cloke, P. and Johnston, R. (eds.) *Spaces of Geographical Thought*, Sage, London.

Griffith, D.A., Doyle, P.G., Wheeler, D.C. and Johnson, D.L. (1998) A tale of two swaths: urban childhood blood-lead levels across Syracuse, New York, *Annals of the Association of American Geographers*, 88, 640–65.

Griffith, J., Riggan, W.B., Duncan, R.C. and Pellom, A.C. (1989) Cancer mortality in US counties with hazardous waste sites and ground water pollution, *Archives of Environmental Health*, 44, 69–74.

Groleau, D., Soulière, M. and Kirmayer, L.J. (2006) Breastfeeding and the cultural configuration of social space among Vietnamese immigrant woman, *Health & Place*, 12, 516–26.

de Gruijl, F.R. and van der Leun, J.C. (2000) Environment and health: 3. Ozone depletion and ultraviolet radiation, *Canadian Medical Association Journal*, 163, 851–5.

Gunnell, D.J., Peters, T.J., Kammerling, R.M. and Brooks, J. (1995) Relation between parasuicide, suicide, psychiatric admissions, and socioeconomic deprivation, *British Medical Journal*, 311, 226–30.

Guo, H.-R. (2004) Arsenic level in drinking water and mortality of lung cancer (Taiwan), *Cancer Causes and Control*, 15, 171–7.

Haalboom, B., Elliott, S.J., Eyles, J. and Muggah, H. (2006) The risk society at work in the Sydney "Tar Ponds," *The Canadian Geographer*, 50, 227–41.

Hadorn, D.C. (2000) Setting priorities for waiting lists: defining our terms, *Canadian Medical Association Journal*, 163, 857–60.

Haggett, P. (1994) Geographical aspects of the emergence of infectious diseases, *Geografiska Annaler*, 76B, 91–104.

Haggett, P. (2000) *The Geographical Structure of Epidemics*, Oxford University Press, Oxford.

Haggett, P. (2001) *Geography: A Global Synthesis,* Prentice-Hall, London.

Haines, A., McMichael, A.J. and Epstein, P.R. (2000) Environment and health: 2. Global climate change and health, *Canadian Medical Association Journal*, 163, 729–34.

Haines, A., Kovats, R.S., Campbell-Lendrum, D., and Corvalan, C. (2006) Climate change and human health: impacts, vulnerability, and mitigation, the *Lancet*, 367, 2101–9.

Haining, R. (2003) *Spatial Data Analysis: Theory and Practice*, Cambridge University Press, Cambridge.

Hales, S., de Wet, N., Maindonald, J. and Woodward, A. (2002) Potential effect of population and climate changes on global distribution of dengue fever: an empirical model, the *Lancet*, 360, 830–834.

Hamilton, J.T. and Viscusi, W.K. (1999) How costly is "clean"? An analysis of the benefits and costs of Superfund site remediations, *Journal of Policy Analysis and Management*, 18, 2–27.

Hanchette, C.L. (2008) The political ecology of lead poisoning in eastern North Carolina, *Health & Place*, 14, 209–16.

Harding, S. and Maxwell, R. (1997) Differences in mortality of migrants, in Drever, F. and Whitehead, M. (eds.) *Health Inequalities*, Office for National Statistics, London.

Hartwig, G.W. and Patterson, K.D., eds. (1978) *Disease in African History: an Introductory Survey and Case Studies*, Durham, North Carolina.

Hatch, M.C., Beyea, J., Nieves, J.W. and Susser, M. (1990) Cancer near the Three Mile Island nuclear plant: radiation emissions, *American Journal of Epidemiology*, 132, 397–412.

Hatch, M.C., Wallenstein, S., Beyea, J., Nieves, J.W. and Susser, M. (1991) Cancer rates after the Three Mile Island nuclear accident and proximity of residence to the plant, *American Journal of Public Health*, 81, 719–24.

Hayashi, L.C., Tamiya, N. and Yano, E. (1998) Correlation between UVB radiation and the proportion of cataract: an epidemiological study based on a nationwide patient survey in Japan, *Industrial Health*, 36, 354–60.

Haynes, R. and Bentham, G. (1982) The effects of accessibility on general practitioner consultations, out-patient attendances and inpatient admissions in Norfolk, England, *Social Science & Medicine*, 16, 561–9.

Haynes, R.B., Sackett, D.L., Taylor, D.W., Gibson, E.S. and Johnson, A.L. (1978) Increased absenteeism from work after detection and labeling of hypertensive patients, *The New England Journal of Medicine*, 299, 741–4.

Head, J., Kivimäki, M., Martikainen, P., Vahtera, J., Ferrie, J.E. and Marmot, M.G. (2006). Influence of change in psychosocial work characteristics on sickness absence: the Whitehall II study, *Journal of Epidemiology and Community Health*, 60, 55–61.

Henry, D. and Lexchin, J. (2002) The pharmaceutical industry as a medicines provider, the *Lancet*, 360, 1590–5.

Hertzman, C. and Frank, J. (2006) Biological pathways linking the social environment, Heymann, J., Hertzman, C., Barer, M.L. and Evans, R.G. (eds.) *Healthier Societies: From Analysis to Action*, Oxford University Press, Oxford.

Hertzman, C. and Power, C. (2006) A life course approach to health and human development, in Heyman, B., ed. (2006) *Risk, Health and Health Care*, Arnold, London.

Heymann, J., Hertzman, C., Barer, M.L. and Evans, R.G., eds. (2006) *Healthier Societies: From Analysis to Action*, Oxford University Press, Oxford.

Hodgson, M.J. (1988) An hierarchical location-allocation model for primary health care delivery in a developing area, *Social Science & Medicine*, 26, 153–61.

Hoffman-Goettz, L. and Donelle, L. (2007) Chat room computer-mediated support on health issues for aboriginal women, *Health Care for Women International, 28*, 397–418.

Holtz C. (2008) *Global Health Care: Issues and Policies*, Jones and Bartlett Publishers, Sudbury, Mass.

Howe, G.M. (1972) *Man, Environment and Disease in Britain: a Medical Geography of Britain Through the Ages*, David and Charles, Newton Abbott.

Hrudey, S.E., Huck, P.M., Payment, P., Gillham, R.W. and Hrudey, E.J. (2002) Walkerton: lessons learned in comparison with waterborne outbreaks in the developed world, *Journal of Environmental Engineering and Science*, 1, 397–407.

Hunter, L.M., Logan, J., Goulet, J.G. and Barton, S. (2006) Aboriginal healing: regaining balance and culture, *Transcultural Nursing*, 17, 13–22.

Hunter, M. (2007) The changing political economy of sex in South Africa: the significance of unemployment and inequalities to the scale of the AIDS pandemic, *Social Science & Medicine*, 64, 689–700.

Hunter, P.R. (1997) *Waterborne Disease: Epidemiology and Ecology*, John Wiley, Chichester.

Hutton, G. and Bartram, J. (2008) Global costs of attaining the Millennium Development Goal for water supply and sanitation, *Bulletin of the World Health Organization*, 86, 13–19.

Jacobs, L. and Morone, J. (2005) *Healthy, Wealthy and Fair: Health Care and the Good Society*, Oxford University Press, Oxford.

Jacqmin, H., Commenges, D., Letenneur, L., Barberger-Gateau, P. and Dartigues, J.-F. (1994) Components of drinking water and risk of cognitive impairment in the elderly, *American Journal of Epidemiology*, 139, 48–57.

James, P.T., Leach, R., Kalamara, E. and Shayeghi, M. (2001) The worldwide obesity epidemic, *Obesity Research*, 9, S228–S33.

Jetten, T.H. and Focks, D.A. (1997) Potential changes in the distribution of dengue transmission under climate warming, *American Journal of Tropical Medicine and Hygiene*, 57, 285–97.

Jin, Y., Ma, X., Chen, X., et al. (2006) Exposure to indoor air pollution from household energy use in rural China: the interactions of technology, behavior, and knowledge in health risk management, *Social Science & Medicine*, 62, 3161–76.

Jochelson, K., Mothibeli, M. and Leger, J.-P. (1991) Human immunodeficiency virus and migrant labour in South Africa, *International Journal of Health Services*, 21, 157–73.

Johansson, L.M., Sundquist, J., Johansson, S.-E., Bergman, B., Qvist, J. and Traskman-Bendz, L. (1997) Suicide among foreign-born minorities and native Swedes: an epidemiological follow-up study of a defined population, *Social Science & Medicine*, 44, 181–7.

Johnsen, S., Cloke, P. and May, J. (2005) Transitory spaces of care: serving homeless people on the street, *Health & Place*, 11, 323–36.

Johnson, P.D., Goldberg, S.J., Mays, M.Z. and Dawson, B.V. (2003) Threshold of trichloroethylene contamination in maternal drinking waters affecting fetal heart development in the rat, *Environmental Health Perspectives*, 111, 289–92.

Jones, A.P. and Bentham, G. (1995) Emergency medical service accessibility and outcome from road traffic accidents, *Public Health*, 109, 169–77.

Jones, C.M., Taylor, G.O., Whittle, J.G., Evans, D. and Trotter, D.P. (1997) Water fluoridation, tooth decay in 5 year olds, and social deprivation measured by the Jarman score: analysis of data from British dental surveys, *British Medical Journal*, 315, 514–7.

Jones, K. (1991) Multilevel models for geographical research, *Concepts and Techniques in Modern Geography*, Environmental Publications, Norwich.

Jones, K. and Duncan, C. (1995) Individuals and their ecologies: analysing the geography of chronic illness within a multilevel modelling framework, *Health & Place*, 1, 27–40.

Jones, K. and Moon, G. (1987) *Health, Disease and Society: An Introduction to Medical Geography*, Routledge, London.

Jones, K.E., Patel, N.G., Levy, M.A., et al.. (2008) Global trends in emerging infectious diseases, *Nature*, 451, 990–3.

Joseph, A.E. and Hallman, B.C. (1998) Over the hill and far away: distance as a barrier to the provision of assistance to elderly relatives, *Social Science & Medicine*, 46, 631–9.

Joseph, A.E. and Phillips, D.R. (1984) *Access and Utilization: Geographical Perspectives on Health Care Delivery*, Harper and Row, London.

Kalipeni, E. and Oppong, J. (1998) The refugee crisis in Africa and implications for health and disease: a political ecology approach, *Social Science & Medicine*, 46, 1637–53.

Kalkenstein, L.S. (1993) Health and climate change: direct impacts in cities, the *Lancet*, 342, 1397–9.

Kampa, M. and Castanas, E. (2008) Human health effects of air pollution, *Environmental Pollution*, 151, 362–7.

Kanaiaupuni, S.M. and Donato, K.M. (1999) Migradollars and mortality: the effects of migration on infant survival in Mexico, *Demography*, 36, 339–53.

Kaplan, G.A. (1996) People and places: contrasting perspectives on the association between social class and health, *International Journal of Health Services*, 26, 507–19.

Kawachi, I. and Berkman, L.F., eds. (2003) Neighborhoods and Health, Oxford University Press, New York.

Kawachi, I. and Kennedy, B., eds. (2002) *The Health of Nations: Inequality is Harmful to Your Health*, New Press, New York.

Kawachi. I. and Wamala, S., eds. (2007) *Globalization and Health*, Oxford University Press, Oxford.

Kawachi, I., Colditz, G.A., Aschemo, A., et al. (1996) A prospective study of social networks in relation to total mortality and cardiovascular disease in men in the USA, *Journal of Epidemiology and Community Health*, 50, 245–51.

Kawachi, I., Kennedy, B.P., Lochner, K. and Prothrow-Stich, D. (1997) Social capital, income inequality, and mortality, *American Journal of Public Health*, 87, 1491–8.

Kazmi, J.H. and Pandit, K. (2001) Disease and dislocation: the impact of refugee movements on the geography of malaria in NWFP, Pakistan, *Social Science & Medicine*, 52, 1043–55.

Kearns, R. and Dyck, I. (1996) Cultural safety, biculturalism and nursing education in Aotearoa/New Zealand, *Health and Social Care in the Community*, 4 (6), 371–380.

Kearns, R.A. and Gesler, W.M., eds. (1998) *Putting Health into Place: Landscape Identity and Wellbeing*, Syracuse University Press, New York.

Keller-Olaman, S., Eyles, J., Elliott, S., Wilson, K., Dostrovsky, N. and Jerrett, M. (2005) Individual and neighbourhood characteristics associated with environmental exposure: exploring relationships at home and work in a Canadian city, *Environment and Behaviour*, 37, 441–64.

Kessel, E. (1999) Access to essential drugs in poor countries, *Journal of the American Medical Association*, 282, 630–1.

Kiecolt-Glaser, J.K., McGuire, L., Robles, T.F. and Glaser, R. (2002) Emotions, morbidity, and mortality: new perspectives from psychoneuroimmunology, *Annual Review of Psychology*, 53, 83–107.

Kim, Y.-E., Gatrell, A.C. and Francis, B.J. (2000) The geography of survival after surgery for colorectal cancer in southern England, *Social Science & Medicine*, 50, 1099–1107.

Kington, R., Carlisle, D., McCaffrey, D., Myers, H. and Allen, W. (1998) Racial differences in functional status among elderly US migrants from the South, *Social Science & Medicine*, 47, 831–40.

Kinlen, L.J., Clarke, K. and Hudson, C. (1990) Evidence from population mixing in British New Towns 1946–85 of an infective basis for childhood leukaemia, the *Lancet*, 336, 577–82.

Kinlen, L.J., O'Brien, F., Clarke, K., Balkwill, A. and Matthews, F. (1993) Rural population mixing and childhood leukaemia: effects of the North Sea oil industry in Scotland, including the area near Dounreay nuclear site, *British Medical Journal*, 306, 743–8.

Kirkham, S.R., Smye, V., Tang, S., et al. (2002) Rethinking cultural safety while waiting to do fieldwork: methodological implications for nursing research, *Research in Nursing and Health*, 25, 222–32.

Kitron, U. and Kazmierczak, J.J. (1997) Spatial analysis of the distribution of Lyme disease in Wisconsin, *American Journal of Epidemiology*, 145, 558–66.

Kjellstrom, T. and Hinde, S. (2007) Car culture, transport policy, and public health, in Kawachi. I. and Wamala, S. (eds.) *Globalization and Health*, Oxford University Press, Oxford.

Kloos, H., Higashi, G.I., Cattani, J.A., Schlinski, V.D., Mansour, N.S. and Murrell, K.D. (1983) Water contact behaviour and schistosomiasis in an Upper Egyptian village, *Social Science & Medicine*, 17, 545–62.

Kloos, H., Fulford, A.J.C., Butterworth, A.E., et al. (1997) Spatial patterns of human water contact and *Schistosoma mansoni* transmission and infection in four rural areas in Machakos District, Kenya, *Social Science & Medicine*, 44, 949–68.

Knapp, K.K. and Hardwick, K. (2000) The availability and distribution of dentists in rural ZIP codes and primary health care professional shortage areas (PC-HPSA) ZIP codes: comparison with primary care providers, *Journal of Public Health Dentistry*, 60, 43–8.

Kralj, B. (2001) Physician distribution and physician shortage intensity in Ontario, *Canadian Public Policy*, 27, 167–78.

Kravdal, O. (2006) Does place matter for cancer survival in Norway? A multilevel analysis of the importance of hospital affiliation and municipality socio-economic resources, *Health & Place*, 12, 527–37.

Krenichyn, K. (2006) "The only place to go and be in the city": women talk about exercise, being outdoors, and the meanings of a large urban park, *Health & Place*, 12, 631–43.

Krieger, N., ed. (2005) *Embodying Inequality: Epidemiologic Perspectives*, Baywood Publishing Co., Amityville, NY,

Krieger, N., Löwy, I., Aronowitz, R., Bigby, J., et al. (2005) Hormone replacement therapy, cancer, controversies, and women's health: historical epidemiological, biological, clinical, and advocacy perspectives, *Journal of Epidemiology and Community Health*, 59, 740–8.

Kuh, D. and Ben-Shlomo, Y., eds. (1997) *A Life Course Approach to Chronic Disease Epidemiology*, Oxford University Press, Oxford.

Kunitz, S.J. (2007) *The Health of Populations: General Theories and Particular Realities*, Oxford University Press, Oxford.

Kunst, A. (1997) *Cross-National Comparisons of Socio-Economic Differences in Mortality*, Department of Public Health, Erasmus University, Rotterdam, Netherlands.

Kuper, H. and Marmot, M. (2003) Job strain, job demands, decision latitude, and risk of coronary heart disease within the Whitehall II study, *Journal of Epidemiology and Community Health*, 57, 147–53.

Kwan, M.-P. (1999) Gender, the home-work link, and space-time patterns of nonemployment activities, *Economic Geography*, 75, 370–94.

Laden, F., Schwartz, J., Speizer, F.E. and Dockery, D.W. (2006) Reduction in fine particulate air pollution and mortality, *American Journal of Respiratory and Critical Care Medicine*, 173, 667–72.

Lagakos, S.W., Wessen, B.J. and Zelen, M. (1986) An analysis of contaminated well water and health effects in Woburn, Massachusetts, *Journal of the American Statistical Association*, 81, 583–96.

Laing, R., Waning, B., Gray, A. and Ford, N. (2003) Twenty-five years of the WHO essential medicines lists: progress and challenges, the *Lancet*, 361, 1723–9.

Landale, N.S., Gorman, B.K., and Oropesa, R.S. (2006) Selective migration and infant mortality among Puerto Ricans, *Maternal and Child Health Journal*, 10, 351–60.

Landale, N.S., Oropesa, R.S. and Gorman, B.K. (2000) Migration and infant death: assimilation or selective migration among Puerto Ricans? *American Sociological Review*, 65, 888–909.

Langford, I.H. (1991) Childhood leukaemia mortality and population change in England and Wales 1969–73, *Social Science & Medicine*, 33, 435–40.

Langford, I.H., Bentham, G. and McDonald, A.-L. (1998) Mortality from non-Hodgkin lymphoma and UV exposure in the European Community, *Health & Place*, 4, 355–64.

Langlois, A. and Kitchen, P. (2001) Identifying and measuring dimensions of urban deprivation in Montreal: an analysis of the 1996 census data, *Urban Studies*, 38, 119–39.

Lanska, D.J. (1997) Geographic distribution of stroke mortality among immigrants to the United States, *Stroke*, 28, 53–7.

Lantz, P.M., House, J.S., Lepkowski, J.M., Williams, D.R., Mero, R.P. and Chen, J. (1998) Socioeconomic factors, health behaviors, and mortality: results from a nationally representative prospective study of US adults, *Journal of the American Medical Association*, 279, 1703–8.

Larson, E.H., Hart, L.G. and Rosenblatt, R.A. (1997) Is non-metropolitan residence a risk factor for poor birth outcome in the US, *Social Science & Medicine*, 45, 171–88.

Law, M., Wilson, K., Eyles, J., et al. (2005) Meeting health need, accessing health care: the role of neighbourhood, *Health & Place*, 11, 367–77.

Learmonth, A. (1988) *Disease Ecology: An Introduction*, Blackwell, Oxford.

Lee, K. and Zwi, A. (2003) A global political economy approach to AIDS: ideology, interests and implications, in Lee, K. (ed.) *Health Impacts of Globalization*, Palgrave Macmillan, London.

Legator, M.S., Harper, B.L. and Scott, M.L. (eds)(1985) *The Health Detective's Handbook*, Johns Hopkins University Press, Baltimore, USA.

Lehoux, P., Poland, B. and Daudelin, G. (2006) Focus group research and the "patient's view," *Social Science & Medicine*, 63, 2091–2104.

Lens, M.B. and Dawes, M. (2004) Global perspectives of contemporary epidemiological trends of cutaneous malignant melanoma, *British Journal of Dermatology*, 150, 179–85.

Levine, R.J., Khan, M.R., D'Souza, S. and Nalin, D.R. (1976) Cholera transmission near a cholera hospital, the *Lancet*, ii, 84–86.

Lewis, N.D. and Kieffer, E. (1994) The health of women: beyond maternal and child health, in Phillips, D.R. and Verhasselt, Y. (eds.) *Health and Development*, Routledge, London.

Lilienfeld, D.E. (1991) The silence: the asbestos industry and early occupational cancer research – a case study, *American Journal of Public Health*, 81, 791–800.

Lincoln Y. and Guba E. (1985) *Naturalistic Inquiry*, Sage, London.

Lindsay, S.W. and Birley, M.H. (1996) Climate change and malaria transmission, *Annals of Tropical Medicine and Parasitology*, 90, 573–88.

Little, M.A. and Baker, P.T. (1988) Migration and adaptation, in Mascie-Taylor, C.G.N. and Lasker, G.W. (eds.) *Biological Aspects of Human Migration*, Cambridge University Press, Cambridge.

Litva, A. and Eyles, J. (1995) "Coming out": exposing social theory in medical geography, *Health & Place*, 1, 5–14.

Lock, M., Nguyen, V.-K. and Zarowsky, C. (2006) Global and local perspectives on population health, in Heymann, J., Hertzman, C., Barer, M.L. and Evans, R.G. (eds.) *Healthier Societies: From Analysis to Action*, Oxford University Press, Oxford.

Loeb, M., Elliott, S.J., Gibson, B., et al. (2005) Protective behavior and West Nile Virus risk, *Emerging Infectious Diseases*, 11, 1433–6.

Loevinsohn, M. (1994) Climatic warming and increased malaria incidence in Rwanda, the *Lancet*, 343, 714–8.

Loffler, W. and Hafner, H. (1999) Ecological pattern of first admitted schizophrenics in two German cities over 25 years, *Social Science & Medicine*, 49, 93–108.

Longhurst, R. (2001) *Bodies: Exploring Fluid Boundaries*, Routledge, London.

Lopez, A.C., Mathers, C.D., Ezzati, M., Jamison, D.T. and Murray, C.J.L. (2006) Global and regional burden of disease and risk factors, 2001: systematic analysis of population health data, the *Lancet*, 367, 1747–57.

Love, D. and Lindquist, P. (1995) The geographical accessibility of hospitals to the aged: a Geographic Information Systems analysis within Illinois, *Health Services Research*, 29, 627–51.

Lovett, A.A. and Gatrell, A.C. (1988) The geography of spina bifida in England and Wales, *Transactions of the Institute of British Geographers*, 13, 288–302.

Lucas, R.M., Repacholi, M.H. and McMichael, A.J. (2006) Is the current public health message on UV exposure correct? *Bulletin of the World Health Organization*, 84, 485–91.

Luginaah, I.N., Jerrett, M., Elliott, S.J., et al. (2001) Health profiles of Hamilton: spatial characterisation of neighbourhoods for health, *GeoJournal*, 55, 135–47.

Lynch, J., Due P., Muntaner, C. and Smith, G.D. (2000) Social capital – is it a good investment strategy for public health? *Journal of Epidemiology and Community Health*, 54, 404–8.

Maantay, J. (2007) Asthma and air pollution in the Bronx: methodological and data considerations in using GIS for environmental justice and health research, *Health & Place*, 13, 32–56.

McConway, K., ed. (1994) *Studying Health and Disease*, Open University Press, Buckingham.

McDonagh, M.S., Whiting, P.F., Wilson, P.M., et al. (2000) Systematic review of water fluoridation, *British Medical Journal*, 321, 855–9.

Macintyre, S. (1998) Social inequalities and health in the contemporary world: comparative overview, in Strickland, S.S. and Shetty, P.S. (eds.) *Human Biology and Social Inequality*, Cambridge University Press, Cambridge.

Macintyre, S., Ellaway, A., and Cummins, S. (2002) Place effects on health: how can we conceptualise, operationalise and measure them? *Social Science & Medicine*, 55, 125–39.

Macintyre, S., MacIver, S. and Sooman, A. (1993) Area, class and health: should we be focusing on places or people? *Journal of Social Policy*, 22, 213–34.

Mackenzie, W.R., Hoxie, N.J., Proctor, M.E., et al. (1994) A massive outbreak in Milwaukee of *Cryptosporidium* infection transmitted through the public water supply, *New England Journal of Medicine*, 331, 161–7.

MacKian, S. (2000) Contours of coping: mapping the subject world of long-term illness, *Health & Place*, 6, 95–104.

McLafferty, S. (1982) Neighborhood characteristics and hospital closure, *Social Science & Medicine*, 16, 1667–74.

McLaren, L. and Gauvin, L. (2003) Does the "average size" of women in the neighbourhood influence a woman's likelihood of body dissatisfaction? *Health & Place*, 9, 327–35.

McMichael, A.J. (1998) The influence of historical and global changes upon the patterns of infectious diseases, in Greenwood, B. and de Cock, K. (eds.) *New and Resurgent Infections: Prediction, Detection and Management of Tomorrow's Epidemics*, John Wiley, Chichester.

McMichael, A.J. (1999) Widening the frame: environment as "habitat," not mere repository of "hazard," in Jedrychowski, W., Vena, J. and Maugeri, U. (eds.) *Challenges to Epidemiology in Changing Europe*, Polish Society for Environmental Epidemiology, Krakow, Poland.

McMichael, A.J. and Haines, A. (1997) Global climate change: the potential effects on health, *British Medical Journal*, 315, 805–9.

McNally, N.J., Williams, H.C., Phillips, D.R., et al. (1998) Atopic eczema and domestic water hardness, the *Lancet*, 352, 527–31.

McNeil, D. (1996) *Epidemiologic Research Methods*, John Wiley, Chichester.

Madronich, S. (1992) Implications of recent total atmospheric ozone measurements for biologically active ultraviolet radiation reaching the earth's surface, *Geophysical Research Letters*, 19, 37–40.

Malone, J.B., Abdel-Rahman, M.S., El Bahy, M.M., Huh, O.K., Shafik M. and Bavia, M. (1997) Geographic information systems and the distribution of *Schistosoma mansoni* in the Nile Delta, *Parasitology Today*, 13, 112–9.

Marcin, J.P., Ellis, J., Mawis, R., Nagrampa, E., Nesbitt, T.S. and Dimand, R.J. (2004) Using telemedicine to provide pediatric subspecialty care to children with special health care needs in an underserved rural community, *Pediatrics*, 113, 1–6.

Marmot, M.G., Davey Smith, G., Stansfield, S., et al. (1991) Health inequalities among British civil servants: The Whitehall II study, the *Lancet*, 337, 1387–93.

Marmot, M.G., Rose, G., Shipley, M. and Hamilton, P.J.S. (1978) Employment grade and coronary heart disease in British civil servants, *Journal of Epidemiology and Community Health*, 3, 244–9.

Marmot, M.G. and Wilkinson, R.G., eds. (1999) *Social Determinants of Health*, Oxford University Press, Oxford.

Martens, P. (1998) *Health and Climate Change: Modelling the Impacts of Global Warming and Ozone Depletion*, Earthscan Publications, London.

Martens, P. and McMichael, A.J. (eds) (2002) *Environmental Change, Climate and Health: Issues and Research Methods*, Cambridge University Press, Cambridge.

Martin, C., Curtis, B., Fraser, C. and Sharp, B. (2002) The use of a GIS-based malaria information system for malaria research and control in South Africa, *Health & Place*, 8, 227–36.

Martyn, C.N., Barker, D.J.P. and Osmond, C. (1989) Geographical relationship between Alzheimer's disease and aluminium in drinking water, the *Lancet*, i, 59–62.

Mascie-Taylor, C.G.N. and Lasker, G.W., eds. (1988) *Biological Aspects of Human Migration*, Cambridge University Press, Cambridge.

Mason, J. (2002) *Qualitative Researching*, 2nd edition, Sage, London.

Massey, D. and Jess, P. (1995) *A Place in the World?* Oxford University Press, Oxford.

Mather, F.J., Chen, V.W., Morgan, L.H. et al. (2006) Hierarchical modeling and other spatial analyses in prostate cancer incidence data, *American Journal of Preventive Medicine*, 30, S88–S100.

Mathers, C.D. and Loncar, D. (2006) Projections of global mortality and burden of disease from 2002 to 2030, *PLoS Medicine*, 3, 2011–30.

Mawby, T.V. and Lovett, A.A. (1998) The public health risks of Lyme disease in Breckland, UK: an investigation of environmental and social factors, *Social Science & Medicine*, 46, 719–27.

Mayer, J.D. (2000) Geography, ecology and emerging infectious diseases, *Social Science & Medicine*, 50, 937–52.

Meade, M. and Earickson, R. (2000) *Medical Geography*, Guilford Press, New York.

Medina-Ramón, M., Zanobetti, A. and Schwartz, J. (2006) The effect of ozone and PM_{10} on hospital admissions for pneumonia and chronic obstructive pulmonary disease: a national multicity study, *American Journal of Epidemiology*, 163, 579–88.

Mendis, S., Fukino, K., Cameron, A., et al. (2007) The availability and affordability of selected essential medicines for chronic diseases in six low- and middle-income countries, *Bulletin of the World Health Organization*, 85, 279–88.

Middleton, N., Sterne, J.A.C. and Gunnell, D. (2006) The geography of despair among 15–44 year old men in England and Wales: putting suicide on the map, *Journal of Epidemiology and Community Health*, 60, 1040–47.

Miles M. and Huberman A. (1994) *Qualitative Data Analysis*, Sage, London.

Miller, K.A., Siscovick, D.S., Sheppard, L., et al. (2007) Long-term exposure to air pollution and incidence of cardiovascular events in women, *New England Journal of Medicine*, 356, 447–58.

Milligan, C. (2001) *Geographies of Care: Space, Place and the Voluntary Sector*, Ashgate, Aldershot.

Milligan, C. (2006) Caring for older people in the 21st century: "notes from a small island," *Health & Place*, 12, 320–31.

Milligan, C. (2007) Restoration or risk? Exploring the place of the common place, in Williams, A. (ed.) *Therapeutic Landscapes*, Ashgate, Aldershot.

Milligan, C., Gatrell, A. and Bingley, A. (2004) "Cultivating health:" therapeutic landscapes and older people in northern England, *Social Science & Medicine*, 58, 1781–93.

Mintz, E.D., Tauxe, R.V. and Levine, M.M. (1998) The global resurgence of cholera, in Noah, N. and O'Mahony, M. (eds.) *Communicable Disease Epidemiology and Control*, John Wiley, Chichester.

Miringoff, M. and Miringoff, M.-L. (1999) *The Social Health of the Nation: How America is Really Doing*, Oxford University Press, Oxford.

Mitchell, H., Kearns, R. and Collins, D. (2006) Nuances of neighbourhood: children's perceptions of the space between home and school in Auckland, New Zealand, *Geoforum*, 38, 614–627.

Miyake, Y., Yokoyama, T., Yura, A., Iki, M. and Shmizu, T. (2004) Ecological association of water hardness with prevalence of childhood atopic dermatitis in a Japanese urban area, *Environmental Research*, 94, 33–7.

Moffatt, S., Phillimore, P., Bhopal, R. and Foy, C. (1995) "If this is what it's doing to our washing, what is it doing to our lungs?" Industrial air pollution and public understanding in north-east England, *Social Science & Medicine*, 41, 883–91.

Mohan, J. (1998) Explaining geographies of health: a critique, *Health & Place*, 4, 113–24.

Mohan, J., Twigg, L., Barnard, S. and Jones, K. (2005) Social capital, geography and health: a small-area analysis for England, *Social Science & Medicine*, 60, 1267–83.

Molarius, A. (2003) The contribution of lifestyle factors to socioeconomic differences in obesity in men and women – a population-based study in Sweden, *European Journal of Epidemiology*, 18, 227–9.

Monarca, S., Donato, F., Zerbini, I., Calderon, R.L. and Craun, G.F. (2006) Review of epidemiological studies on drinking water hardness and cardiovascular diseases, *European Journal of Cardiovascular Prevention and Rehabilitation*, 13, 495–506.

Monbiot, G. (2007) *Heat: How to Stop the Planet from Burning*, Random House, Toronto, Ontario.

Monique, A. (2008) Mechanisms underlying developmental programming of elevated blood pressure and vascular dysfunction: evidence from human studies and experimental animal models, *Clinical Science*, 114, 1–17.

Moon, G. and Brown, T. (1998) Place, space, and health service reform, in Kearns, R.A. and Gesler, W.M. (eds.) *Putting Health into Place: Landscape Identity and Wellbeing*, Syracuse University Press, New York.

Moon, G., Quarendon, G., Barnard, S., Twigg, L. and Blyth, B. (2007) Fat nation: deciphering the distinctive geographies of obesity in England, *Social Science & Medicine*, 65, 20–31.

Moorin, R.E., Holman, C.D.J., Garfield, C. and Brameld, K.J. (2006) Health related migration: evidence of reduced "urban-drift," *Health & Place*, 12, 131–40.

Moorman, J.E., Rudd, R.A., Johnson, C.A., et al. (2007) National surveillance for asthma: United States, 1980–2004, *MMWR*, 56, 1–14; 18–54.

Morone, J.A. and Jacobs, L.R. (2005) *Healthy, Wealthy and Fair: Health Care and the Good Society*, Oxford University Press, Oxford.

Morris, J.A., Butler, R., Flowerdew, R. and Gatrell, A.C. (1993) Retinoblastoma in children of former residents of Seascale, *British Medical Journal*, 306, 650.

Mostashari, F., Kulldorff, M., Hartman, J.J., Miller, J.R. and Kulasekera, V. (2003) Dead bird clusters as an early warning system for West Nile virus activity, *Emerging Infectious Diseases*, 9, 641–6.

Mukerjee, M. (1995) Persistently toxic: the Union Carbide accident in Bhopal continues to harm, *Scientific American*, 252, June 8, 10.

Muller, I., Smith, T., Mellor, S., Rare, L. and Genton, B. (1998) The effect of distance from home on attendance at a small rural health centre in Papua New Guinea, *International Journal of Epidemiology*, 27, 878–84.

Muntaner, C. and Lynch, J. (1999) Income inequality, social cohesion, and class relations: a critique of Wilkinson's neo-Durkheimian research program, *International Journal of Health Services*, 29, 59–81.

Murray, C.J.L. and Lopez, A.D., eds. (1996) *Quantifying Global Burden Health Risks: The Burden of Disease Attributable to Selected Risk Factors*, Harvard University Press, Cambridge, Mass.

Murray, C.J.L. and Lopez, A.D. (1997) Global mortality, disability, and the contribution of risk factors: Global Burden of Disease study, the *Lancet*, 349, 1436–42.

Nast, H.J. and Pile, S., eds. (1998) *Places Through the Body*, Routledge, London.

National Research Council (1991) *Environmental Epidemiology: Public Health and Hazardous Wastes*, National Academy Press, Washington.

Navarro, V. (2002) A critique of social capital, *International Journal of Health Services*, 32, 423–32.

Nazroo, J. (1998) Genetic, cultural or socio-economic vulnerability? Explaining ethnic inequalities in health, in Bartley, M., Blane, D. and Davey Smith, G. (eds.) *The Sociology of Health Inequalities*, Blackwell, Oxford.

Nemet, G.F. and Bailey, A.J. (2000) Distance and health care utilization among the rural elderly, *Social Science & Medicine*, 50, 1197–1208.

Newbold, K.B. (1998) Problems in search of solutions: health and Canadian aboriginals, *Journal of Community Health*, 23, 59–74.

Newbold, K.B. (2005) Self-rated health within the Canadian immigrant population: risk and the healthy immigrant effect, *Social Science & Medicine*, 60, 1359–70.

Nicholls, N. (1993) El Niño-southern Oscillation and vector-borne disease, the *Lancet*, 342, 1284–5.

Noor, A.M., Amin, A.A., Gething, P.W., Atkinson, P.M., Hay, S.I. and Snow, R.W. (2006) Modeling distances traveled to government health services in Kenya, *Tropical Medicine and International Health*, 11, 188–96.

Norman, P., Boyle, P. and Rees, P. (2005) Selective migration, health and deprivation: a longitudinal analysis, *Social Science & Medicine*, 60, 2755–71.

Nuyt A. (2008) Mechanisms underlying development programming of elevated blood pressure and vascular dysfunction, *Clinical Science*, 114, 1–17.

Oberlander, J., Marmor, T. and Jacobs, L. (2001) Rationing medical care: rhetoric and reality in the Oregon health plan, *Canadian Medical Association Journal*, 164, 1583–7.

O'Campo, P. (2003) Invited commentary: advancing theory and methods for multilevel models of residential neighborhoods and health, *American Journal of Epidemiology*, 157, 9–13.

O'Campo, P., Xue, X., Wang, M.-C., and Caughy, M. (1997) Neighborhood risk factors for low birth-weight in Baltimore: a multilevel analysis, *American Journal of Public Health*, 87, 1113–8.

Office for National Statistics (1997) *Health in England 1996: What People Know, What People Think, What People Do*, Stationery Office, London.

Oliver, L.N. and Hayes, M.V. (2005) Neighborhood socio-economic status and the prevalence of overweight Canadian children and youth, *Canadian Journal of Public Health*, 96, 515–20.

Opar, A., Pfaff, A., Seddique, A.A., Ahmed, K.M., Graziano, J.H. and van Geen, A. (2007) Responses of 6500 households to arsenic mitigation in Araihazar, Bangladesh, *Health & Place*, 13, 164–72.

Openshaw, S. (1983) The modifiable areal unit problem, *Concepts and Techniques in Modern Geography*, Environmental Publications, Norwich.

Openshaw, S., Craft, A.W., Charlton, M. and Birch, J.M. (1988) Investigation of leukaemia clusters by use of a geographical analysis machine, the *Lancet*, i, 272–3.

Packard, R.M. (1989) *White Plague, Black Labor: Tuberculosis and the Political Economy of Health and Disease in South Africa*, University of California Press, Berkeley, California.

Paigen, B., Goldman, L.R., Magnant, M.M., Highland, J.H. and Steegmann, A.T. (1987) Growth of children living near the hazardous waste site, Love Canal, *Human Biology*, 59, 489–508.

Paneth, N., Vinten-Johansen, P., Brody, H. and Rip, M. (1998) A rivalry of foulness: official and unofficial investigations of the London cholera epidemic of 1854, *American Journal of Public Health*, 88, 1545–53.

Parr, H. (1998) The politics of methodology in "post-medical geography:" mental health research and the interview, *Health & Place*, 4, 341–54.

Parr, H. (2002a) New body-geographies: the embodied spaces of health and medical information on the Internet, *Environment and Planning D: Society and Space*, 20, 73–95.

Parr, H. (2002b) Medical geography: diagnosing the body in medical and health geography, 1999–2000, *Progress in Human Geography*, 26, 240–51.

Parr, H. (2003) Medical geography: care and caring, *Progress in Human Geography*, 27, 212–21.

Parry, M.L. and Rosenzweig, C. (1993) Food supply and the risk of hunger, the *Lancet*, 342, 1345–7.

Pearce, J., Barnett, R. and Kingham, S. (2006) Slip! Slap! Slop! Cutaneous malignant melanoma incidence and social status in New Zealand, 1995–2000, *Health & Place*, 12, 239–52.

Pécoul, B., Chirac, P., Trouiller, P. and Pinel, J. (1999) Access to essential drugs in poor countries: a lost battle? *Journal of the American Medical Association*, 281, 361–7.

Perry, B. and Gesler, W. (2000) Physical access to primary health care in Andean Bolivia, *Social Science & Medicine*, 50, 1177–88.

Petersen, A. and Lupton, D. (1996) *The New Public Health: Health and Self in the Age of Risk*, Sage, London.

Phillimore, P. and Moffatt, S. (1994) Discounted knowledge: local experience, environmental pollution and health, in Popay, J. and Williams, G. (eds.) *Researching the People's Health*, Routledge, London.

Phillips, D.R. (1990) *Health and Health Care in the Third World*, Longman, Harlow.

Phillips, D.R. (1991) Problems and potential of researching epidemiological transition: examples from southeast Asia, *Social Science & Medicine*, 33, 395–404.

Phillips, D.R. (1994) Epidemiological transition: implications for health and health care provision, *Geografiska Annaler*, 76B, 71–89.

Phillips, D.R. and Verhasselt, Y., eds. (1994) *Health and Development*, Routledge, London.

Philo, C. (1989) "Enough to drive one mad:" the organization of space in 19th-century lunatic asylums, in Wolch, J. and Dear, M. (eds.) *The Power of Geography: How Territory Shapes Social Life*, Unwin Hyman, Boston.

Philo, C. (1996) Staying in? Invited comments on "Coming out: exposing social theory in medical geography," *Health & Place*, 2, 35–40.

Philo, C. (1997) Across the water: reviewing geographical studies of asylums and other mental health facilities, *Health & Place*, 3, 73–89.

Pickett, K.E., Collins, J.W., Masi, C.M. and Wilkinson, R.G. (2005) The effects of racial density and income incongruity on pregnancy outcomes, *Social Science & Medicine*, 60, 2229–38.

Pocock, S.J., Cook, D.G. and Shaper, A.G. (1982) Analysing geographic variation in cardiovascular mortality: methods and results, *Journal of the Royal Statistical Society, Series A*, 145, 313–41.

Pocock, S.J., Shaper, A.G., Cook, D.G., et al. (1980) British Regional Heart Study: geographic variations in cardiovascular mortality, and the role of water quality, *British Medical Journal*, 280, 1243–8.

Poland, B.D. (1998) Smoking, stigma, and the purification of public space, in Kearns, R.A. and Gesler, W.M. (eds.) *Putting Health into Place: Landscape Identity and Wellbeing*, Syracuse University Press, New York.

Popay, J., Thomas, C., Williams, G., Bennett, S., Gatrell, A.C. and Bostock, L. (2003) A proper place to live: health inequalities, agency and the normative dimensions of place, *Social Science & Medicine*, 57, 55–69.

Pope, C.A. (1989) Respiratory disease associated with community air pollution and a steel mill, Utah Valley, *American Journal of Public Health*, 79, 623–8.

Pope, C., Ziebland, S. and Mays, N. (2006) Analyzing qualitative data, in Pope, C. and Mays, N. (eds.) *Qualitative Research in Health Care*, 3rd edition, Blackwell, London.

Porsch-Oezcueruemez, M., Bilgin, Y., Wollny, M., et al., (1999) Prevalence of risk factors of coronary heart disease in Turks living in Germany: the Giessen study, *Atherosclerosis*, 144, 185–98.

Potvin, L., Richard, L. and Edwards, A.C. (2000) Knowledge of cardiovascular disease risk factors among the Canadian population: relationships with indicators of socioeconomic status, *Canadian Medical Association Journal*, 162, S5–S24.

Powell, D. and Leiss, W. (1997) *Mad Cows and Mother's Milk: The Perils of Poor Risk Communication*, McGill-Queen's University Press, Montreal.

Power, C., Manor, O. and Matthews, S. (1999) The duration and timing of exposure: effects of socio-economic environment on adult health, *American Journal of Public Health*, 89, 1059–65.

Prothero, R.M. (1965) *Migrants and Malaria*, Longmans, London.

Prothero, R.M. (1977) Disease and mobility: a neglected factor in epidemiology, *International Journal of Epidemiology*, 6, 259–67.

Pyle, G.F. (1969) The diffusion of cholera in the United States in the nineteenth century, *Geographical Analysis*, 1, 59–75.

Rajaratnam, J.K., Burke, J.G. and O'Campo, P. (2006) Maternal and child health and neighborhood context: the selection and construction of area-level variables, *Health & Place*, 12, 547–56.

Reijneveld, S.A. (1998) Reported health, lifestyles, and use of health care of first generation immigrants in the Netherlands: do socioeconomic factors explain their adverse position? *Journal of Epidemiology and Community Health*, 52, 298–304.

Remennick, L.I. (1999) Preventive behavior among recent immigrants: Russian-speaking women and cancer screening in Israel, *Social Science & Medicine*, 48, 1669–84.

Richardson, J.L., Langholz, B., Bernstein, L., Burciaga, C., Danley, K. and Ross, R.K. (1992) Stage and delay in breast cancer diagnosis by race, socioeconomic status, age and year, *British Journal of Cancer*, 65, 922–6.

Rigby, J.E. and Gatrell, A.C. (2000) Spatial patterns in breast cancer incidence in north-west Lancashire, *Area*, 32, 71–8.

Riise, T., Grønning, M., Klauber, M.R., Barrett-Connor, E., Nyland, H. and Albrektsen, G. (1991) Clustering of residence of multiple sclerosis patients at age 13 to 20 years in Hordaland, Norway, *American Journal of Epidemiology*, 133, 932–9.

Roberts, E.M. (1997) Neighborhood social environments and the distribution of low birthweight in Chicago, *American Journal of Public Health*, 87, 597–603.

Robinson, K., Elliott, S.J., Driedger, S.M., et al. (2005) Using linking systems to build capacity and enhance dissemination in heart health promotion: a Canadian multiple-case study, *Health Education Research*, 20, 499–513.

Rogers, A., ed. (1992) *Elderly Migration and Population Redistribution: A Comparative Study*, Belhaven Press, London.

Rogers, A., Hassell, K., Noyce, P. and Harris, J. (1998) Advice-giving in community pharmacy: variations between pharmacies in different locations, *Health & Place*, 4, 365–74.

Rogers, D.J. and Williams, B.G. (1993) Monitoring trypanosomiasis in space and time, *Parasitology*, 106, S77–S92.

Roos, N.P., Brownell, M. and Menec, V. (2006) Universal medical care and health inequalities: right objectives, insufficient tools, in Heymann, J., Hertzman, C., Barer, M. L. and Evans, R.G. (eds.) *Healthier Societies: From Analysis to Action*, Oxford University Press, Oxford.

Rose, G. (1993) *Feminism and Geography: The Limits of Geographical Knowledge*, Polity Press, Cambridge.

Rose, G. (1995) Place and identity: A sense of place, in Massey, D. and Jess, P. (eds.) *A Place in the World?*, Open University Press, Milton Keynes.

Rosner D. and Markowitz G. (2002) *Deceit and Denial: The Deadly Politics of Industrial Pollution*, University of California Press, Los Angeles.

Roy, A. (1999) *The Cost of Living*, Flamingo, London.

Rudant, J., Baccaïni, B., Ripert, M., et al. (2006) Population-mixing at the place of residence at the time of birth and incidence of childhood leukaemia in France, *European Journal of Cancer*, 42, 927–33

Ruger, J.P. and Kim, H.-J. (2006) Global health inequalities: an international comparison, *Journal of Epidemiology and Community Health*, 60, 928–36.

Rushton, G. and Lolonis, P. (1996) Exploratory spatial analysis of birth defects in an urban population, *Statistics in Medicine*, 15, 717–26.

Sabel, C.E., Gatrell, A.C., Löytönen, M., Maasilta, P. and Jokelainen, M. (2000) Modelling exposure opportunities: estimating relative risk for motor neurone disease in Finland, *Social Science & Medicine*, 50, 1121–37.

Salmond, C., Crampton, P., Hales, S., Lewis, S. and Pearce, N. (1999) Asthma prevalence and deprivation: a small area analysis, *Journal of Epidemiology and Community Health*, 53, 476–80.

Salmond, C., Crampton, P. and Sutton, F. (1998) NZDep91: a New Zealand index of deprivation, *Australian and New Zealand Journal of Public Health*, 22, 95–7.

Schabas, R. (2002) Public health: what is to be done? *Journal de L'Association Médicale Canadienne*, 166, 1282–3.

Schærstrom, A. (1996) *Pathogenic Paths? A Time Geographical Approach in Medical Geography*, Lund University Press, Lund.

Schrecker, T. and Labonte, R. (2007) What's politics got to do with it? Health, the G8, and the global economy, in Kawachi, I. and Wamala, S. (eds.) *Globalization and Health*, Oxford University Press, Oxford.

Schrijvers, C.T.M. and Mackenbach, J.P. (1994) Cancer patient survival by socioeconomic status in the Netherlands: a review for six common cancer sites, *Journal of Epidemiology and Community Health*, 48, 441–6.

Scott, P.A., Temovsky, C.J., Lawrence, K., Gudaitis, E. and Lowell, M.J. (1998) Analysis of Canadian population with potential geographic access to intravenous thrombolysis for acute ischaemic stroke, *Stroke*, 29, 2304–10.

Seale, C. and Pattison, S., eds. (1994) *Medical Knowledge: Doubt and Certainty*, Open University Press, Buckingham.

Sellstrom, E., Arnoldsson, G., Bremberg, S. and Hjern, A. (2008) The neighbourhood they live in: does it matter to women's smoking habits during pregnancy? *Health & Place*, 14, 155–66.

Senior, M.L. (1991) Deprivation payments to GPs: not what the doctor ordered, *Environment and Planning C*, 9, 79–94

Senior, M.L., New, S.J., Gatrell, A.C. and Francis, B.J. (1993) Geographic influences on the uptake of infant immunisations: 2. Disaggregate analyses, *Environment and Planning A*, 25, 467–79.

Senior, M.L., Williams, H.C.W.L. and Higgs, G. (1998) Spatial and temporal variation of mortality and deprivation 2: Statistical modelling, *Environment and Planning A*, 30, 1815–34.

Sharp, B.L. and le Sueur, D. (1996) Malaria in South Africa: the past, the present and selected implications for the future, *South African Medical Journal*, 86, 83–9.

Shaw, M., Dorling, D., Gordon, D. and Davey Smith, G. (1999) *The Widening Gap: Health Inequalities and Policy in Britain*, the Policy Press, Bristol.

Shaw, M., Dorling, D. and Mitchell, R. (2002) *Health, Place and Society*, Pearson Education Ltd., Harlow.

Shilling, C. (2005) *The Body in Culture, Technology and Society*, Sage, London.

Shrivistava, P. (1992) *Bhopal: Anatomy of a Crisis*, 2nd edition, Paul Chapman, London.

Singh, G.K. (2003) Area deprivation and widening inequalities in US mortality, 1969–1998, *American Journal of Public Health*, 93, 1137–43.

Slack, R., Ferguson, B. and Ryder, S. (1997) Analysis of hospitalization rates by electoral ward: relationship to accessibility and deprivation data, *Health Services Management Research*, 10, 24–31.

Slaper, H. Velders, G.J.M., Daniel, J.S., de Gruijl, F.R. and van der Leun, J.C. (1996) Estimates of ozone depletion and skin cancer incidence to examine the Vienna Convention achievements, *Nature*, 384, 256–8.

Smallman-Raynor, M. and Phillips, D. (1999) Late stages of epidemiological transition: health status in the developed world, *Health & Place*, 5, 209–22.

Smallman-Raynor, M, Muir, K.R. and Smith, S.J. (1998) The geographical assignment of cancer units: patient accessibility as an optimal location problem, *Public Health*, 112, 379–83.

Smith, A.H., Lingas, E.O. and Rahman, M. (2000) Contamination of drinking water by arsenic in Bangladesh: a public health emergency, *Bulletin of the World Health Organization*, 78, 1093–1103.

Smith, B.E. (1981) Black lung: the social production of disease, *International Journal of Health Services*, 11, 343–59.

Snow, J. (1854) *On the Mode of Communication of Cholera*, John Churchill, London.

Snow, R.W., Craig, M., Deichmann, U. and Marsh, K. (1999) Estimating mortality, morbidity and disability due to malaria among Africa's non-pregnant population, *Bulletin of the World Health Organization*, 77, 624–40.

Sooman, A. and Macintyre, S. (1995) Health and perceptions of the local environment in socially contrasting neighbourhoods in Glasgow, *Health & Place*, 1, 15–26.

Sorlie, P.D., Backlund, E. and Keller, J.B. (1995) US mortality by economic, demographic, and social characteristics: the National Longitudinal Mortality Study, *American Journal of Public Health*, 85, 949–56.

Squires, B.P. (2000) Cardiovascular disease and socioeconomic status, *Canadian Medical Association Journal*, 162, S3–S24.

Stainton Rogers, W. (1991) *Explaining Health and Illness*, Harvester Wheatsheaf, Hemel Hempstead.

Staples, J.A., Ponsonby, A.-L., Lim, L.L.-Y. and McMichael, A.J. (2003) Ecologic analysis of some immune-related disorders, including Type 1 diabetes, in Australia: latitude, regional ultraviolet radiation, and disease prevalence, *Environmental Health Perspectives*, 111, 518–23.

Stead, M., MacAskill, S., MacKintosh, A.M., Reece, J. and Eadie, D. (2001) "It's as if you're locked in:" qualitative explanations for area effects on smoking in disadvantaged communities, *Health & Place*, 7, 333–43.

Stiller, C.A. and Boyle, P.J. (1996) Effect of population mixing and socioeconomic status in England and Wales, 1979–85, on lymphoblastic leukaemia in children, *British Medical Journal*, 313, 1297–1300.

Streuning, K. (2007) Do government sponsored marriage promotion policies place undue pressure on individual rights? *Policy Sciences*, 40, 241–59.

Sun, W.Y., Sangweni, B., Butts, G. and Merlino, M. (1998) Comparisons of immunisation accessibility between non-US born and US-born children in New York City, *Public Health*, 112, 405–8.

Sundquist, K. and Ahlen, H. (2006) Neighborhood income and mental health: a multilevel follow-up study of psychiatric hospital admissions among 4.5 million women and men, *Health & Place*, 12, 594–602.

Sundquist, J. and Johansson, S.E. (1997) Long-term illness among indigenous and foreign-born people in Sweden, *Social Science & Medicine*, 44, 189–98.

Sundquist, K., Malmström, M. and Johansson, S.-E. (2004) Neighbourhood deprivation and incidence of coronary health disease: a multilevel study of 2.6 million women and men in Sweden, *Journal of Epidemiology and Community Health*, 58, 71–7.

Tanser, F., Sharp, B. and le Sueur, D. (2003) The potential effect of climate change on malaria transmission in Africa, the *Lancet*, 362, 1792–1798.

Taylor, H.R. (1995) Ocular effects of UV-B exposure, *Documenta Ophthalmologica*, 88, 285–93.

Tchuenté, L.A. (2006) Control of schistosomiasis: challenge and prospects for the 21st century, *Bulletin de la Société de Pathologie Exotique*, 99, 372–6.

Thamer, M., Richard, C., Casebeer, A.W. and Ray, N.F. (1997) Health insurance coverage among foreign-born US residents: the impact of race, ethnicity, and length of residence, *American Journal of Public Health*, 87, 96–102.

Thiel de Bocanegra, H. and Gany, F. (1997) Providing health services to immigrant and refugee populations in New York, in Ugalde, A. and Cardenas, G. (eds.) *Health and Social Services among International Labor Migrants*, CMAS Books, University of Texas Press, Austin, USA.

Thomas, C. (1999) *Female Forms*, Open University Press, London.

Thomas, C. (2007) *Sociologies of Disability and Illness: Contested Ideas in Disability Studies and Medical Sociology*, Palgrave Macmillan, Basingstoke.

Thomas, J.C. and Thomas, K.K. (1999) Things ain't what they ought to be: social forces underlying racial disparities in rates of sexually transmitted diseases in a rural North Carolina county, *Social Science & Medicine*, 49, 1075–84.

Thomas, R.W. (1992) *Geomedical Systems: Intervention and Control*, Routledge, London.

Thomson, M.C., Connor, S.J., Milligan, P.J.M. and Flasse, S.P. (1996) The ecology of malaria – as seen from earth-observation satellites, *Annals of Tropical Medicine and Parasitology*, 90, 243–64.

Timander, L. and McLafferty, S. (1998) Breast cancer in West Islip, NY: a spatial clustering analysis with covariates, *Social Science & Medicine*, 46, 1623–35.

Tonne, C., Melly, S., Mittleman, M., Coull, B., Goldberg, R. and Schwatz, J. (2007) A case-control analysis of exposure to traffic and acute myocardial infarction, *Environmental Health Perspectives*, 115, 53–7.

Townsend, J. (1995) The burden of smoking, in Benzeval, M. Judge, K., and Whitehead, M. (eds.) *Tackling Health Inequalities: An Agenda for Action*, King's Fund, London.

Trevelyan, B., Smallman-Raynor, M. and Cliff, A.D. (2005) The spatial dynamics of poliomyelitis in the United States: from epidemic emergence to vaccine-induced retreat, *Annals of the Association of American Geographers*, 95, 269–93.

Tudor Hart, J. (1971) The inverse care law, the *Lancet*, 27 February, 407–12.

Turshen, M. (1984) *The Political Ecology of Disease in Tanzania*, Rutgers University Press, New Brunswick, NJ.

Turshen, M. (1999) *Privatizing Health Services in Africa*, Rutgers University Press, New Brunswick, NJ.

Ugalde, A. (1997) Health and health services utilization in Spain among labor immigrants from developing countries, in Ugalde, A. and Cardenas, G. (eds.) *Health and Social Services among International Labor Migrants*, CMAS Books, University of Texas Press, Austin, USA.

UK Photochemical Oxidants Review Group (1993) *Ozone in the United Kingdom; Third Report*, Department of the Environment, London.

United Nations (2005) *The Millennium Development Goals Report*, United Nations Publications, Geneva.

United Nations (2006) *Water: a Shared Responsibility, Second UN World Water Development Report*, UNESCO Publishing and Berghahn Books.

United Nations (2007) *International Panel on Climate Change: Climate Change 2007: Impacts, Adaptation and Vulnerability*, Cambridge University Press, Cambridge.

Urry, J. (2007) *Mobilities*, Polity Press, Cambridge.

Valberg, P.A., van Deventer, T.E. and Repacholi, M.H. (2007) Workgroup report: base stations and wireless networks – radiofrequency (RF) exposures and health consequences, *Environmental Health Perspectives*, 115, 416–24.

Valentine, G. (2003) Geography and ethics: in pursuit of social justice – ethics and emotions in geographies of health and disability research, *Progress in Human Geography*, 27, 375–80.

Valjus, J., Hongisto, M., Verkasalo, P., Jarvinen, P., Heikkila, K. and Koskenvuo, M. (1995) Residential exposure to magnetic fields generated by 100–400kV power lines in Finland, *Bioelectromagnetics*, 16, 365–76.

Vanasse, A., Demers, M., Hemiari, A. and Courteau, J. (2006) Obesity in Canada: where and how many? *International Journal of Obesity*, 30, 677–83.

Varia, M., Wilson, S., Sarwal, S., et al. (2003) Investigation of a nosocomial outbreak of severe acute respiratory syndrome (SARS) in Toronto, Canada, *Canadian Medical Association Journal*, 169, 285–92.

Veenstra, G. (2007) Social space, social class and Bourdieu: health inequalities in British Columbia, Canada, *Health & Place*, 13, 14–31.

Veenstra, G., Luginaah, I., Wakefield, S., Birch, S., Eyles, J. and Elliott, S.J. (2005) Who you know, where you live: social capital, neighbourhood and health, *Social Science & Medicine*, 60, 2799–2818.

Verheij, R.A., de Bakker, D.H. and Groenewegen, P.P. (1999) Is there a geography of alternative medical treatment in The Netherlands? *Health & Place*, 5, 83–98.

Verrept, H. and Louckx, F. (1997) Health advocates in Belgian health care, in Ugalde, A. and Cardenas, G. (eds.) *Health and Social Services among International Labor Migrants*, CMAS Books, University of Texas Press, Austin, USA.

Vianna, N.J. and Polan, A.K. (1984) Incidence of low birthweight among Love Canal residents, *Science*, 226, 1217–9.

Von Reichert, C., McBroom, W.H., Reed, F.W. and Wilson, P.B. (1995) Access to health care and travel for birthing: Native American-white differentials in Montana, *Geoforum*, 26, 297–308.

Vrijheid, M. (2000) Health effects of residence near hazardous waste landfill sites: a review of epidemiological literature, *Environmental Health Perspectives*, 108, S101–S12.

Waitzkin, H. (2003) The social origins of illness: a neglected history, in Krieger, N. (ed.) (2005) *Embodying Inequality: Epidemiologic Perspectives*, Baywood Publishing Co., Amityville, NY.

Walberg, P., McKee, M., Shkolnikov, V., Chenet, L. and Leon, D.A. (1998) Economic change, crime, and mortality crisis in Russia: regional analysis, *British Medical Journal*, 1998, 317, 312–8.

Walters, N.M., Zietsman, H.L. and Bhagwandin, N. (1998) The geographical distribution of diagnostic medical and dental X-ray services in south Africa, *South African Medical Journal*, 88, 383–9.

Wang, T.J. and Stafford, R.S. (1998) National patterns and predictors of β-blocker use in patients with coronary artery disease, *Archives of Internal Medicine*, 158, 1901–6.

Warfa, N., Bhui, K., Craig, T., et al. (2006) Post-migration geographical mobility, mental health and health service utilisation among Somali refugees in the UK: a qualitative study, *Health & Place*, 12, 503–15.

Warin, M., Baum, F., Kalucy, E., Murray, C. and Veale, B. (2000) The power of place: space and time in women's and community health centres in South Australia, *Social Science & Medicine*, 50, 1863–75.

Warnes, A.M., King, R., Williams, A.M. and Patterson, G. (1999) The well-being of British expatriate retirees in southern Europe, *Ageing and Society*, 19, 717–40.

Watson, P. (1995) Explaining rising mortality among men in Eastern Europe, *Social Science & Medicine*, 41, 923–34.

Watson, P. (2006) Stress and social change in Poland, *Health & Place*, 12, 372–82.

Watts, S. (1997) *Epidemics and History: Disease, Power and Imperialism*, Yale University Press, New Haven.

Watts, S., Khallaayoune, K., Bensefia, R., Laamrani, H. and Gryseels, B. (1998) The study of human behavior and *Schistosomiasis* transmission in an irrigated area in Morocco, *Social Science & Medicine*, 46, 755–65.

Wei, M., Valdez, R.A., Mitchell, B.D., Haffner, S.M., Stern, M.P. and Hazuda, H.P. (1996) Migration status, socioeconomic status, and mortality rates in Mexican Americans and non-Hispanic whites: the San Antonio heart study, *Annals of Epidemiology*, 6, 307–13.

Wells, B.L. and Horm, J.W. (1992) Stage at diagnosis in breast cancer: race and socio-economic factors, *American Journal of Public Health*, 82, 1383–5.

Wellstood, K., Wilson, K. and Eyles, J. (2006) "Reasonable access" to primary care: assessing the role of individual and system characteristics, *Health & Place*, 12, 121–30.

Whincup, P. and Cook, D. (1997) Blood pressure and hypertension, in Kuh, D. and Ben-Shlomo, Y. (eds.) *A Life Course Approach to Chronic Disease Epidemiology*, Oxford University Press, Oxford.

Whiteis, D.G. (1997) Unhealthy cities: corporate medicine, community economic underdevelopment and public health, *International Journal of Health Services*, 27, 227–42.

Whiteis, D.G. (1998) Third world medicine in first world cities: capital accumulation, uneven development and public health, *Social Science & Medicine*, 47, 795–808.

Whitley, E., Gunnell, D., Dorling, D. and Davey Smith, G. (1999) Ecological study of social fragmentation, poverty, and suicide, *British Medical Journal*, 319, 1034–7.

Wickrama, K.A. and Kaspar, V. (2007) Family context of mental health risk in tsunami-exposed adolescents: findings from a pilot study in Sri Lanka, *Social Science & Medicine*, 64, 713–23.

Wiles, J. and Rosenberg, M.W. (2001) "Gentle caring experience." Seeking alternative health care in Canada, *Health & Place*, 7, 209–24.

Wilkinson, P., ed. (2006) *Environmental Epidemiology*, Open University Press, Maidenhead, UK.

Wilkinson, R. and Marmot, M., eds. (2003) *Social Determinants of Health: the solid facts*, 2nd edition, World Health Organization: Denmark.

Wilkinson, R.G. (1996) *Unhealthy Societies: The Afflictions of Inequality*, Routledge, London.

Wilkinson, R.G. and Pickett, K.E. (2006) Income inequality and population health: a review and explanation of the evidence, *Social Science & Medicine*, 62, 1768–84.

Williams, A., ed. (2007) *Therapeutic Landscapes*, Ashgate, Aldershot.

Williams, G. and Popay, J. (1994) Researching the people's health: dilemmas and opportunities for social scientists, in Popay, J. and Williams, G. (eds.) *Researching the People's Health*, Routledge, London.

Wilson, K. (2003) Therapeutic landscapes and First Nations peoples: an exploration of culture, health and place, *Health & Place*, 9, 83–93.

Wilson, K., Elliott, S.J., Keller-Olaman, S., Law, M. and Eyles, J. (2004) Linking perceptions of neighborhood to health in Hamilton, Canada, *Journal of Epidemiology and Community Health*, 58, 192–8.

Wilson, K., Eyles, J., Keller-Olaman, S., Elliott, S. and Devcic, D. Linking social exclusion and health: explorations in contrasting neighborhoods in Hamilton, Ontario, *Canadian Journal of Urban Research*, in press.

Wilton, R.D. (1996) Diminished worlds? The geography of everyday life with HIV/AIDS, *Health & Place*, 2, 69–83.

Wilton, R.D. (2003) Poverty and mental health: a qualitative study of residential care facility tenants, *Community Mental Health Journal*, 39, 139–56.

Wilton, R.D. (2004a) From flexibility to accommodation? Disabled people and the reinvention of paid work, *Transactions of the Institute of British Geography*, 29, 420–32.

Wilton, R.D. (2004b) Keeping your distance: balancing political engagement and scientific autonomy with a psychiatric consumer/survivor group, in Fuller, D. and Kitchin, R. (eds.) *Radical Theory/ Critical Praxis: Making a Difference Beyond the Academy?* Praxis Press, British Columbia.

Wilton, R.D. and DeVerteuil, G. (2006) Spaces of sobriety/sites of power: examining social model alcohol recovery programs as therapeutic landscapes, *Social Science & Medicine*, 63, 649–61.

Wing, S., Richardson, D., Armstrong, D. and Crawford-Brown, D. (1997) A re-evaluation of cancer incidence near the Three Mile Island nuclear plant: the collision of evidence and assumptions, *Environmental Health Perspectives*, 105, 52–7.

Wingate, M.S. and Alexander, G.R. (2006) The healthy migrant theory: variations in pregnancy outcomes among US-born migrants, *Social Science & Medicine*, 62, 491–8.

Winter, H., Cheng, K.K., Cummins, C., Maric, R., Silcocks, P. and Varghese, C. (1999) Cancer incidence in the south Asian population of England (1990–92), *British Journal of Cancer*, 79, 645–54.

Wisner, B. (1988) GOBI versus PHC? Some dangers of selective primary health care, *Social Science & Medicine*, 26, 963–9.

Wood, D.J., Clark, D. and Gatrell, A.C. (2004) Equity of access to adult hospice inpatient care within north-west England, *Palliative Medicine*, 18, 543–9.

Wood, E., Yip, B., Gataric, N., et al. (2000) Determinants of geographic mobility among participants in a population-based HIV/AIDS drug treatment program, *Health & Place*, 6, 33–40.

Woods, L.M., Rachet, B., Riga, M., Stone, N., Shah, A. and Coleman, M.P. (2005) Geographical variation in life expectancy at birth in England and Wales is largely explained by deprivation, *Journal of Epidemiology and Community Health*, 59, 115–20.

Woodward, A., Guest, C., Steer, K., et al. (1995) Tropospheric ozone: respiratory effects and Australian air quality goals, *Journal of Epidemiology and Community Health*, 49, 401–7.

World Health Organization (1957) *Measurement of Levels of Health: Report of the WHO Study Group*, WHO, Geneva.

World Health Organization (2000) *The World Health Report 2000, Health Systems: Improving Performance*, WHO, Geneva.

World Health Organization (2005) http://www.afro.who.int/home/countires/fact_sheets/zimbabwe.pdf

World Health Organization (2006) Cholera 2005, *Weekly Epidemiological Record*, 81, 297–308.

World Health Organization (2007) *Schistosomiasis, Fact Sheet Nº115*: www.who.int/mediacentre/factsheets/fs115/en/

World Health Organization (2008) WHO Statistical Information System http://www.who.int/whosis/en/index.html

Yamada, S., Caballero, J., Matsunaga, D.S., Agustin, G. and Magana, M. (1999) Attitudes regarding tuberculosis in immigrants from the Philippines to the United States, *Family Medicine*, 31, 477–82.

Yantzi, N.M., Rosenberg, M.W. and McKeever, P. (2006) Getting out of the house: the challenges mothers face when their children have long-term care needs, *Health and Social Care in the Community*, 15, 45–55.

Yarnell, J., Yu, S., McCrum, E., et al. (2005) Education, socioeconomic and lifestyle factors, and risk of coronary heart disease: the PRIME study, *International Journal of Epidemiology*, 34, 268–75.

Yi-Xin, H. and Manderson, L. (2005) The social and economic context and determinants of schistosomiasis japonica, *Acta Tropica*, 96, 223–31.

Zhang, Y., Huang, W., London, S.J., et al. (2006) Ozone and daily mortality in Shanghai, China, *Environmental Health Perspectives*, 114, 1227–32.

Appendix: Web-based Resources for the Geographies of Health

Introduction

The World Wide Web offers an unrivalled resource for the health researcher. Almost certainly, anyone picking up this book will be well used to surfing the net in order to extract research and other material, and may well have visited some of the following sites in order to obtain information and perhaps data. We can guarantee that there are sites, especially those relating to particular countries, or diseases and disabilities, that are missing from the abbreviated list given below. But we think we can also guarantee that at least some of the sites listed provide adequate links to other, unlisted, sites that will provide additional, valuable material. In the last resort, using one of the available search engines, along with a judicious choice of keyword(s), will usually lead to something of value.

This appendix is structured into two main sections. The first offers some sites that provide gateways into the research literature and good, general links, while the second lists some more specific sites of relevance to health geographers, including software.

A number of caveats are in order. First, we make no claims to completeness or comprehensiveness. Second, although we have visited all sites listed, the time lag between delivery of the manuscript to the publisher and its appearance as the book you are reading means that some sites may now be out of date; certainly, the information they contain will itself have been updated in many cases. You should therefore treat these sites as a set of signposts and routes rather than as a complete and accurate map. Third, although the pedigree of the sites listed here is quite impeccable, careful scrutiny of the small print will reveal that sources of data are of variable quality.

General Sources

Accessing relevant literature

There are a number of on-line bibliographic search tools available, of which PubMed is a most useful source. Bookmark the following site, which will be useful for obtaining abstracts of research papers:

www.ncbi.nlm.nih.gov/sites/entrez?db=PubMed

Some journals permit access on-line to full-length research papers. In some cases this involves an institution's library subscribing to the relevant journals.

Major health sites

Cross-country comparisons   A good starting point is the World Health Organization's main site:

www.who.int

From this site you can navigate to a wide range of sources and resources. In particular, their "Statistical Information System" (www.who.int/whosis/) has a wealth of data on health indicators, outcomes, diseases, health service provision, and so on. This site also has maps of global disease distribution.

There are similar sites for regional offices of WHO. For example, data on avoidable mortality in countries of Europe (at the regional level) are available from the Danish (Copenhagen) office at:

www.euro.who.int/

The United Nations have developed eight Millennium Development Goals ranging from halving the rates of extreme poverty worldwide, to halting the spread of HIV/AIDS, and providing universal primary education by the target date of 2015. Further details and updates on the status of this effort can be found at:

www.un.org/millenniumgoals/

A vast amount of information is available for individual countries at UNICEF's web-site:

www.unicef.org

For example, data relating to its Progress of Nations program (2000), and its report on the State of the World's Children (2007), both of which contain numerous health indicators, may be found at:

www.unicef.org/pon00 and www.unicef.org/sowc07

For further information about the Global Burden of Disease research outlined in Chapter 4, see:

www.hsph.harvard.edu/organizations/bdu/index.html

and the International Burden of Disease Network at:

www.ibdn.net/

USA   The best starting point is the Centers for Disease Control, based in Atlanta:

www.cdc.gov

From here you can navigate to numerous relevant sites, including the National Center for Health Statistics; this collects, analyses and disseminates health statistics, including those on a state-by-state basis. The introductory page for NCHS is:

www.cdc.gov/nchs

For data on cancer incidence and mortality the American Cancer Society has a wealth of valuable material. Visit:

www.cancer.org/docroot/stt/stt_0.asp

See further sections of this appendix for information relating to the Atlas of Cancer Mortality.

A wealth of data on the social, political and economic organization of the United States, including data on health services, insurance coverage and service provision may be obtained from the Statistical Abstract of the United States:

www.census.gov/compendia/statab/

The Morbidity and Mortality Weekly Report from the Centers for Disease Control has disease data at the state level as well as reports that may be downloaded:

www.cdc.gov/mmwr/

UK   Although having only limited data disaggregated by geographical area there is useful material from the Office for National Statistics:

www.statistics.gov.uk

The Department of Health site is also worth visiting for national data:

www.dh.gov.uk/

Regional Public Health Observatories contain a wealth of data and information relating to public health. You can navigate to specific regions from the main site:

www.apho.org.uk/

Canada   In addition to conducting a census every five years, Statistics Canada conducts approximately 350 active surveys on virtually all aspects of the Canadian population, including statistics relating to health, the environment and crime:

www.statcan.ca/

See for example:

National Population Health Survey   www.statcan.ca/english/concepts/nphs/
Canadian Health Measures Survey   www.statcan.ca/english/concepts/hs/measures.htm
Aboriginal Peoples Survey   www12.statcan.ca/english/profil01aps/home.cfm

Other Countries   Australia: the Department of Health and Ageing site: www.health.gov.au has numerous links to Australian health information, as well as helpful related links not listed here. The Australian Health Map: www.abc.net.au/health/healthmap/ provides a state-by-state guide to health statistics, resources, and health information.

New Zealand: www.moh.govt.nz/moh.nsf/ links to the New Zealand Ministry of health site. Further health information, statistics related to the performance of the health sector, and links to numerous other New Zealand health sites are available from: www.stats.govt.nz/people/health/default.htm

European Union: EuroStat: www.epp.eurostat.ec.europa.eu, gathers and analyses data and figures from the different European statistics offices in order to provide comparable and harmonized data to define, implement and analyze Community policies. Public Health Europa has many interesting and helpful links on varying health topics, as well as information on environmental conditions, and risk assessments available from: www.ec.europa.eu/health/index_en.htm

Specific diseases

On-line searches will reveal information – and possibly data – on particular diseases. Research literature on particular diseases can be obtained via particular on-line databases, such as PubMed, though some care is needed; for example, asking for all papers on "cancer" will be counter-productive! On the other hand, searching for papers on breast cancer survival in the USA will generate a more manageable list.

Again, WHO has numerous relevant sites, including those relating to tropical diseases such as malaria, and dengue fever. The schistosomiasis site, for example, is:

www.who.int/topics/schistosomiasis/en/

WHO also produces epidemiological fact sheets, for example on HIV and AIDS, which contain data and maps for particular countries. Visit:

www.who.int/GlobalAtlas/predefinedReports/EFS2006/index.asp

There are also numerous country-specific sites that link to important data on obesity trends:

USA   www.cdc.gov/overweight
Canada www.hc-sc.gc.ca/index_e.html or www.statcan.ca/ each have useful links
UK   www.dh.gov.uk/en/Policyandguidance/Healthandsocialcaretopics/Obesity/index.htm

The geography of health

One extremely useful site, with numerous links (including those to many listed here) has been devised by health geographers at the University of Portsmouth, UK. See:

www.envf.port.ac.uk/geo/research/health/links/weblinks.htm

and

www.port.ac.uk/research/ardsu/usefullinks/

See also the following:

geography.about.com/science/geography/msub19.htm

Margie Roswell at the University of Maryland has provided much useful information at the following site:

www.chpdm.org/

For a site relating to the historical work of Dr John Snow (see Chapter 8), visit:

www.ph.ucla.edu/epi/snow.html

There are a growing number of lists containing news of events, meetings, publications, job opportunities, and so on. For example, in Britain the Geography of Health Study Group of the Royal Geographical Society (with the Institute of British Geographers) has a regular newsletter. Visit the RGS web-site at:

www.rgs.org

Also see the American (Association of American Geographers) and Canadian equivalents at:

www.aag.org and www.cag-acg.ca/en/index.html

Electronic atlases and associated sources

Several sites contain maps or entire atlases of health-related maps, in some cases giving access to the data underlying such maps. A superb example is the Atlas of Cancer Mortality in the United States, 1950–94. See:

www3.cancer.gov/atlasplus/

Examine the "HealthVis" system available at Pennsylvania State University; this allows the user to dynamically link different "views" of the data, such as a map with a scatterplot of rates and socio-demographic data:

www.geovista.psu.edu/software/software.jsp

The following site covers the use of remote sensing in monitoring human and animal health. It contains a bibliography and much other valuable material:

geo.arc.nasa.gov/sge/health/sensor/sensor.html

For a comprehensive overview of health in Europe, the *Atlas of Health in* Europe is available from the Atlas of Leading and Avoidable Causes of Death in Countries of Central and Eastern Europe, including data for 14 countries by administrative subdivision, see:

www.euro.who.int/InformationSources/Publications/Catalogue/20030630_1

Increasing amounts of data and accompanying maps relating to environmental quality are now available. The US Environmental Protection Agency web-site carries data on air quality and pollution, with some data provided in real-time, and collections of maps on a state-by-state basis. See:

www.epa.gov

Data are available here on hazardous waste ("Superfund") sites. See the maps at:

www.epa.gov/superfund/sites/npl/npl.htm

The Environment Agency in England and Wales also makes available reports and data:

www.environment-agency.gov.uk

The Canadian Surveillance and Risk Assessment Division of the Public Health Agency of Canada provides numerous links to major chronic disease mortality maps online at:

www.dsol-smed.phac-aspc.gc.ca/dsol-smed/mcd-smcm/index_e.html

Software

ESRI markets a number of GIS products, including ArcGIS and the desktop package ArcView. Most academic institutions will have licences for these products, as will many health organizations. For some sample applications of GIS in the health field, look at the following page on ESRI's web-site:

www.esri.com/industries.html

A variety of software products are becoming available for spatial analysis. Although devised for use in the study of crime, rather than health, CrimeStat may be used for performing some of the spatial statistical analyses discussed in Chapter 3. It is described in, and available from:

www.icpsr.umich.edu/CRIMESTAT/

GeoDa, a program with a user-friendly graphical interface, may be used to explore and model spatial data. It is available free of charge from:

www.geoda.uiuc.edu/

Index